Karsten Wendt
Multi-Objective Optimization utilizing Cluster Analy
applied to Dimensional Transposed Problems

**TUD**_press_

Karsten Wendt

# Multi-Objective Optimization utilizing Cluster Analysis applied to Dimensional Transposed Problems

**TUD**press

2016

Die vorliegende Arbeit wurde am 27. Februar 2015 an der Fakultät Elektrotechnik und Informationstechnik der Technischen Universität Dresden als Dissertation eingereicht und am 23. November 2015 verteidigt.

Vorsitzender:
Prof. Dr.-Ing. Dr. h.c. Karlheinz Bock, Technische Universität Dresden

Gutachter:
Prof. Dr.-Ing. habil. René Schüffny, Technische Universität Dresden
Prof. Dr.-Ing. habil. Wilfried Krug, DUALIS GmbH IT Solution Dresden

Bibliografische Information der Deutschen Nationalbibliothek
Die Deutsche Nationalbibliothek verzeichnet diese Publikation in der Deutschen Nationalbibliografie; detaillierte bibliografische Daten sind im Internet über http://dnb.d-nb.de abrufbar.

Bibliographic information published by the Deutsche Nationalbibliothek
The Deutsche Nationalbibliothek lists this publication in the Deutsche Nationalbibliografie; detailed bibliographic data are available in the Internet at http://dnb.d-nb.de.

ISBN 978-3-95908-042-2


© 2016 w.e.b. Universitätsverlag & Buchhandel
Eckhard Richter & Co. OHG
Bergstr. 70 | D-01069 Dresden
Tel.: 0351/47 96 97 20 | Fax: 0351/47 96 08 19
http://www.tudpress.de

Technische Universität Dresden

# Multi-Objective Optimization utilizing Cluster Analysis applied to Dimensional Transposed Problems

**Dipl.-Ing. Karsten Wendt**

der Fakultät Elektrotechnik und Informationstechnik der Technischen Universität Dresden

zur Erlangung des akademischen Grades

**Doktoringenieur**

(Dr.-Ing.)

genehmigte Dissertation

Tag der Einreichung: 27.02.2015
Tag der Verteidigung: 23.11.2015

Vorsitzender: Prof. Dr.-Ing. Dr. h.c. Karlheinz Bock, Technische Universität Dresden

Gutachter: Prof. Dr.-Ing. habil. René Schüffny, Technische Universität Dresden
Prof. Dr.-Ing. habil. Wilfried Krug, DUALIS GmbH IT Solution Dresden

## Acknowledgments

Over the past six years I have received support and encouragements from a great number of individuals, making this work a thoughtful and rewarding journey.

First of all I would like to thank my doctoral supervisor Prof. Dr.-Ing. habil. René Schüffny for the continuous support of my Ph.D study, for the opportunities to be part of several very challenging international research projects, for his patience and his many helpful advices. He and his guidance made it possible to pursue such a project of researching and writing over this long time.

Besides my advisor, I would like to thank the rest of my thesis committee: Prof. Dr.-Ing. habil. Wilfried Krug and Prof. Dr.-Ing. habil. Uwe Petersohn for the insightful discussions and their encouragements, but also for the hard questions which inspired me to widen my research from various perspectives.

In particular I would like to thank Dr. Matthias Ehrlich for the long-term cooperation as well as his motivation and knowledge to discuss and support my scientific work generously. Without him, this thesis would not have been possible. I am glad to got to know him as a colleague and as a friend.

Furthermore my thanks go to the entire group of the HPSN chair for the stimulating team work, for all the discussions and for the experiences we shared among each other. I would also like to thank the team of the DUALIS GmbH IT Solution Dresden, and in particular Mrs. Heike Wilson, which has welcomed me for many years and gave me the chance to gather knowledge and inspirations in industrial contexts.

Last but not least I would like to thank my family, i.e. my parents, my grandparents and especially my beloved wife and children for the moral support and their great patience.

This work is dedicated to my grandfather, who died in January 2016.

## Abstract

With respect to the importance of multi-objective optimization in the context of the today's information processing and analysis, as well as the limitation of current approaches to treat large and complex tasks in practical time and little adjustment costs, this work proposes a novel optimization concept, based on data model transformations and subsequent cluster analyses to solve multi-objective problems. The approach abstracts the transposition of large, high-dimensional and diverse data models to low-dimensional uniform equivalents within an independent framework, which is optimized regarding data similarity conservation, i.e. the relations of the data items to each other are preserved, and low runtime complexity, i.e. linearly increasing model sizes also cause only linearly growing runtimes in spite of considering all data relations. The cluster analysis step is represented by an enhanced version of the k-Means algorithm, which is designed to group large numbers of data items to large numbers of clusters with also linear time complexity. Applying and adapting these components to two representative multi-objective optimization problems illustrate and prove the usability of the proposed concept, by solving these tasks with high qualities of results and low runtimes with virtually linear time complexities. The abstracted components, as well as their application extensions are tested and analyzed utilizing artificial, scalable databases, to determine valid parameter ranges, qualities of results and runtimes, and to ensure repeatable and comparable tests.

## Kurzfassung

Mit Hinblick auf die Bedeutung der multi-kriteriellen Optimierung im Kontext der heutigen Datenverarbeitung und -analyse, sowie auf die Grenzen herkömmlicher Verfahren bei der Behandlung großer und komplexer Aufgaben in akzeptabler Zeit mit möglichst geringem Anpassungsaufwand, stellt die vorliegende Arbeit ein neuartiges Optimierungsverfahren vor, welches auf Datenmodell-Transformation und anschließender Clusteranalyse basiert. Der Ansatz abstrahiert die Transposition von großen, hochdimensionalen und diversiven Datenmodellen zu niedrig-dimensionalen, uniformen Äquivalenten in einem unabhängigen Framework. Dessen eingebettete Algorithmen sind hinsichtlich Ähnlichkeitserhaltung und niedriger Zeitkomplexität optimiert, d.h. zum Einen verbleiben Relationen der Datenelemente zueinander unverändert und zum Anderen erhöht sich die Rechenzeit bei linear wachsender Datengröße trotz Betrachtung aller Datenbeziehungen auch nur linear. Der Schritt der nachfolgenden Clusteranalyse wird durch eine weiterentwickelte Form des k-Means Algorithmus vertreten, welche für die Verarbeitung großer Datenmengen in vielen Zielclustern mit ebenso nur linearer Laufzeiterhöhung entwickelt wurde. Die Adaption und Anwendung dieser beiden Komponenten auf zwei repräsentative Beispiele zur multi-kriteriellen Optimierung illustrieren und validieren die Brauchbarkeit des vorgeschlagenen Konzepts durch hohe Ergebnisqualitäten und niedrige Laufzeiten mit nahezu linearer Komplexität. Test und Analyse der abstrahierten Komponenten, genauso wie deren Anwendungserweiterungen erfolgten auf Basis künstlicher, skalierbarer, Datenmodelle zur Evaluierung geeigneter Parameterbereiche, der Ergebnisqualitäten und Laufzeiten, und um wiederholbare und vergleichbare Tests sicher zu stellen.

# Contents

**1  Introduction**                                                                        **5**

  1.1  Multi-Objective Optimization . . . . . . . . . . . . . . . . . . . . . . . . . . .   6

  1.2  Current Approaches to the Multi-Objective Optimization  . . . . . . . . . . . . .   7

  1.3  Current Challenges of the Multi-Objective Optimization  . . . . . . . . . . . .   9

  1.4  Motivation . . . . . . . . . . . . . . . . . . . . . . . . . . . . . . . . . . . .  11

**2  Dimensional Transposition Framework**                                                 **15**

  2.1  Introduction to the Data Analysis and Transposition . . . . . . . . . . . . . . .  15

    2.1.1  Current Challenges of the Cluster Analysis  . . . . . . . . . . . . . . .  15

    2.1.2  Current Developments with Respect to the Challenges  . . . . . . . . . .  17

      2.1.2.1  Processing Large Data Sets . . . . . . . . . . . . . . . . . . .  17

      2.1.2.2  Address High Dimensionality . . . . . . . . . . . . . . . . . .  18

      2.1.2.3  Treating Complex Metrics . . . . . . . . . . . . . . . . . . . .  23

    2.1.3  Approach of the Dimensional Transposition . . . . . . . . . . . . . . . .  24

      2.1.3.1  Motivation for the Data Transposition . . . . . . . . . . . . .  24

      2.1.3.2  Base Concept of the Dimensional Transposition Framework . . . . .  25

  2.2  Methods and Materials  . . . . . . . . . . . . . . . . . . . . . . . . . . . . . .  27

    2.2.1  Algorithm Analysis . . . . . . . . . . . . . . . . . . . . . . . . . . . .  27

    2.2.2  Measuring the Algorithm's Results and Performance  . . . . . . . . . . .  28

    2.2.3  Parameters of the Data Models and Algorithm Configuration . . . . . . . .  29

    2.2.4  Exploring the Parameter Space . . . . . . . . . . . . . . . . . . . . . .  30

    2.2.5  Algorithm Development Procedure  . . . . . . . . . . . . . . . . . . . .  32

  2.3  Results . . . . . . . . . . . . . . . . . . . . . . . . . . . . . . . . . . . . . .  32

    2.3.1  Runtime Profiling . . . . . . . . . . . . . . . . . . . . . . . . . . . .  33

    2.3.2  The Dimensional Transposition Framework in Detail . . . . . . . . . . . .  34

      2.3.2.1  Property Comparison as Computing Base . . . . . . . . . . . . .  35

      2.3.2.2  The Equilibration Algorithm as Core of the Framework  . . . . . .  38

      2.3.2.3  Internal Tree Building . . . . . . . . . . . . . . . . . . . . .  45

      2.3.2.4  Maximal Leaf Distance Calculation . . . . . . . . . . . . . . .  49

      2.3.2.5  Hierarchical Branch Representation as Meta Heuristic . . . . . . .  54

      2.3.2.6  Tree Based Equilibration Algorithm  . . . . . . . . . . . . . .  56

      2.3.2.7  Algorithm Enhancement by Tree Maintenance . . . . . . . . . . .  61

      2.3.2.8  Algorithm Enhancement by Substitution . . . . . . . . . . . . .  65

      2.3.2.9  Algorithm Enhancement by Insertion . . . . . . . . . . . . . . .  65

      2.3.2.10  The Final Dimensional Transposition Algorithm . . . . . . . . .  70

2.3.3    Result and Performance Measurements  . . . . . . . . . . . . . . . . . . . .   73

       2.3.3.1    Transposition Quality Dependency of the Simulated Annealing Factor   74

       2.3.3.2    Transposition Quality Dependency of the Branch Size  . . . . . . . .   75

       2.3.3.3    Transposition Quality Dependency of the Dimension Count  . . . . . .   77

       2.3.3.4    Runtime Dependency of the Data Model Size  . . . . . . . . . . . . .   78

       2.3.3.5    Runtime Dependency of Large Scale Data Models  . . . . . . . . . .   80

2.3.4    Summary of the Insights regarding the Dimensional Transposition Framework  .   81

       2.3.4.1    The Dimensional Transposition Framework as a Novel MDS Technique   81

       2.3.4.2    Result and Performance Measurements  . . . . . . . . . . . . . . . .   82

       2.3.4.3    Framework Application and Comparison  . . . . . . . . . . . . . . .   83

2.4    Discussion . . . . . . . . . . . . . . . . . . . . . . . . . . . . . . . . . . . . . . .   84

    2.4.1    Advantages and Disadvantages . . . . . . . . . . . . . . . . . . . . . . . . .   85

    2.4.2    Potential Applications  . . . . . . . . . . . . . . . . . . . . . . . . . . . . . .   86

    2.4.3    Further Enhancements and Improvements  . . . . . . . . . . . . . . . . . . .   86

**3   Tree Fusion k-Means Algorithm**                                               **89**

3.1    Introduction to the k-Means based Cluster Analysis and its Enhancements  . . . . . . . .   89

    3.1.1    The Meaning of Cluster Analysis  . . . . . . . . . . . . . . . . . . . . . . .   89

    3.1.2    The Standard k-Means Algorithm  . . . . . . . . . . . . . . . . . . . . . . .   91

       3.1.2.1    Base Algorithm Scheme of the k-Means Algorithm  . . . . . . . . .   91

       3.1.2.2    Fields of Application for k-Means Variants . . . . . . . . . . . . . .   92

       3.1.2.3    Advantages and Disadvantages  . . . . . . . . . . . . . . . . . . .   92

       3.1.2.4    Available Enhancements . . . . . . . . . . . . . . . . . . . . . . .   93

    3.1.3    The Tree Fusion k-Means Approach  . . . . . . . . . . . . . . . . . . . . .   95

       3.1.3.1    Motivation for a new k-Means Enhancement . . . . . . . . . . . . .   95

       3.1.3.2    Base Concept for the Tree Fusion k-Means Algorithm  . . . . . . . .   96

3.2    Methods and Materials  . . . . . . . . . . . . . . . . . . . . . . . . . . . . . . . .   98

    3.2.1    Algorithm Comparison . . . . . . . . . . . . . . . . . . . . . . . . . . . . .   98

    3.2.2    Measuring the Algorithm's Results and Performance . . . . . . . . . . . . . .   98

    3.2.3    Parameters of the Data Models and Algorithm Configuration . . . . . . . . . .   99

    3.2.4    Exploring the Configuration Space . . . . . . . . . . . . . . . . . . . . . .   100

    3.2.5    Algorithm Development Procedure  . . . . . . . . . . . . . . . . . . . . .   102

3.3    Results . . . . . . . . . . . . . . . . . . . . . . . . . . . . . . . . . . . . . . . .   102

    3.3.1    Runtime Profiling . . . . . . . . . . . . . . . . . . . . . . . . . . . . . . .   102

    3.3.2    Tree Fusion k-Means in Detail  . . . . . . . . . . . . . . . . . . . . . . . .   104

       3.3.2.1    Overview of *Tree Fusion - k-Means* Scheme  . . . . . . . . . . . . .   104

       3.3.2.2    Tree Building  . . . . . . . . . . . . . . . . . . . . . . . . . . .   106

       3.3.2.3    Maximal Leaf Distance Calculation . . . . . . . . . . . . . . . . .   107

       3.3.2.4    Algorithm Enhancement by Cluster Distance Estimation . . . . . . .   108

       3.3.2.5    Minimal Leaf Interspace Filtering . . . . . . . . . . . . . . . . . .   109

       3.3.2.6    Algorithm Enhancement by Tree Updating . . . . . . . . . . . . . .   112

       3.3.2.7    Tree Fusion / Element - Cluster Assignment Swaps  . . . . . . . . .   114

3.3.3    Result and Performance Measurements . . . . . . . . . . . . . . . . . . . . . 118

3.3.3.1    Runtime Dependency of the Data Model Type and Size . . . . . . . 119

3.3.3.2    Speedup Dependency of the Data Model Type and Size . . . . . . . 121

3.3.3.3    Speedup and Runtime Slope Dependency of Large Scale Data Models  125

3.3.3.4    Speedup Dependency of the Dimension Count . . . . . . . . . . . 126

3.3.3.5    Speedup Dependency of the Branch Size . . . . . . . . . . . . . . 128

3.3.3.6    Efficiency Dependency of the Data Model Size . . . . . . . . . . . 129

3.3.4    Summary of the Insights regarding the new Clustering Algorithm . . . . . . . 131

3.3.4.1    Algorithm Scheme . . . . . . . . . . . . . . . . . . . . . . . . . 131

3.3.4.2    Measurement Analysis . . . . . . . . . . . . . . . . . . . . . . . 131

3.3.4.3    Performance Comparisons . . . . . . . . . . . . . . . . . . . . . 132

3.4    Discussion . . . . . . . . . . . . . . . . . . . . . . . . . . . . . . . . . . . . . 133

3.4.1    Advantages and Disadvantages . . . . . . . . . . . . . . . . . . . . . . . 133

3.4.2    Applications . . . . . . . . . . . . . . . . . . . . . . . . . . . . . . . . 133

3.4.3    Further Enhancements and Improvements . . . . . . . . . . . . . . . . . . 134

**4    Applications                                                                            135**

4.1    Introduction to the new Approach's Application . . . . . . . . . . . . . . . . . . . 135

4.1.1    Segmentation as Neuron Placement . . . . . . . . . . . . . . . . . . . . . 136

4.1.2    Pattern Recognition as Peak Detection . . . . . . . . . . . . . . . . . . . . 140

4.2    Application to the Segmentation Issue as Neuron Placement . . . . . . . . . . . . . 143

4.2.1    Methods and Materials . . . . . . . . . . . . . . . . . . . . . . . . . . . 143

4.2.1.1    Proof of Concept . . . . . . . . . . . . . . . . . . . . . . . . . . 143

4.2.1.2    Data Models and Algorithm Configuration . . . . . . . . . . . . . 144

4.2.1.3    Measuring Results and Performance . . . . . . . . . . . . . . . . 148

4.2.1.4    Performance Evaluation . . . . . . . . . . . . . . . . . . . . . . . 149

4.2.2    Results . . . . . . . . . . . . . . . . . . . . . . . . . . . . . . . . . . . 150

4.2.2.1    The Segmentation Algorithm . . . . . . . . . . . . . . . . . . . . 150

4.2.2.2    Measurement Results . . . . . . . . . . . . . . . . . . . . . . . . 156

4.2.2.3    Results Summary . . . . . . . . . . . . . . . . . . . . . . . . . . 159

4.3    Application to the Pattern Recognition Issue as Peak Detection . . . . . . . . . . . 160

4.3.1    Methods and Materials . . . . . . . . . . . . . . . . . . . . . . . . . . . 160

4.3.1.1    Proof of Concept . . . . . . . . . . . . . . . . . . . . . . . . . . 160

4.3.1.2    Data Models and Algorithm Configuration . . . . . . . . . . . . . 160

4.3.1.3    Measuring Results and Performance . . . . . . . . . . . . . . . . 163

4.3.1.4    Performance Evaluation . . . . . . . . . . . . . . . . . . . . . . . 164

4.3.2    Results . . . . . . . . . . . . . . . . . . . . . . . . . . . . . . . . . . . 165

4.3.2.1    The Peak Detection Algorithm . . . . . . . . . . . . . . . . . . . 165

4.3.2.2    Measurement Results . . . . . . . . . . . . . . . . . . . . . . . . 169

4.3.2.3    Results Summary . . . . . . . . . . . . . . . . . . . . . . . . . . 172

4.4    Discussion . . . . . . . . . . . . . . . . . . . . . . . . . . . . . . . . . . . . . 173

**5    Summary and Final Discussion                                                             177**

**6 Appendix**                                                                      **181**

  6.1  Dimensional Transposition Framework . . . . . . . . . . . . . . . . . . . . . 181

      6.1.1  Profiling Results . . . . . . . . . . . . . . . . . . . . . . . . . . 181

      6.1.2  Measurement Results . . . . . . . . . . . . . . . . . . . . . . . 182

            6.1.2.1  Transposition Quality Dependency of the Simulated Annealing Factor  182

            6.1.2.2  Transposition Quality Dependency of the Branch Size . . . . . . . . 183

            6.1.2.3  Transposition Quality Dependency of the Dimension Count . . . . . 184

            6.1.2.4  Runtime Dependency of the Data Model Size . . . . . . . . . . . 187

  6.2  *Tree Fusion k-Means* Algorithm . . . . . . . . . . . . . . . . . . . . . 188

      6.2.1  Profiling Results . . . . . . . . . . . . . . . . . . . . . . . . . . 188

      6.2.2  Measurement Results . . . . . . . . . . . . . . . . . . . . . . . 191

            6.2.2.1  Runtime Dependency of the Data Model Type and Size . . . . . . . 191

            6.2.2.2  Speedup Dependency of the Data Model Type and Size . . . . . . . 194

            6.2.2.3  Speedup and Runtime Slope Dependency of Large Scale Data Models  200

            6.2.2.4  Speedup Dependency of the Dimension Count . . . . . . . . . . . 200

            6.2.2.5  Speedup Dependency of the Branch Size . . . . . . . . . . . . . 203

            6.2.2.6  Efficiency Dependency of the Data Model Size . . . . . . . . . . . 204

  6.3  Applications . . . . . . . . . . . . . . . . . . . . . . . . . . . . . . . 206

      6.3.1  Segmentation . . . . . . . . . . . . . . . . . . . . . . . . . . . 206

            6.3.1.1  Data Models and Algorithm Configuration . . . . . . . . . . . . . 206

**Acronyms**                                                                        **207**

**Bibliography**                                                                    **209**

# 1 Introduction

Nowadays virtually all domains, which apply information technologies, comprise optimization tasks with several, partly contradictory objectives, involving increasingly larger and more complex data sets. This development already stresses conventional optimization approaches to the limit of their performance. This is especially relevant for future oriented fields of application, beginning with the analysis and evaluation of market data, including consumer information and behavior to optimize products and marketing strategies. This focus can be enlarged to general process and logistic optimizations of manufactures, national or international production chains, or entire economies. With regard to their monetary policy of, for example, central banks, these organizations have to find proper tradeoffs, balancing competing objectives such as inflation, employment states and trade deficits among other issues [Deb, 2001]. Furthermore, the multi-objective requirements are extendable to challenges of the applied sciences to find ideal machine and method specifications, along with optimized schedules and a minimal consumption of resources [Marler and Arora, 2004]. The fields of application also encapsulate core issues of the computer science itself, e.g. efficient distributions of computing capacity, network optimizations or data security as well as data processing and analysis [Deb, 2001]. Novel computer architectures such as the neural hardware have to be configured by so called mapping processes, transforming complex neural networks to hardware-specific equivalents, while considering different objectives like topology specifications or parameter and capacity constraints [Brüderle et al., 2011]. Also, last but not least, modern types of energy production and consumption require flexible, so called smart networks, which have to be optimized on demand with respect to varying and competing objectives [Pohekar and Ramachandran, 2004].

All these tasks can be summarized under the term *Multi-Objective Optimization (MOO)*, which is equivalent to the determination of a large number of variables with regard to many, competing objectives with different priorities. In spite of the ongoing hardware development and increase of computing power, the majority of the *MOO* problems exhibit non-linear time complexities, thus the required computation effort grows faster than the available capacities. Reflecting the current development of data creation and aggregation, it can be anticipated that the solving of future *MOO* problems requires novel approaches of data processing and algorithms.

For this reason the present work elaborates a novel *MOO* concept, which takes into account the important role of the *Cluster Analysis (CA)* in data investigation and data mining, processing large amounts of information efficiently. Utilizing its similarity to the *MOO* in general with regard to the objectives' requirements, it is reasonable to combine existing proven *MOO* approaches with state-of-the-art data analysis and transformation techniques. The resulting methods should be able to approach to global *MOO* optima from other directions than the conventional techniques, while exhibiting practical runtime behaviors.

The definition of the *MOO* is described in Section 1.1 providing a formal base for these issues. Although a lot of approaches exist to solve *MOO* problems in a general or specific manner, these techniques become impractical or the qualities of results decrease, if they are applied to large and/or complex data sets. The state-of-the-art of *MOO* methods is summarized in Section 1.2, followed by considerations about the current challenges regarding recent optimization tasks in Section 1.3. The introduction is completed by the formulation of the theses, e.g. the motivation in Section 1.4 to propose a novel concept to transform, analyze, and reinterpret, hence solve *MOO* tasks.

The entire work is divided as follows: Section 2 presents a technique to transform large and complex models to an uniform equivalent, which is more suitable to be processed by *CA* algorithms or other data mining methods. Subsequently an enhanced version of the *k-Means* cluster algorithm is introduced in Section 3, specialized in the processing of large numbers of data items and target clusters. Both these components form the base of the proposed *MOO* approach and are applied to selected examples in Section 4 to prove the concept and its efficiency. Finally the obtained results, measurements and insights are discussed in Section 5.

## 1.1 Multi-Objective Optimization

Formally the *MOO* is defined as follows [Miettinen, 1999]: It is given a feasible set $X$ of decision vectors in a decision space $\mathbb{R}^n$, as well as an objective vector of at least two target functions $\vec{f}(\vec{x}) = (f_1(\vec{x}), f_2(\vec{x}), .., f_k(\vec{x}))^T$, which projects a vector $\vec{x} \in \mathbb{R}^n$ into a target space $\mathbb{R}^k$, $Y$. Thus, the resulting mapping $\vec{f}$ can be expressed as:

$$\vec{f} : \mathbb{R}^n \mapsto \mathbb{R}^k \tag{1.1}$$

which is also illustrated in Figure 1.1. The task of the *MOO* is to find a vector $\vec{x}^*$, which minimizes the target function vector $\vec{f}(\vec{x})$.

Where the optimization of only a single objective allows an unambiguous solution $\vec{x}^*$, the *MOO* encapsulates a set of solutions $\overrightarrow{(\vec{x}^*)}$, representing compromises with regard to the different optimization targets. The simultaneous optimization of more than one objective is referred to as *MOO*.

By the introduction of *pareto dominance* as a criterion of order [Ehrgott, 2005], the solution alternatives become comparable. A solution vector $\vec{y}^* = \vec{f}(\vec{x}^*)$ in $\mathbb{R}^k$ is *pareto-dominant* ($\prec_{pareto}$) in comparison to another vector $\vec{y} = \vec{f}(\vec{x})$, if the following applies:

$$\forall i \in \{1, .., k\} : y_i^* \leq y_i$$
$$and \tag{1.2}$$
$$\exists i \in \{1, .., k\} : y_i^* < y_i$$

Such a solution $\vec{x}^*$ is referred to as *pareto-optimal* in $\mathbb{R}^n$ or as non-dominated point, if no other solution $\vec{x} \in X$ exists, for which applies $\vec{f}(\vec{x}) \prec_{pareto} \vec{f}(\vec{x}^*)$. The task of *MOO* methods is to find and identify such pareto-optimal solutions firstly and secondly to select a suitable result. Figure 1.1 shows the mapping $\vec{f}$ as described above in combination with the introduced pareto-optimal front, or short the pareto front.

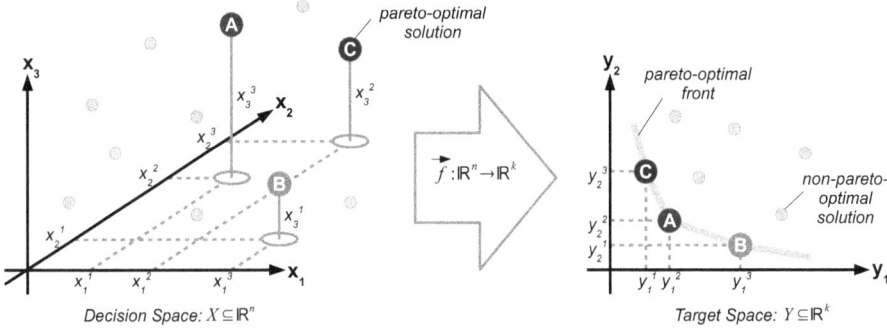

*Decision Space:* $X \subseteq \mathbb{R}^n$                     *Target Space:* $Y \subseteq \mathbb{R}^k$

**Figure 1.1:** *Illustration of Pareto-Optimal Solutions and the Pareto Front*; Out of the feasible set $X$ in $\mathbb{R}^3$ of the decision variables (left side) the target functions $\vec{f}(\vec{x})$ create a mapping $Y = \vec{f}(X)$ onto the target space in $\mathbb{R}^2$ (right side). Shown are three pareto-optimal solutions $A$, $B$ and $C$, and non-pareto-optimal solutions (gray circles), which contain a vector $(x_1, x_2, x_3)$ each in the decision space. Mapped to the target space, each solution refers to a vector $(y_1, y_2)$, presenting the optimization objectives. The pareto-optimal solution alternatives form the pareto front, which to find and to select from is the task of the *MOO*.

## 1.2 Current Approaches to the Multi-Objective Optimization

The conventional approaches to solve *MOO* problems can be separated into two groups:

On the one hand the objective vector $\vec{f}$ (Formula (1.1)) can be compressed to an integrating function $f_S(\vec{f})$, e.g. by linear combination, thus transforming the multi-objective task to a *single-objective* problem [Miettinen, 1999]. This so called *Scalarizing* enables the utilization of the broad spectrum of conventional (numerical) methods to treat single objective optimization problems, and furthermore, the determination of pareto-optimal solutions by varying the scaling parameters. However, as adding these parameters enlarges the target space, it may have an adverse effect onto the computation time or, if chosen inconveniently, decrease the quality of the achievable results.

On the other hand alternative methods first detect representative solutions within the pareto front [Ehrgott, 2005; Coello et al., 2007] and subsequently a so called *decision maker* (an algorithm or a human) selects the most suitable pareto-optimal solution(s). Hence the selection of the final solution is not based on previous objective parametrizations, but is performed when a sufficient number of comparable solutions are available, thus avoiding distortions caused by inconvenient initial assumptions. The determination of representative sets of pareto-optimal solutions requires comparable much computing power and thus is comparably slow in contrast to the scalarizing approaches. In addition the decision makers need extra information to select the best solutions.

Regardless of the general approaches described above, the techniques to determine pareto-optimal solutions are dividable into three types:

- *Mathematical Optimization (MAO)* [Williams, 2013]: Starting at a (random) position within the decision space, these algorithms determine solutions by the iterative improvement of preliminary results, e.g. by *gradient descent/ascent*[1] [Snyman, 2005]. These methods are comparably fast, but vulnerable to local optima, i.e. once a local minimum or maximum is found, it is difficult to recognize it as such to resume the search. In consequence these algorithms are applied several

---

[1]iterative technique to improve a solution by following the steepest gradient in the target space

times, starting at different positions and with varying parameter values. The general scheme is shown in Figure 1.2.

- *Evolutionary Algorithms* (*EVA*) [Deb, 2001]: Initialized at multiple positions, which are as distant as possible, these methods try to cover the entire decision space and in general create a set of different pareto-optimal solutions, requiring only a single algorithm run. The techniques of the *EVA* are comparably slow and due to their global focus, they may miss optimal results, if evaluation differences are too small with regard to the overall context (no fine granularity). The general scheme is also shown in Figure 1.2.

- *Combinations of EVA and MAO* [Miettinen et al., 2008]: The best results are achieved by hybrid methods, which determine an initial set of pareto-optimal solutions, utilizing *EVA*s, and subsequently refine these results by *MAO*s. In consequence the major drawback are the longest runtimes in comparison to other approaches.

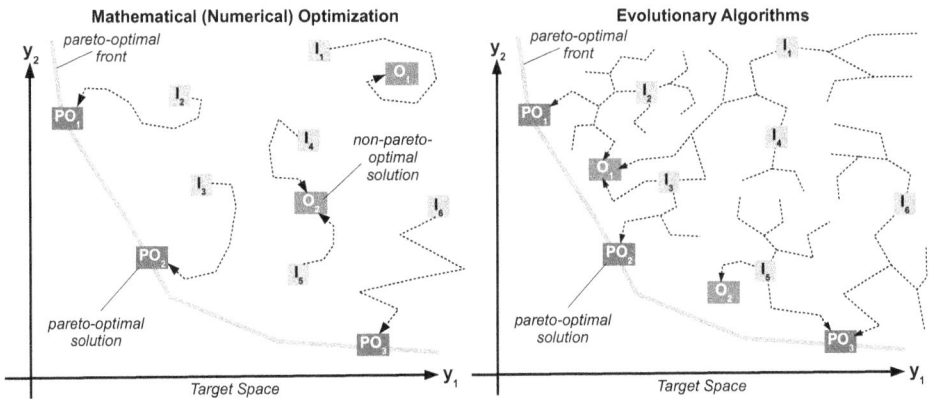

**Figure 1.2:** *Visualization of the principles of the MAO (left side) and the EVA (right side) to solve MOO problems;*
Shown is the target space with $k = 2$ and the pareto-optimal front, which to find or to select from is the task of the *MOO*. Given different starting positions $I_1..I_6$, the algorithms of both approaches try to locate elements of the unknown pareto front.
**Left (*MAO*):** Starting from $I_1$ the search, e.g. a gradient descent, approaches to the local optimum $O_1$ and is unable to pass it. Hence the search is started several times at $I_2$ and $I_3$, with the result, that this time two better, pareto-optimal solutions $PO_1$ and $PO_2$ are found. The subsequent two search runs, starting at $I_4$ and $I_5$, yield the solution $O_2$, which can also be classified as non-pareto-optimal. Only the start from $I_6$ provides the third pareto-optimal solution $PO_3$. The pareto front can be refined further by more search iterations and the adaptation of the search algorithm's parameters.
**Right (*EVA*):** In contrast the evolutionary based strategies start simultaneously at the different initial positions $I_1..I_6$. At the expense of more computation time a set of solutions is tracked at the same time, allowing for refinement of promising paths and to discard less potential candidates. The start from $I_2$ for example develops three different branches, but only the left one leads to the pareto-optimal solution $PO_1$. The others are discarded. It is obvious that this approach covers the entire target space better than the *MAO*, but is less precise, i.e. in this example it stops at the local optima $O_1$ and $O_2$ without refining them to the actual pareto front.

All presented *MOO* techniques are based on the principle to improve already existing, non-optimal solutions. The improvement is done e.g. by the gradually adaptation of variables in direction of the closest optima (*MAO* - gradient descent/ascent [Snyman, 2005]), by the stochastic search within the current neighborhood of the decision space (*MAO - Simulated Annealing (SA)* [Kirkpatrick, 1984]), or by random changes or recombinations of solution components (*EVA - Mutation*[2] and *Crossover*[3] operations [Coello et al., 2007]).

## 1.3 Current Challenges of the Multi-Objective Optimization

First of all the current challenges of *MOO* are determined by the characteristics of the tasks to solve, which can be summarized as follows:

- *Large Number of Variables*: The complexity of current *MOO* problems is determined by the large number of variables, thus complicating efficient searches in the decision space, due to its exponential growing [Ehrgott, 2005].

- *Complex Variable Dependencies and Secondary Conditions*: Beside the target functions, *MOO* tasks are often complemented by additional constraints, e.g. *if $x_i > 0 \Rightarrow x_j < 0$*, which are dividable in the types *hard*[4] and *soft*[5], making the creation of valid and optimized solutions more expensive [Caramia, 2006].

- *Large Number of and Expensive Target Functions*: Not only the growing number of target functions increases the computing effort per algorithm iteration, but also their complexity caused by, for example, multiplications, exponentiations, logarithms or further extensive operations [Marler and Arora, 2004].

- *Limited Evaluations*: The application of heuristic methods provides no or only limited information about the distance of the current solution(s) to the pareto front, which complicates the decision to continue or abort the current search [Coello et al., 2007].

As consequences of the characteristics listed above, the following challenges arise during the application of *MOO* techniques: As illustrated in Section 1.2 most of the algorithms are based on the iterative improvement of already existing, valid, but not optimal solutions. Due to the large sizes of the decision and target space in combination with complex target functions and secondary conditions (see also Section 1.1), the creation of new, valid solutions on the base of currently available ones, becomes increasingly costly. The modification of a single variable implies the consideration of all relevant secondary conditions, usually the evaluation of all target functions and the estimation of the remaining optimization potential. Thus, the investigation of the current solutions' entire environment (within $X$ and $Y$) requires a comparably high computing effort. In consequence methods of the *MAO* and the *EVA*, or combinations of them require significant longer runtimes, or, in the case of limited computation time, produce results of lower quality.

Figure 1.3 illustrates the search within the environments of preliminary solutions in the decision and target space for their iterative improvement.

---

[2] single random change of an element within a solution
[3] combination of fractions of multiple solutions to a new solution
[4] crucial criteria to retain the solution valid
[5] optional criteria to improve the solution

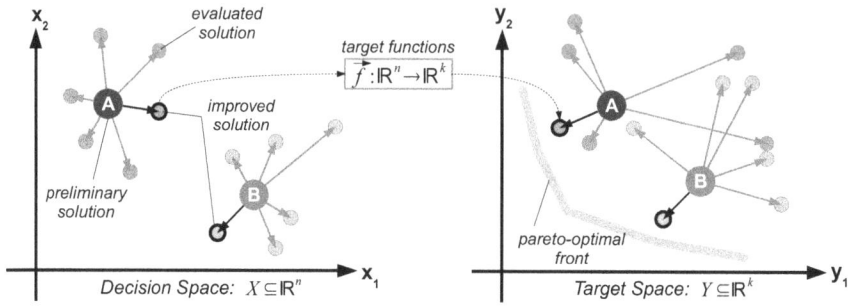

**Figure 1.3:** *Example of MOO by Solution Environment Investigation and Iterative Improvement.* Shown are the preliminary solutions $A$ and $B$ within the decision and the target space, both with two dimensions. To approach the sought pareto front, the *MOO* techniques of the *MAO* and/or the *EVA* have to create valid solutions (small colored circles) by investigating the environment of the provided preliminary solutions $A$ and $B$ in $X$ and evaluate them in $Y$, utilizing the mapping functions $\vec{f}$. The solutions which are closest to the pareto front are accepted and provide the starting points for the next iteration. It is obvious that the larger both spaces, i.e. the larger $n$ and $k$, and the more secondary conditions are given, the more neighbor solutions exist and the more expensive it is, to map the solution from $X$ to $Y$ for their evaluation. These effects slow down traditional *MOO* approaches and/or may reduce their qualities of results respectively.

As the pareto front is the optimization objective and thus unknown, the increasingly large and complex *MOO* tasks make it increasingly difficult to estimate the remaining optimization potential, i.e. to evaluate authoritative information about the current distances to the pareto front and to decide whether to improve a solution further, to discard it or to declare it as final.

Beside the task-inflicted challenges, the following issues arise from the commonly utilized algorithm techniques: Because most of the current *MOO* algorithms aim at a general use, supportive enhancements are difficult to apply, as they depend mainly on the specific problem characteristics. For example the implementation of a $\vec{\delta}$-functionality for costly target functions, i.e. the calculation of the difference of a target function $f_i(\vec{x})$, if $x_j \in \vec{x}$ is altered by the value of $\vec{\delta}$, to accelerate the investigation of the current solution's environment $(\vec{x} + \vec{\delta})$, requires detailed knowledge about $f_i(\vec{x})$, and thus is problem specific.

Furthermore, partial solutions are often not evaluable, as the target functions cover the entire decision space and thus a solution can only be mapped to the target space, if the majority of all variables is determined and valid. Especially for combinatorial problems this challenge slows down the algorithms, as the search often implies the partial or even the entire creation of new solutions. For example, the *knapsack problem* [Kellerer et al., 2004] as *MOO* task to assign items to a set of containers with different capacities, while not exceeding the containers' capacities and optimizing set of target function, is affected by this issue: Given a valid solution, the investigation of its environment is equivalent to the evaluation of all possible swaps of an item to another container. The hard secondary condition to not over-occupy containers may require additional subsequent items swaps, until no capacity is exceeded. In consequence for each swap a resulting valid solution is necessary to evaluate the swap in the target space, slowing down the investigation of all neighbor solutions and thus the overall optimization process.

## 1.4 Motivation

Reflecting the importance of *MOO* with regard to the wide fields of application (see the introduction to this chapter), the prevalent conventional approaches, i.e. the *MAO*, the *EVA* and mixed forms of them (Section 1.2), and the present challenges inflicted by the processed problems or by the utilized techniques (Section 1.3), arises the question for novel approaches to treat the increasingly difficult optimization tasks.

Prior to the development of new concepts and their combination with refined existing techniques to design a novel *MOO* approach, the general target objectives have to be defined. The focus of the new approach is on the following aspects, treating the challenges as introduced in Section 1.3:

- *Problem Size Reduction* - As described above a significant portion of the challenges is caused by the *MOO* problem itself, i.e. the requirements with regard to the data model processing can be separated as follows:

  - *Variable Count Reduction*: The number of variables, which define the size of the decision space $X$ (Section 1.1), determines the amount of solutions mainly, and thus have to be investigated. Decreasing the variable count may reduce the solution space, and hence the solution complexity significantly.

  - *Target Function Reduction*: Analogue, the size of $Y$, the target space (Section 1.1), is defined by the number of target functions, hence determining the space in which the sought pareto-optimal solutions are or the entire pareto front is located. Reducing that space in the same manner as carried out on $X$ can facilitate the evaluation of solution candidates.

  - *Function and Constraint Encapsulation*: The solutions' validation and mapping from the decision to the target space is described by the secondary conditions and the target functions respectively. Large numbers and high complexities of both of them imply extensive and complicate validation and mapping processes. Thus an efficient encapsulation of these descriptions, i.e. their transition to more efficient expressions, with regard to the reduced space sizes may enable an accelerated overall process.

- *Performance Enhancement* - As introduced above the conventional approaches have several inherent limitations that makes them inappropriate to treat *MOO* tasks of large sizes and complexities. The resulting requirements for the processing behavior of the novel approach can be summarized as follows:

  - *Fine Granular Coverage of the Entire Search Space*: Based on the utilized approach or applied (meta) heuristics, many common techniques tend either to cover the entire solution space at the cost of fine granularity or to refine found solutions to their (local) optima at the risk of missing globally better results. The novel approach should aim for both, i.e. a complete search space coverage as well as a fine granular refinement.

  - *Runtime Complexity Reduction*: As the general increase of the variable and target function counts causes an exponential growth of the resulting decision and target spaces, an inconstant runtime increment, and thus a non-linear time complexity of the conventional methods can be expected. This may lead to impractical runtimes for problems beyond a certain size, depending on the processing system. Hence a crucial criterion is a low, or in the best case, a linear time complexity of the new *MOO* approach.

  – *Redundant Computing Minimization*: It is assumed, that a considerable amount of computing effort is redundant and thus, can be isolated to be performed only once, saving computing power.

- *Extendability* - Beside the requirements regarding data processing and solution detection, the following objectives aim at the practical application of the novel *MOO* approach:

  – *Scalability*: Because of the expected growth of *MOO* problems, it is desirable to support the scalability of the new methods, i.e. they should be capable of utilizing expanded hardware resources, e.g. additional computation cores for parallel processing.

  – *Compatibility*: With respect to the development of specialized techniques to treat certain *MOO* tasks, hybrid forms, i.e. combinations of proven concepts and the new approach should be considered. In consequence a concept of general interfaces is preferable.

The novel approach to comply with these requirements, is based on techniques of the *CA*, which can be considered as a subfield of the *MOO* [Ehrgott, 2005; Miettinen, 1999], as it aims to create a small number of dense and meaningful groups of data items (clusters). Conversely, this means that a subset of all *MOO* problems, maybe the majority, can be modeled as *CA* tasks [Pulido and Coello Coello, 2004; Taboada and Coit, 2007].

Hence, assuming that provided *MOO* problems can be transformed to suitable equivalents, methods of the *CA* can be utilized to structure the data in way, that the resulting groups are interpretable as the sought pareto-optimal solutions or elements of the pareto front, thus providing an efficient *MOO* technique. As theses that presumption can be formulated as follows:

*Thesis 1:* *MOO* problems, i.e. their decision and target spaces as well as their mapping functions and secondary conditions, are transposable to expressions, which enable *CA* techniques to process them efficiently.

*Thesis 2:* The *CA* results of a transformed *MOO* problem can be transposed reversely, i.e. the clusters and their contents can be interpreted as pareto-optimal solutions. Clusters of optimized and comparable values regarding the (transposed) target functions represent the pareto front.

*Thesis 3:* The forward transposition of a specific *MOO*, the subsequent *CA* and the final interpretation, i.e. the partial backward transposition, feature a low time complexity in comparison to conventional approaches.

*Thesis 4:* The achievable quality of result is comparable to or better than the outcome of the conventional methods.

The proposed general approach to solve *MOO* problems is referred to as *Transposition and Cluster Analysis based Multi-Objective Optimization* (TRACS-MOO). Figure 1.4 illustrates a schematic example of the TRACS-MOO, transposing a *MOO* problem to a *CA* task. According to Figure 1.1 the heterogeneous decision and target spaces $X$ and $Y$, as well as the mapping $\vec{f}$ can be transposed to an uniform

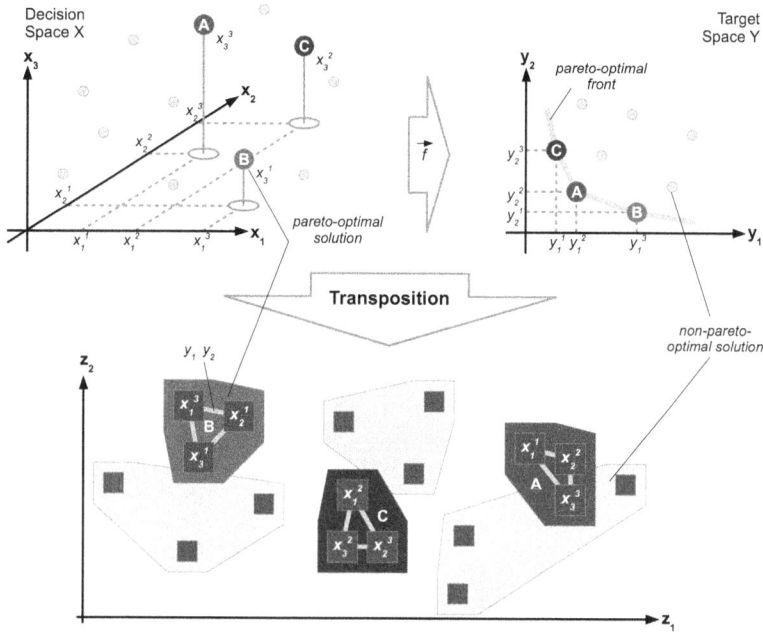

**Figure 1.4:** *Visualization of the* TRACS-MOO *concept as a MOO Problem Transposition to the CA Domain;*
Shown are the decision and the target space, including the mapping vector $\vec{f}$ of a *MOO* example
and their transformation to another uniform space for clustering.
The solutions $A$, $B$, and $C$ each consist of a vector $(x_1, x_2, x_3)$ and are pareto-optimal, i.e. in this
example they minimize the target functions $y_1$ and $y_2$. Non-pareto-optimal solutions are visualized as
small gray circles. The transposition of the single vector elements of each solution $(x_1^1, x_1^2, .., x_3^n)$, and
thus the entire *MOO* problem, to a space $Z$ is performed in such a way, that the resulting distance of
the elements (gray squares) corresponds with values of the target functions, i.e. the closer an element
pair, the smaller $y_1$ and $y_2$ are. A subsequent *CA* creates dense clusters, so that each cluster contains
exactly one element of each dimension in $X$, thus optimizing the target function implicitly. The most
dense clusters are equivalent to the pareto-optimal solution, representing the pareto front in $Y$.

space $Z$, so that the utilization of adapted *CA* methods and the resulting clusters are equivalent to the
original *MOO*. In this way TRACS-MOO has the potential to comply with the approach's general target
objectives, introduced above.

Figure 1.5 shows the coarse data scheme of the TRACS-MOO concept. A *MOO* task is transposed
to an equivalent expression, processed by an adapted *CA* technique and interpreted, i.e. partially re-
transposed to the original spaces. This scheme represents the utilized base concept for the proposed
*MOO* technique as elaborated in the following.

In the next chapters the conceptual design, the results of the development and the performance measurements of the two key components (*transposition* and *cluster analysis*) of the TRACS-MOO approach
as well as its application are presented. Chapter 2 describes the transposition component as general technique to transpose *MOO* problems to preferably low-dimensional and uniform equivalents. Chapter 3
introduces an efficient, adaptable cluster technique as *CA* component within the overall context. Finally
Chapter 4 presents two examples the TRACS-MOO's application to current *MOO* tasks.

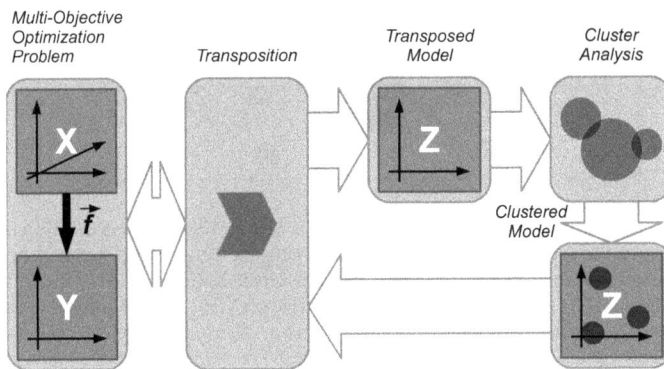

**Figure 1.5:** *Scheme of the novel* TRACS-MOO *approach to solve MOO problems.* Shown is the general algorithm's data scheme. Starting with a *MOO* problem as described in Section 1.1 the first step is to transpose the data model, i.e. the decision space $X$, the target space $Y$ and the mapping $\vec{f}$ (the target functions), as well as the secondary conditions (not in the figure), to a new space $Z$. This uniform space embeds $\vec{f}$ in distances between the elements of $X$, thus implicit solving the *MOO* task partially. The second step performs a *CA*, so that the sought for elements within the former space $Y$, i.e. the pareto-optimal solutions are determined as clusters of predefined characteristics. Finally the results are transferred back to the original spaces, forming the final *MOO* solution(s).

# 2 Dimensional Transposition Framework

The following section introduces the first component of the TRACS-MOO approach (see also Section 1.4) to transpose a provided large and complex data model to an uniform equivalent, which is more suitable for subsequent general purpose and efficient data processing. Section 2.1 represents the importance of the automated data analysis, along with the current challenges and the base concept of this first abstracted TRACS-MOO component. To evaluate and explore its fields of application and performance, Section 2.2 describes the applied testing and analysis strategy. Subsequently the obtained insights, i.e. the developed algorithm components and the measurement results are presented in Section 2.3. Finally the results' meaning with respect to the proposed *MOO* approach and to data preparation in general are discussed in Section 2.4.

## 2.1 Introduction to the Data Analysis and Transposition

First, the importance of the data analysis, in particular the *CA*, is reflected in Section 2.1.1 with regard to the occurring challenges while processing tasks of current sizes and complexities. Subsequently conventional and modern approaches to address these challenges are summarized in Section 2.1.2. The resulting requirements for new approaches, along with the proposed data transposition method are finally introduced in Section 2.1.3.

### 2.1.1 Current Challenges of the Cluster Analysis

As shown in Section 1, *MOO* and *CA* as its sub field become increasingly more important to investigate and analyze large data sets, known as *big data* [1]. Applying methods of the *CA* to real world problems, such as *Exploratory Pattern Analysis*, *Grouping*, *Decision Making*, *Data Mining*, *Document Retrieval*, *Image Segmentation*, and *Pattern Classification* in fields of *Web Analysis*, *Customer Relationship Management*, *Marketing*, *Medical Diagnostics*, *Computational Biology* and *Computer* or *Social Science* [Berkhin, 2006; Soni and Ganatra, 2012], reveal various upcoming issues the traditional approaches are unable to handle [McCallum et al., 2000; Can and Ozkarahan, 1990; Agrawal et al., 2005; Jiang et al., 2004].

The current challenges can be separated into three categories:

1. the increasing size of data sets [Berkhin, 2006; Jain, 2008; Andrews and Fox, 2007; Willett, 1988; Lukashin et al., 2003; Philbin, 2007; Steinbach et al., 2003],

2. the increasing dimensionality of the data items [Indyk and Motwani, 1998; Kriegel et al., 2009; Andrews and Fox, 2007; Lukashin et al., 2003; Philbin, 2007; Steinbach et al., 2003] and

---

[1] common term to describe data sets of sufficient complexity and size, exceeding the human's capability to analyze them manually

3. the increasing diversity of the data items' features regarding their value domains, scaling factors and offsets, completeness and dynamic behaviors [Arkhangel'skii, 1990; Jaeger, 2008; Jain, 2008; Steinbach et al., 2003].

In general occurring effects in one category amplify effects of the other ones, e.g. processing high-dimensional data is even more difficult in combination with large data sets and/or many heterogeneous value domains and ranges. Figure 2.1 illustrates the complexity of current *CA* problems, including the three categories introduced above.

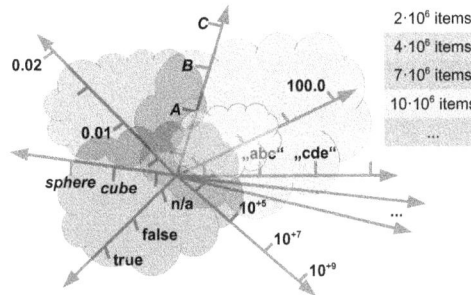

**Figure 2.1:** *Illustration of a Large and Complex Data Model to cluster*; shown are various orthogonal axes, containing different *continuous* value ranges ([0.005..0.02], [25.0..100.0] and [$10^{+5}$..$10^{+9}$]), *categorical* values ([*n/a, true, false*] and [*A, B, C*]), *string-based* values (["abc"], ["cde"]) and *complex* values ([*sphere*], [*cube*], which may encapsulate inferior features). Clouds of different colors symbolize concentrations of up to $10 \cdot 10^6$ items, which overlap and are embedded within the axes system. The *CA*'s task is to find the significant item clusters within the clouds, while minimizing the number of groups and their values' variance while ignoring noise and outsiders with regard to the user-defined preferences.

While comparing the items ($n$) to the clusters ($k$) several times ($l$), the time complexity never falls below $O(nkl)$ for unsupervised approaches [Berkhin, 2006; Jain et al., 1999] such as *k-means* by Lloyd [1982] or *ISODATA* by Ball and Hall [1965]. This ignores the fact that $k$ is unknown and thus has to be determined during multiple runs, whose qualities of results are also dependent on the initial seeds. This means the total runtime of these techniques is a multiple of $O(nkl)$ and thus scales non-linearly [Arthur and Vassilvitskii, 2007; Arthur et al., 2009]. More advanced techniques, such as *PROCLUS* by Aggarwal et al. [1999], *DBSCAN* by Ester et al. [1996], *CURE* by Guha et al. [1998], *k-prototypes* by Huang [1997], *CLARANS* by Ng and Han [2002] or *BIRCH* by Zhang et al. [1996], have a time complexity significantly higher than $O(nk)$, often $O(n^2k)$. As consequence the runtimes of most of the known *CA* techniques become impractical if the data sets' size (data item count, feature count per data item and feature complexities) exceeds certain limits [2], which excludes the known *CA* methods for processing large data sets.

Secondly, the adaptation of common cluster algorithms or their available frameworks (see above) to current problems becomes even more expensive, due to the data models' complexities [Estivill-Castro, 2002]. This involves the diversity of the data items' features regarding value domains, ranges and availabilities in combination with useful metrics (similarity functions) to compare them [Arkhangel'skii,

---

[2]limit sizes are strongly dependent on the computing machines and the given contexts (real time decisions of air plane software vs. archeological analyzing tools)

1990]. To achieve the requirement to handle large numbers of data items with features of different kind efficiently, metrics have to be optimized or transformed to more efficient but equivalent expressions [Schwartz and Bilsky, 1990; Jaeger, 2008].

Besides the non-linear time complexity, also the memory costs and their time behavior influence the resulting runtime [Jain et al., 1999], especially if the data models exceed the available RAM and have to be swapped to the storage memory [Zhang et al., 1996; Ester et al., 1996].

As a result of the high dimensionality of current data models the *curse of dimensionality* decreases the achieved qualities of clusters or causes outcomes to be more sensible to noise or outliers [Indyk and Motwani, 1998]. Also, last but not least, the increasing variety and complexity of available data models as well as *CA* objectives complicate the meaningful comparison of cluster algorithms and thus the decision which technique is promising for which problem class [Estivill-Castro, 2002].

## 2.1.2 Current Developments with Respect to the Challenges

In the following the representative approaches for the three major challenges (data model size, data dimensionality and data type diversity) of the *MOO* or the *CA* respectively (Section 2.1.1) are summarized.

### 2.1.2.1 Processing Large Data Sets

Beside the large number of clustering techniques for different purposes [Berkhin, 2006; Jain et al., 1999; Jain, 2008; Soni and Ganatra, 2012] (see also Section 3.1.1), some approaches focus the issue to process large data sets in comparable short time with compromises regarding the quality of results. Following Berkhin [2006]; Jain [2008] the attempts can be divided into:

- *Incremental Clustering*: This approach [Hartigan, 1975; Jain et al., 1999] scans the data set usually a single time and assigns the data items to existing clusters while simultaneously updating them (e.g. split, merge or discard). As an example *DIGNET* by Thomopoulos et al. [1995] utilizes *k-means* to process data items during a single scan, while updating the centroid[3] priorities depending on the assigned data items. The cluster quality is strongly influenced by the scan order of the data items, thus may require multiple runs with different seeds.

- *Data Squashing*: To reduce the data amount during runtime, data items can be summarized in temporary clusters, which provide statistical information about their substitutes. Typically these temporary clusters are organized in a hierarchy, which is constructed during a single scan. Subsequent phases optimize this tree (e.g. balancing, detecting outliers, ..) and finally cluster it by applying common techniques like *k-means* to the compressed data structure. Popular examples are the *BIRCH* algorithm by Zhang et al. [1996], relying on a tree of *cluster features*, which is rebuild several times and the *k-prototypes* algorithm by Huang [1997], which is based on continuously updated representative *prototypes* for entire data set. Similar to the incremental clustering the compression quality and thus also the final quality of clusters depends on the ordering of the data items. Restructuring the substituted data improves the quality, but is time expensive [Berkhin, 2006]. Other techniques to compress data are the utilization of *minimal spanning trees* [Kruskal, 1956] or histograms of *visual words* [Jain, 2008].

---

[3]one of the most common cluster type definition, where the cluster center is the average of all cluster member

- *Reliable Sampling*: As an intuitive approach the sampling utilizes the fact, that a random fraction of the complete data set is usually sufficient to identify underlying structures and thus the clusters. The *CURE* algorithm by Guha et al. [1998] defines randomly a subset of the data model, on which the cluster structure is constructed and to which the remaining data items are assigned subsequently. As the processing of data models larger than the available RAM may cause significant runtime increases, as parts of the data model has to be swapped to the slower storage memory, random sampling is an appropriate method to reduce the data model sizes and to enable the *CA* techniques to investigate them efficiently [Jain et al., 1999].

- *Efficient Nearest Neighbor Search*: One of the basic operations of each cluster algorithm is to decide the cluster affiliation for each data item, hence efficient methods to detect the nearest (cluster) neighbors accelerates the algorithms' execution [Jain, 2008]. One of the most common approach is to utilize the triangular inequality [Inequality, 2014] on vector space based data models, improving centroid-based algorithms [Elkan, 2003].

- *Parallel Computing*: To address the challenge of non-linear runtime complexities, algorithms can be designed and implemented for parallel data processing which can speed up the execution time up to a factor equal to the number of cores on a multi-core system [Dhillon and Modha, 2000; Bae et al., 2012].

The majority of current algorithms utilize a combination of these approaches, for example the *ISO-DATA* by Ball and Hall [1965], the *DBSCAN* by Ester et al. [1996] or the *CLARANS* by Ng and Han [2002].

### 2.1.2.2 Address High Dimensionality

In addition to the data amount and its handling (Sections 2.1.1 and 2.1.2.1), current *CA*s have to treat high dimensional data items, containing large numbers of features. According to Kriegel et al. [2009] the processing of such data models is accompanied by four issues:

- *Curse of Dimensionality*: This term describes the effect that the volume of the (decision) space, which is spanned by the axes, increases faster (exponentially) with additional axes than the actual amount of data (linearly) [Bellman, 1957; Indyk and Motwani, 1998]. This leads to a data sparseness in relation to the describing axes system, which is problematic for all methods based on statistics, such as the *CA*, as arithmetic centers or standard deviations become imprecise. Given an $n$-dimensional space, its volume $v^n$ can be expressed as:

$$v^n = \prod_{i=0}^{n} (a_{max}^i - a_{min}^i) = a_{range}^n \tag{2.1}$$

where $a_{max}^i$ and $a_{min}^i$ are the maximal and minimal occurring values of the axis $a^i$. Assuming an average value range $a_{range}$ for all axes, it becomes apparent, that the volume grows exponentially with $n$. If the amount of data items does not also grow with a constant factor for each additional feature, the model becomes sparse in $n$.

- *Distance Discrimination*: As single axes play only minor roles in high-dimensional systems, the metrics (distances or affinities) of different data items converge to average values and thus become meaningless. Given the position or feature vectors

$$\vec{pos}_x = \begin{pmatrix} p_x^0 \\ .. \\ p_x^k \\ .. \\ p_x^n \end{pmatrix} \qquad \vec{pos}_y = \begin{pmatrix} p_y^0 \\ .. \\ p_y^k \\ .. \\ p_y^n \end{pmatrix} \tag{2.2}$$

of two data items $x$ and $y$ in $n$ dimensions, the influence of a single value pair $p_x^k$ and $p_y^k$ of axis $k$ on the distance $dist_{x,y}$ decreases with growing $n$:

$$\begin{aligned} dist_{x,y}^2 &= \sum_{i=0}^{n} (p_x^i - p_y^i)^2 \\ &= \sum_{i=0, i \neq k}^{n} (p_x^i - p_y^i)^2 + (p_x^k - p_y^k)^2 \\ &\cong \sum_{i=0, i \neq k}^{n} (p_x^i - p_y^i)^2 \quad \textit{for large } n \end{aligned} \tag{2.3}$$

Assuming a large $n$ and that the data of all axes is not correlated except in $k$, the distance is dominated by noise and thus not suitable any more to express the data items' similarity.

- *Local Feature Relevance*: Large numbers of features may form uncorrelated clusters, in consequence global analyses can conceal smaller clusters in subspaces. An example can be found in Figure 2.2.

- *Arbitrary Orientation*: Correlations are not necessarily oriented parallel to axes, they can occur arbitrarily oriented, which makes it more difficult to detect them. An example can be found in Figure 2.2.

To address these issues various approaches were introduced [Berkhin, 2006; Kriegel et al., 2009]: namely 1) the *High-Dimensional Clustering*, such as 1.*a*) the *Projected Clustering* and 1.*b*) the *Subspace Clustering* and 2) the *Data Model Preprocessing*, as 2.*a*) the *Feature Processing*, 2.*b*) the *Principal Component Analysis* (PCA) and 2.*c*) the *Multi-Dimensional Scaling* (MDS). In the following these methods are introduced briefly, each accompanied by representative examples:

1. *High-Dimensional Clustering*: Algorithms of this class aim to handle the large number of axes by assuming inherent correlations of subsets in all dimensions:

    a) *Projected Clustering*: These methods analyze the axes step by step to detect structured regions. This assumes that clusters exists in different subspaces, i.e. are defined by different and overlapping sets of features [Kriegel et al., 2009]. In this way the axes and their statistics

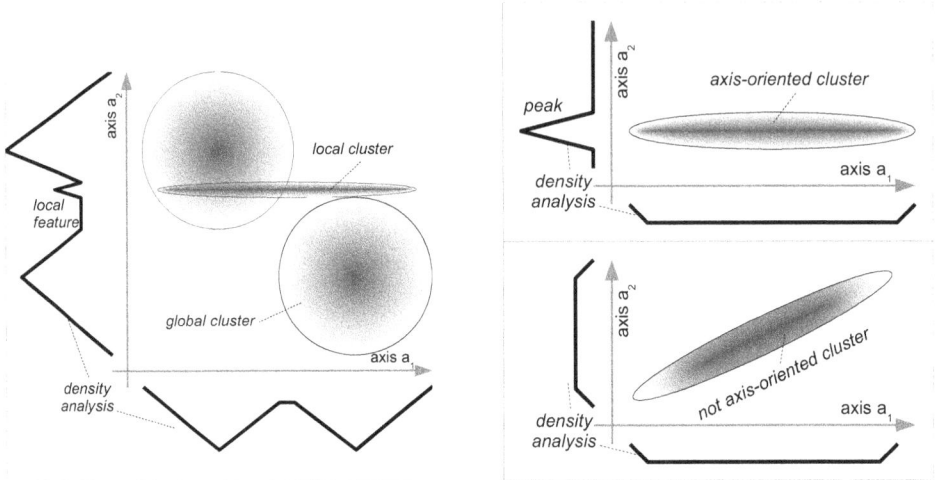

**Figure 2.2:** *High Dimensionality Challenges Examples*; For the purpose of clarity only two dimensions / axes are visualized.
**Local Feature Relevance (left):** Shown are two global clusters, which have density peaks in all axes, and a local cluster, which is stretched evenly across all axes, except one ($a_2$). The density analysis of all axes reveals peaks in all axes for the global clusters; the local one appears only in a single axis. As the most common *CA* techniques for high dimensionality aim to combine peaks to clusters, uncorrelated and/or small single peaks of density may be classified as irrelevant or noise.
**Arbitrary Orientation (right):** Compared are an axis-oriented cluster (top), and a not axis-oriented cluster (bottom). Where the density analysis reveals significant peaks in several axes for the axis-oriented cluster, it fails to identify the not axis-oriented one, due to its inconvenient orientation. In spite of its same size, shape and density distribution, the second one is more difficult to detect, than the first one.

are utilized to detect the characteristics of the sought clusters from different perspectives. The approach of Aggarwal et al. [1999] determines a set of clusters randomly and minimizes the standard deviation to identify their subspaces by assigning the data items in a *k-means* manner. Bad initialized clusters are replaced as long as the cluster quality increases. The algorithm produces clusters of spherical shape and has to be executed multiple times due to its seed dependency. In contrast Boehm et al. [2004] extends the *DBSCAN* algorithm by Ester et al. [1996] by utilizing an adaptive (learning) distance function, which reflects the axes' relevances for the current cluster. As a drawback this method utilizes a set of user-defined parameters, which are difficult to initialize with appropriate values.

b) *Subspace Clustering*: In contrast to the projected clustering, this approach tries to identify clusters as combinations of relevant (dense) regions of different single axes. In a first step the data items are projected to small subsets of all axes, similar to the *Projected Clustering*, and analyzed statistically [Kriegel et al., 2009]. Subsequently algorithms combine significant single-axis peaks to clusters (bottom-up) or divide existing clusters to smaller ones along the largest gaps (top-down) [Parsons et al., 2004]. The pioneer algorithm *CLIQUE* by Agrawal et al. [2005] divides each axis into fractions of equal size and combines only fractions, whose resulting density exceeds a given threshold, thus implementing a bottom-up strategy. On the other side top-down algorithms like *SUBCLU* by Kailing et al. [2004]; Rakesh et al. [2005] extend the approach of *DBSCAN* by Ester et al. [1996] by utilizing density-connected sets as metrics for the tree inserting mechanism. Furthermore, other extensions of the *k-means* algorithm introduce a relevance per axis, which are adapted iteratively to identify meaningful combinations of dimensions [de Amorim and Mirkin, 2012].

While being comparable successful in detecting distinct axis-parallel clusters, these methods however fail to identify arbitrary oriented structures, which results in cluster results of lower quality or no results at all [Berkhin, 2006; Kriegel et al., 2009]. A second general issue are the dependencies of parameters for thresholds, quantizations, adaptation strengths and so on, which influence the quality of clusters and runtime behavior greatly [Berkhin, 2006; Kriegel et al., 2009; Parsons et al., 2004]. Due to the gradual transition of both high-dimensional clustering categories, the assignment of available cluster techniques is handled differently by many authors.

2. *Data Model Preprocessing*: As a second track these techniques aim to transform the data model to an equivalent representation, reducing dimensionality, suppressing irrelevancies and noise, and uniforming feature value ranges and domains. For example, the approach of McCallum et al. [2000] divides the data coarsely into overlapping preclusters, called "canopies", before applying known clustering algorithms, which process only data items of common canopies. In this way the data model is transferred to a preclustered state, thus accelerating the overall clustering, but implying a quality of results dependence of the precluster strategy. Another technique is based on local linear embeddings by Roweis and Saul [2000], producing low-dimensional representations of high-dimensional data while preserving the data items' local neighborhood and allowing the application of common cluster algorithms. In the following the three major groups of data model preprocessing approaches are introduced:

    a) *Feature Processing*: These comparable light methods, which process the existing axes of a data model, can be divided in [Kriegel et al., 2009; Parsons et al., 2004]:

        • *Feature Transformation*: This class of techniques can be described as the clustering of the features themselves, applying the same algorithms as for the data clustering, which search for regions of small variance, also known as *co-clustering* [Hartigan, 1975; Kriegel et al., 2009]. In this way the dimensionality can be compressed, combining redundant axes and detecting relevant regions. As a drawback these methods are sensitive to many irrelevant axes and due to the compressed feature space, the final cluster results are hard to interpret [Parsons et al., 2004].

        • *Feature Selection*: The second approach seeks to identify irrelevant axes in advance and to remove them completely. Usually these methods operate entropy-based and follow different strategies, hence it is strongly parameter dependent which features are classified as relevant for clustering [Berkhin, 2006; Parsons et al., 2004].

    Recent preprocessing methods implement both of these strategies, as it was shown that the less the data models contain redundancies and irrelevancies, the less the curse of dimensionality affects the qualities of results [Houle et al., 2010].

b) *Principal Component Analysis (PCA)*: These often utilized methods aim not only to reduce the data model system, but also to transform it to a low-dimensional equivalent. Given a data model in $\mathbb{R}^p$ the purpose of the *PCA* is to project the data model to a space $\mathbb{R}^q$, where $q <<$ $p$, while preserving the metrics between the data points to avoid data loss or distortions [Pearson, 1901; Jolliffe, 2005]. The projection is carried out via the calculation of the covariance matrix of the data, its eigenvalues and normalized eigenvectors respectively. This set of vectors can be interpreted as axes of an ellipsoid embedding the original data [ter Braak, 1986]. Each eigenvalue's fraction of the sum of all eigenvalues expresses how well the according vector represents the underlying data. A representative vector subset as a sufficient projected data model system, can either be selected by choosing those $q$ vectors of the smallest variance or optimized by minimizing the total variance prior to the selection [Hilbert, 1904; Wilkinson, 1965]. The application of this technique was presented by Hestenes [1958] and Lathauwer et al. [2000]. On the one hand the *PCA* exhibits great potential to find meaningful low-dimensional data representations, on the other hand it is limited to features of value domains and suggests a time complexity of at least $O(n^2)$. If it is not combined with further (heuristic) optimizations, it is not well suited for large data sets [Jolliffe, 2005].

c) *Multi-Dimensional Scaling (MDS)*: Similar to the *PCA* this approach seeks to transpose a given high-dimensional data model to a low-dimensional equivalent. Unlike the methods above, it is based on processing the data items' affinities rather than their single feature values, and thus is of more general nature. The techniques can be divided in:

- *Classical* and *Metric MDS*: Also known as the *Togerson Scaling* [Torgerson, 1952] and as the first presentation of *MDS*, it expects an input matrix of dissimilarities (between the data items) and produces a coordinate matrix of chosen dimensionality, minimizing a loss function. As it is based on the eigenvalue decomposition [Wilkinson, 1965], it is similar to the *PCA*. By generalizing the classical approach, metric *MDS* techniques process matrices of weighted data item distances and a set of loss functions [Borg and Groenen, 2005]. For example *Metric MDS* can be applied to the face recognition problem; Bronstein et al. [2005] introduced an even more generalized method to embed high-dimensional surfaces into lower ones while minimizing the distortions.

- *Non-Metric MDS*: These methods extend the *Metric MDS* to avoid affinity-equivalent distance functions, which may distort the results, if defined inappropriately. The correlation of projected distances and affinities assumes a monotonic relation [Kruskal, 1964]: if $aff^{A,B} > aff^{B,C}$ applies for the affinities of the data items $A$, $B$ and $C$, the Euclidean distances within the low-dimensional target system have to satisfy $dist^{A,B} < dist^{B,C}$, otherwise the distances, and thus the position vectors have to be adapted. Most of the approaches determine the projected coordinate vectors in an iterative manner while minimizing the disparities, like the pioneer algorithm by Kruskal [1964]. The *Shepard-Kruskal Algorithmus* [Shepard, 1980] introduced the first general *MDS* method for non-metric data models.

Similar to the *PCA*, the available *MDS* techniques consider all data item pairs during run-time, suggesting a time and memory complexity of $O(n^2)$, excluding their implementation for large data sets without further enhancements. Present approaches utilize sparse data analyzing methods to improve the performance by selecting a subset of the data items, following different strategies. One of the best known method is the *Landmark* algorithm by de Silva and Tenenbaum [2003], which selects a set of "landmarks"[4] as references for the remaining data items, thus reducing the computation time. If the data is available in a graph based representation, other developments utilize the implicit neighbor relations[5] to accelerate the pairwise affinity evaluation [Brandes and Pich, 2007; Yehuda et al., 2002].

A more general approach was presented by Agarwal et al. [2010], introducing a parameter-less framework to handle a large variety of *MDS* problems. It is specialized in transpositions to spherical representations[6] and is proved to converge in every case. The underlying "placecenter" algorithm is based on iterative local improvements, i.e. relocates the data items sequentially to positions, where the user-defined cost functions become minimal. The time complexity of this method is also $O(n^2)$, thus prohibits its application to large data sets.

### 2.1.2.3 Treating Complex Metrics

The issues of *CA*s regarding large data sets (Section 2.1.2.1) and high dimensionality (Section 2.1.2.2) are accompanied by the current trend, that data item features reside in heterogeneous domains of growing complexity [Jain, 2008]. While former *CA* models were mostly located in analysis-friendly spaces of $\mathbb{R}^p$ [Lloyd, 1982], present data sets more often exhibit categorical, string or object based features [Can and Ozkarahan, 1990; Huang, 1997].

To enable *CA* algorithms to process such data, domain compatible metrics have to be provided to compare features and thus the data items. The metrics, determining a distance or an affinity of two items, encompass a wide field [Arkhangel'skii, 1990] and have to meet two conditions: $i$) express the items' or groups' similarities in an uniform manner and $ii$) be as fast as possible.

As advanced cluster techniques even compare data items at least a single time, usually more often[7], inefficient metrics could decrease the algorithm's performance significantly due to the common non-linear time complexity [Berkhin, 2006; Jain, 2008]. If metrics are utilized more than once, available advanced buffering techniques have to be utilized, as storing all comparisons conventionally may rapidly exceed the memory as consequence of the $O(n^2)$ size complexity [Jain, 2008].

The algorithm of Schwartz and Bilsky [1990] shows, that it is possible to compare categorical features by applying *MDS* techniques and ranked based metrics. Jaeger [2008] introduced logistic regression based models to overcome complex categorical features.

---

[4]describing preferably distinct and representative positions within the data model, characterizing the data structure in a global scope

[5]starting from a node, all other nodes, which are reachable via maximal $k$ edges can be regarded as neighbor-related, and thus affine, within the radius $k$

[6]the target system resides on the surface of a $n$-dimensional sphere

[7]compare popular methods like *PROCLUS* by Aggarwal et al. [1999], *CLIQUE* by Agrawal et al. [2005], *ISODATA* by Ball and Hall [1965], *DBSCAN* by Ester et al. [1996], *CURE* by Guha et al. [1998], *k-prototypes* by Huang [1997], *SUBCLU* by Kailing et al. [2004], *CLARANS* by Ng and Han [2002], *DIGNET* by Thomopoulos et al. [1995] and *BIRCH* by Zhang et al. [1996]

As conclusion the utilization of expensive metrics has to be minimized to process large and high-dimensional data sets in sufficiently short time, or more preferably, the data models have to be transposed to forms, which provide more efficient metrics (see *PCA* and *MDS* in Section 2.1.2.2) [Jain, 2008].

### 2.1.3 Approach of the Dimensional Transposition

After reflecting the current challenges of the *CA* in Section 2.1.1 and existing approaches to address them in Section 2.1.2, the following Section 2.1.3.1 summarizes the motivation for a novel *MDS* technique, whose base concept is illustrated in Section 2.1.3.2.

#### 2.1.3.1 Motivation for the Data Transposition

Considering the introduced issues and current developments (Sections 2.1.1 and 2.1.2), an universal data transformation framework is reasonable, which combines data model representations, available cluster techniques and interpretations in a generalized way, i.e. prepare the data for efficient processing, taking into account the following objectives:

1. Operational:

   - *Universality:* The framework should be as general as possible, i.e. basing on user-defined data item metrics and including all feature domains.

   - *Time Complexity:* It should enable the possibility to process large data sets, i.e. the preparation's time complexity has to be significantly below $O(n^2)$, $O(n)$ in the best case.

   - *Storage:* The results should be storable, as the preparation is usually done only once.

   - *Updating:* With regard to the storage aspect, the results have to offer access for updates, if the original databases change, are extended or the user preferences are altered.

   - *Parameter Dependency:* With regard to the *CA*'s complexity, the number of user-defined *MDS* parameters should be restricted to a meaningful maximal set.

   - *Interface:* To easily apply common cluster techniques and other interpretation methods, the access to the prepared data should be as intuitive as possible. To enable different kind of data models, the data import has to be designed as general as possible.

2. Qualitative:

   - *Equivalence:* The distortions caused by the data transposition have to be minimized, preserving the data's meaningfulness.

   - *Prestructuring:* As many *CA*s support precalculated information, the transposition should produce viable data structures.

   - *Dimensionality:* Due to the reasons given above, the data dimensionality has to be reduced to a minimum with regard to the equivalence aspect.

   - *Memory Efficiency:* Considering the time complexity and storage objectives, the results should be as compact as possible.

With regard to the available preprocessing methods (Section 2.1.2.2), the *Metric MDS* approach is selected for several reasons: It is based on an universal metric system to compare data items (universality and interface aspects), the results, a set of position vectors, are easily storable, exportable as well as efficient (storage, interface and memory efficiency aspects) and occurring data or preference changes can be merged into existing results (updating aspect). The underlying *MDS* algorithms should complement the framework's ability to handle large and complex data sets in practical time spans (time complexity aspect) and should be able to reduce the data model to a meaningful low-dimensional expression with sufficiently less distortions (equivalence and dimensionality aspects). The requirement to minimize the runtime makes preliminary data clustering probable, which can be exported to subsequent *CA* methods (prestructuring aspect).

In the following the framework is referred to as the *Dimensional Transposition Framework* (DT Framework). Figure 2.3 shows the context of the approach, illustrating the DT Framework's role as an interface between challenging data models and available *CA* techniques or other data interpretation methods. Considering the objectives defined above, it provides not only a standardized data process method, but should make numerous *MOO* and *CA* problems processable in the first place.

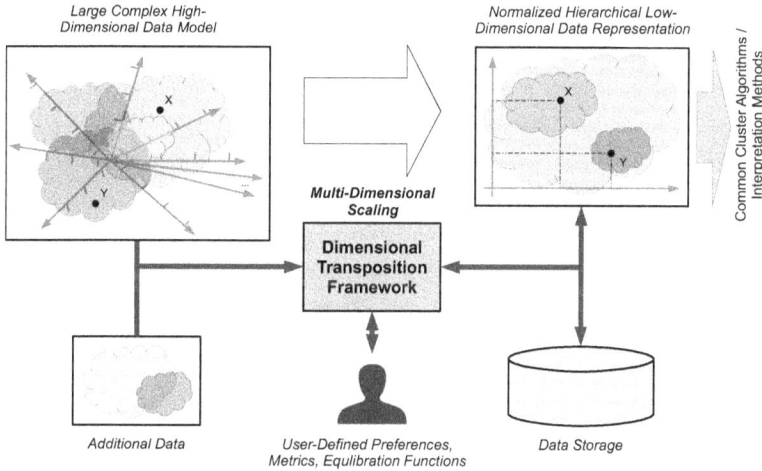

**Figure 2.3:** *Structure of the* DT *Framework;* Given a large and complex high-dimensional data set (top left) the DT Framework transforms the data model to a normalized and (hierarchical) prestructured equivalent of low dimensionality (top right). The method is able to take into account user-defined preferences and definitions, such as metric and equilibration functions (center). The result can be stored and restored, making costly data preparations necessary only once and enabling the import of data changes (bottom right and left). In this way challenging data models can be transformed to a standardized expression, making them processable by common data analysis techniques without adapting these to heterogeneous data model types.

## 2.1.3.2 Base Concept of the Dimensional Transposition Framework

For the core concept and the detailed presentation of the DT Framework in Section 2.3, following terms are defined:

- *Element:* data item, container of properties, cluster content

- *Property:* feature of an element, describing attribute

- *Affinity:* value for the similarity between a pair of elements, basing on their properties

- *Relevance:* value of the affinity's importance with respect to the overall data model

- *Distance:* current Euclidean distance of an element pair within a normalized data model, based on the position vectors

- *Optimal Distance:* optimal distance of an element pair regarding its affinity

- *Transposition Quality:* average deviation of all pairing distances in relation to the optimal distances

- *Weight:* importance of an element with respect to the overall data model

- *Priority:* importance of a property (equal for all elements)

As the eigenvalue based concept (see Section 2.1.2.2) has to be excluded due to its time complexity, the DT Framework utilizes an approach, which initially places the elements randomly and then alters their position vectors iteratively, so that the transposition quality becomes maximal. This repositioning evaluates the affinities of elements to element groups, compares the resulting optimal distance with the current distance and finally applies resulting $\delta$-vectors to the current position vectors. This technique considers the relevances, weights and priorities during each iteration to gradually shift the elements to their final position, so that the applied $\delta$-vectors are in equilibrium, i.e. the Euclidean distances are as close as possible to their optimal values. The underlying precalculated data structure supports the $\delta$-vector evaluation, accelerating the process to be able to handle large data sets and providing subsequent CAs with preclustered structures. As soon as the transposed data representation is found, the original data model can be neglected, only links from the low-dimensional equivalents to their original are preserved, thus reducing the final data amount.

Figure 2.4 shows the base approach of the DT Framework, importing a large complex data model with according property metrics and exporting a standardized data model representation for further processing. The representation is found by applying iteratively affinity based changes, in the following referred to as *forces*, to the position vectors.

**Property Based Data Model**

| | A | B | C |
|---|---|---|---|
| Int Prop. | 99 | 93 | 4 |
| Float Prop. | 0.1 | 0.12 | 0.9 |
| Categorical Prop. | $C_1,C_2$ | $C_2$ | $C_3$ |
| Bool Prop. | true | true | false |
| ... | ... | ... | ... |
| String Prop. | „abc" | „abcd" | „xyz" |

**Dimensional Transposition Framework**

Iterative Metric MDS — Force Based Equilibration

**Affinity Based Data Model**

| | X | Y |
|---|---|---|
| A | 0 | 0 |
| B | 1 | 0 |
| C | 0.5 | 10 |

Cluster Analysis / Interpretation Methods

*Metric Interface*
Property — Distance — Property

*Vector Interface*
Element ↔ Position Vector

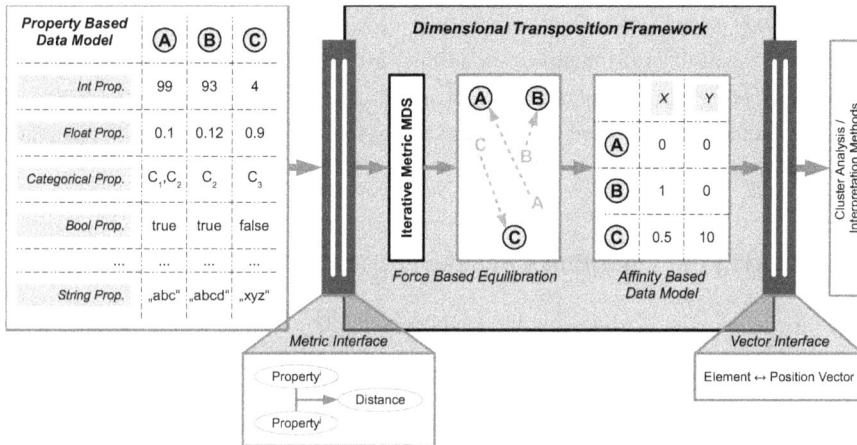

**Figure 2.4:** *Base Approach of the* DT *Framework;* The elements **A**, **B** and **C** contain several properties of different types and ranges. An example data model is given as a *property based* description, providing sets of properties for each data item. The DT Framework requires a *Metric Interface*, which offers relations between a property pair of one kind and an according distance. Basing on this information the *MDS* creates an *affinity based* expression of the given data model, shifting the elements iteratively to their equilibration positions. As shown by the property values, **A** differs only slightly from **B** in contrast to **C**. This leads to resulting position vectors, locating **A** and **B** closely and **C** distantly due to their overall affinities. Subsequent *CA* methods can access the position vectors via the *Vector Interface* as a low-dimensional, property independent data model equivalent.

## 2.2 Methods and Materials

In the following the testing and analysis strategy for the proposed *MDS* technique (Section 2.1.3) is presented. First, Section 2.2.1 describes which algorithms are compared, followed by the introduction of the evaluated measurements in Section 2.2.2. Subsequently Section 2.2.3 defines the parameter ranges during the testing, i.e. the considered data models and algorithm configurations. The last two Sections 2.2.4 and 2.2.5 describe the designed test scenarios for the final performance analysis and the development process respectively.

The DT Framework was implemented in C# .NET 4.5 as a transparent and well developed programming language [C#.NET, 2001].

### 2.2.1 Algorithm Analysis

As presented in Section 2.1 no comparable algorithms or techniques exist to transpose a cluster analysis problem of different value types and ranges to an uniform space. Hence two algorithms within the DT Framework are compared: First the *Equilibration Algorithm* (EQ Algorithm) (Section 2.3.2.2), which compares the given data items in pairs and repositions them gradually to find a proper representation within the targeted low-dimensional space. Although slow in execution this algorithm is able to find solutions close to the optimum, as it will be shown later in the measurement results (Section 2.3.3). Second, the *Dimensional Transposition Algorithm* (DT Algorithm) is tested, extending the technique of the EQ Algorithm by utilizing several heuristic improvements to accelerate the process at the costs of lower transposition quality.

For the measurements of the transposition qualities and runtimes (Section 2.2.2), both algorithms process identical tests, consisting of different data models and algorithm configurations (Section 2.2.3). During each test the data model and configuration parameters were varied within different ranges and domains step-wisely (Section 2.2.4).

## 2.2.2 Measuring the Algorithm's Results and Performance

To detect the transposition qualities and the runtimes of the DT Algorithm in comparison to EQ Algorithm, the following values were measured or evaluated. Table 2.1 shows an overview of the single measurements, detailed explanation can be found below.

| Measurement | Symbol | Description | Unit of Measurement |
|---|---|---|---|
| Runtime | $t_{run}$ | algorithm's execution time | *seconds* |
| Average Quality | $quality_{average}$ | overall transposition quality | - |
| Worst Quality | $quality_{worst}$ | worst single transposition quality | - |
| Quality Std. Deviation | $quality_{SD}$ | transposition quality standard deviation | - |
| Runtime Slope | $slope_{runtime}$ | relative runtime increment | *second/MSI* |

**Table 2.1:** *Measurements / Evaluations during* DT *Framework Testing*; Shown are the names and the symbols of the single measurements in combination with short descriptions and the according units.

The algorithm's runtime $t_{run}$ represents the total execution duration, starting after the data model creation and ending before the transposition quality evaluation. In this way it describes the total time costs of the chosen algorithm.

As an expression of the transposition quality the combination of the average ($quality_{average}$), the worst single ($quality_{worst}$) and the quality's standard deviation ($quality_{SD}$) is utilized:

$$quality_{average} = \frac{\sum_{e^i,e^j,i\neq j} quality^{e^i,e^j}}{n_{element} \cdot (n_{element} - 1)}$$
$$quality_{worst} = min(quality^{e^0,e^1}, .., quality^{e^{n-1},e^n})$$
$$quality_{SD} = \sqrt{\frac{\sum_{e^i,e^j,i\neq j} (quality_{average} - quality^{e^i,e^j})^2}{n_{element} \cdot (n_{element} - 1)}}$$

$$(2.4)$$

where $quality^{e^i,e^j}$ is the single transposition quality of two elements $e^i$ and $e^j$ within the transposed data model (Formula (2.11)) and $n_{element}$ is the number of items to transpose. In this way $quality_{average}$ describes the overall transposition quality and lays in the range of 0 (transposition result does not represent the items' relations) and 1 (transposition result represents the items' relations ideally). On the other side $quality_{worst}$ stands for the worst single transposition, which may result in inappropriate cluster results of subsequent *CA*s. The more it differs from $quality_{average}$, the more a later cluster assignment is probably inappropriate. Similar $quality_{SD}$ describes the mean variation around the average value, i.e. the smaller the standard deviation is, the more $quality_{average}$ represents the overall transposition quality.

To evaluate the time complexity when processing larger data models, the runtime slope is introduced:

$$slope_{runtime} = \frac{1}{2} \left( \frac{t^i_{run} - t^{i-1}_{run}}{size^i - size^{i-1}} + \frac{t^{i+1}_{run} - t^i_{run}}{size^{i+1} - size^i} \right) \quad (2.5)$$

where $size^i$ describes the data model size dependent on a chosen parameter with the index $i$ (e.g. the number of items $n_{element}$) and $t^i_{run}$ stands for the according runtime while processing the data model. In the following chosen size parameter is abstracted to the *Model Size Increment* (*MSI*). Calculating the relative differences between the previous and the next data model size step provides an expression of the runtime increment, while constant values would describe the best possible case (a linear runtime increment at a linear data model size increment). All other cases indicate impractical runtimes above a certain data model size.

### 2.2.3 Parameters of the Data Models and Algorithm Configuration

To detect the quality and runtime sensitivities, several data model and algorithm parameters are varied as shown in Table 2.2. A detailed explanation can be found below.

| Data Model Parameter | Symbol | Description | Range |
|---|---|---|---|
| Element Count | $n_{element}$ | number of elements | $1k..300k$ |
| Property Count | $n_{property}$ | number of properties | 2..10 |
| Dimensionality | $n_{dimension}$ | position vector size | 2..10 |
| Data Model Type | $type_{distribution}$ | property value distribution method | *DVD, CVD* |
| **Algorithm Parameter** | | | |
| Simulated Annealing Factor | $f_{SA}$ | degree of global scaling decrease, maximal transposition range | 0.5..1.0 |
| Equilibration Strategy | $type_{algorithm}$ | chosen equilibration strategy | EQ, DT |
| Optimal Branch Size | $size_{opt}$ | desired count of subbranches per branch (only DT Alg.) | 4..16 |

**Table 2.2:** *Varied Parameters between Algorithm Runs;* Shown are the names and symbols of the single parameters in combination with short descriptions and typical value ranges during the tests.

The data model size is defined by the parameter $n_{element}$, which represents the number of elements to be transposed. Furthermore, $n_{property}$, the number of properties per element, describes the characteristics according to which the transposition takes place. It is desired to handle large numbers of elements in combination with a relatively large numbers of properties.

The dimensionality of the targeted uniform space is defined by $n_{dimension}$ as the number of unitless and orthogonal axes. The more axes reside within the target space, the larger is the degree of freedom granted to the algorithm to find suitable transposition results, but the more limited are the capabilities of subsequent *CA* [Indyk and Motwani, 1998]. Hence a practicable compromise has to be found.

The last data model parameter $type_{distribution}$ defines the value distribution method for single property values. To cover a large field of possible distributions, two corner cases are selected:

- the *Distinct Value Distribution* (*DVD*) type, which divides a property's value range in at least two narrow regions and distributes the values of single properties randomly to one of these regions.

- the *Continuous Value Distribution* (*CVD*) type, which covers the entire value range of a property evenly and distributes the values of single properties randomly within this range.

Table 2.3 shows examples of both value distributions types in comparison to its allowed value range. Assuming an according property describes the degree of coloring, the *DVD* type would generate black and white elements in contrast to the *CVD* type, which would result in elements of every shade of gray.

| Distribution Type | Allowed Range | Distribution Range(s) |
|---|---|---|
| DVD | 0.0..1.0 | 0.0 or 1.0 |
| CVD | 0.0..1.0 | 0.0..1.0 |

**Table 2.3:** *Examples of the Utilized Value Distribution Types*

In this way both types stand for the extrema of conceivable distribution types and it is assumed that if the algorithm is capable to handle them, it can also handle mixed distribution types. Furthermore, the resulting artificial data models of arbitrary sizes are easily reproducible by other authors, thus making methods comparable, which utilize them.

The second group of parameters are for the algorithms. Their configuration is determined mainly by the *SA* factor $f_{SA}$ and the parameter for equilibration strategy $type_{algorithm}$. The *SA* factor defines the accuracy of the algorithm process by setting the dynamic step size and the number of iterations (Formulas (2.18) and (2.23). The closer $f_{SA}$ is to 1, the more accurate the algorithms equilibrate. A value of 1 would mean maximal accuracy in the scope of the target system and infinite runtime.

Furthermore, $type_{algorithm}$ differs between the more accurate, but slow equilibration core algorithm (EQ Algorithm, Section 2.3.2.2) and its fast, but less accurate heuristic variant (DT Algorithm, Section 2.3.2.10) to transpose the given data model.

If the DT Algorithm is selected, an additional parameter, the optimal branch size ($size_{opt}$) of the internal tree becomes relevant. It defines the structure of the preprocessed data model representation and hence the behavior of the algorithm.

## 2.2.4 Exploring the Parameter Space

To test and evaluate the DT Algorithm (Section 2.2.1), multiple tests are performed: Each test consists of a data model and an algorithm parameter set (Section 2.2.3). To compensate initialization dependent scatterings, for each configuration multiple runs with different random seeds are performed. Finally the test measurements (Section 2.2.2) are evaluated and stored for later analysis.

The tests are grouped in several scenarios as shown in Table 2.4. A scenario consists of fixed and varied parameters and includes the evaluation of different measurements. The purpose of each test scenario is to detect or validate performance indicators and their dependencies of given parameters. In some cases scaling behaviors will be investigated if dependencies exist.

The scenarios 1 and 2 are designed to detect possible transposition qualities and sufficient values for the *SA* factor and the branch size parameters. For this purpose, data models of different sizes, property counts and value distribution types are processed by both algorithms, utilizing different *SA* factors. The so far unknown optimal branch sizes are set to constant average values. The comparison of both algorithm results yields a minimal *SA* factor, which is applied during the subsequent tests and referred to as $f_{SA}^{best}$. Ensuring that no quality distortions are caused by dimensional reductions, the axes count $n_{dimension}$ is set to the same value as the property count, providing a sufficient large degree of freedom to the algorithms.

Second, after the determination of a suitable *SA* factor, the branch sizes are altered step-wisely while processing data models with the same configurations. As tests during the development process showed, setting the other tree defining parameters (Section 2.3.2.3) in dependence of the optimal branch size is

| ID: Scenario | Varied Parameter | Step Size | Fixed Parameter | Measurements |
|---|---|---|---|---|
| 1: Optimal *SA* Factor Detection | $n_{element} = 100..1k$<br>$n_{property} = 3..10$<br>$f_{SA} = 0.5..0.95$<br>$type_{algorithm} = EQ / DT$<br>$type_{distribution} = DVD, CVD$ | 100<br>1<br>0.05 | $n_{dimension} = n_{property}$<br>$size_{opt} = 8$ | $quality_{average}$<br>$quality_{worst}$<br>$f_{SA}^{best}$ |
| 2: Optimal Branch Size Detection | $n_{element} = 100..1k$<br>$n_{property} = 3..10$<br>$size_{opt} = 4..16$<br>$type_{algorithm} = DT$<br>$type_{distribution} = DVD, CVD$ | 100<br>1<br>1 | $n_{dimension} = n_{property}$<br>$f_{SA} = f_{SA}^{best}$ | $quality_{average}$<br>$quality_{worst}$<br>$size_{opt}^{best}$ |
| 3: Dimension Reduction Validation | $n_{property} = 3..10$<br>$n_{dimension} = 2..10$<br>$type_{algorithm} = EQ / DT$<br>$type_{distribution} = DVD, CVD$ | 1<br>1 | $n_{element} = 1k$<br>$f_{SA} = f_{SA}^{best}$<br>$size_{opt} = size_{opt}^{best}$ | $quality_{average}$<br>$quality_{worst}$<br>$quality_{SD}$ |
| 4: Runtime Dependency | $n_{element} = 5k..50k$<br>$n_{property} = 3..10$<br>$type_{distribution} = DVD, CVD$ | 5k<br>1 | $n_{dimension} = 3$<br>$type_{algorithm} = DT$<br>$f_{SA} = f_{SA}^{best}$<br>$size_{opt} = size_{opt}^{best}$ | $t_{run}$<br>$slope_{runtime}$ |
| 5: Large Scale Behavior | $n_{element} = 25k..500k$<br>$type_{distribution} = DVD, CVD$ | 1k | $n_{property} = 10$<br>$n_{dimension} = 3$<br>$type_{algorithm} = DT$<br>$f_{SA} = f_{SA}^{best}$<br>$size_{opt} = size_{opt}^{best}$ | $t_{run}$<br>$slope_{runtime}$ |

**Table 2.4:** *Test Scenarios for* DT *Framework*; shown are configurations of the DT Framework test scenarios, to detect suitable parameters and analyze the influence of the parameters described in Section 2.2.3 on the quality and performance indicators as introduced in Section 2.2.2.

a suitable strategy to reduce the search space and avoid misshaped trees[8], hence only $size_{opt}$ is altered. The value determined in this way is applied to the subsequent tests and referred to as $size_{opt}^{best}$.

Scenario 3 considers the issue to ensure sufficient transposition qualities while transferring data models with high property counts to low-dimensional spaces. Utilizing the algorithm configurations found in the scenarios 1 and 2, both algorithms process data models, while varying the number of properties as well as the target dimensions and applying both distribution types. The subsequent analysis and comparison of the results reveal the achievable quality as well as the quality differences and deviation between and of both algorithms.

Finally the scenarios 4 and 5 treat the runtime determination and the runtime scaling behavior of the DT Algorithm. This is realized in scenario 4 by transposing data models of different sizes (regarding element and property counts) and distribution types to a low-dimensional target space, utilizing the former detected suitable algorithm configuration. In contrast scenario 5 sets the number of properties to a fixed value and alters only the number of elements within a large range for both distribution types. The transposition quality evaluation of both these scenarios is avoided, as according to Formula (2.4) the calculation of the average qualities implies quadratically increasing computing times and thus impractical runtimes for the entire tests.

To suppress the impact of randomness onto the results, the number of runs per configuration is always set to five, i.e. each data model and algorithm configuration is instantiated and processed five times with different seeds. The different measurement values serve as base to calculate average result values and

---

[8]this term describes tree characteristics, which are not suitable for the purpose to accelerate the algorithm; in this case too narrow (large tree depth) or too wide branches (small tree depth) would have a negative impact on the algorithm's performance

standard deviations for each configuration set.

The tests were carried out in parallel on a *Windows 7 Professional* workstation, hosting an *Intel Core* $i7-2760QM$ CPU with access to $12$ GB of RAM. The results were collected automatically into detailed logs, which serve as the charts' database for the final analysis in Section 2.3.3.

### 2.2.5 Algorithm Development Procedure

To analyze the algorithm performances during development, a profiling mechanism was introduced, which logs the CPU ticks at every method entry end exit along with the calling method's name. After a test run completion a log provides precise information about the time costs of every code block together with the source of calling. These logs represent the analysis tool for the identification of the performance bottlenecks and the development.

For the step-wise improvement of the EQ the DT Algorithm, an iterative development approach was chosen: As the execution of all tests within the test scenarios (Section 2.2.4) after each progression step would be too extensive, a subset of them is selected and used during the development. The selected test are shown in Table 2.5.

| Scenario | Varied Parameter | Fixed Parameter | Evaluations |
|---|---|---|---|
| Development | $n_{element} = 10k; 20k; 40k;$ $f_{SA} = 0.5 .. 0.9$ $type_{distribution} = DVD / CVD$ $type_{algorithm} = $ EQ / DT | $n_{property} = 3$ $n_{dimension} = 2$ $n_{seed} = 3$ $size_{opt} = 8$ | $t_{run}$ $quality_{average}$ $quality_{SD}$ |

**Table 2.5:** *Test Configurations utilized during the* DT *Framework Development*; Shown are the data models, which were transposed and evaluated during the development process for validation and improvement of the algorithms.

The development tests, containing data models of different sizes as well as property counts and applying both distribution types, are processed by both algorithms with varying accuracy in a comparable short time span, which makes them suitable for repeated execution.

After each development step the DT Framework processed these data models and was analyzed by the profiling method explained above. By comparing the measurements to the results of the previous development iterations, the current bottlenecks were identified. Subsequently an improvement was conceptually developed and implemented, after which a repeated profiling run and check against the results of the EQ Algorithm validates time cost savings and whether or not the results are identical. If the validation was successful, the changes became permanent, otherwise they were discarded.

## 2.3 Results

In the following Section 2.3.1 gives a brief insight in the profiling results as development tool. Subsequently in Section 2.3.2 the actual DT Framework is presented, starting with the approach's bases, introducing the main algorithm and enhancements, right through to the final *MDS* framework. The measurement results of the final software are illustrated in Section 2.3.3 with respect to the testing strategy previously described in Section 2.2. Finally the results are summarized in Section 2.3.4.

## 2.3.1 Runtime Profiling

To give an example of the evaluated data during the development process, in the following truncated runtime profile logs of both the EQ and the final version of the DT Algorithm are shown in the Tables 2.6 and 2.7. As the algorithms process the same transposition task, the tables show the absolute and relative time costs per method and its callers. Extended versions of the tables can be found in the appendix in Section 6.1.1.

| Method Name | Time Costs | Total Tick Count | Ticks / Call | Calling Method | Total Callings | Ratio: Callings |
|---|---|---|---|---|---|---|
| CompareProperties | 49,10 % | 2.683.616.643 | 92,63 | | | |
| | | | | Leaf.CompareProperties | 28.971.000 | 100,00 % |
| CalculateSpecificWeight | 14,67 % | 802.017.010 | 27,68 | | | |
| | | | | Leaf.CalculateSpecificWeight | 28.971.000 | 100,00 % |
| Property.CompareTo | 11,52 % | 629.655.988 | 1,09 | | | |
| | | | | CompareProperties | 579.420.000 | 100,00 % |
| CalculateTotalForce | 9,15 % | 500.228.393 | 17,27 | | | |
| | | | | CalculateForce | 28.971.000 | 100,00 % |
| CalculateSquareDistance | 4,36 % | 238.352.880 | 8,20 | | | |
| | | | | CalculateDistance | 29.062.000 | 100,00 % |
| .. | .. | .. | .. | .. | .. | .. |

**Table 2.6:** EQ *Algorithm Example Profiling Log*;
*Data Model*: 1.000 elements with 20 features, distributed in 10 Gaussian fields of different sizes;
*Algorithm*: EQ Algorithm;
Only the first five entries are shown; the left side represents the time costs of all involved methods, the right side shows the calling distribution of the according method.

The profiling shows that half of the computing time is spent in the method `CompareProperties`, which calculates the affinity of an element pair, based on the user-defined affinity functions. With regard to the fact, that this method only iterates the element features and calls the according external compare functions, it indicates that a significant portion of the runtime is spent outside the framework. The processed element count of 1.000 implies a number of $10^6$ element comparisons per iteration, which leads to almost $30 \cdot 10^6$ affinity calculations in total, if an iteration count of 30 is assumed. The property compare method `CompareTo`, called by `CompareProperties`, reflects with approximately $0.5 \cdot 10^9$ calls the treatment of 20 properties per element. In combination with the weight calculation it can be concluded, that more than 75% of the computing time is required to calculate the relations between the elements.

These insights form the base for the subsequent algorithm enhancements. By investigating these parts of the initial implementation the development of the DT Algorithm was started and iteratively improved, utilizing repeated profiling runs. Table 2.7 illustrates the runtime distribution for the final version of the transposition algorithm.

The table shows, that now approximately one third of the computation time is occupied by the `CompareProperties` method, but with significant lesser calls, i.e. only 2, 3% in comparison to the call count of the original implementation. In the same way the call counts of the other methods are reduced significantly. None of the remaining methods cause obvious runtime peaks with respect to their tasks, i.e. the computation effort is distributed comparable equally, providing only little potential for further enhancements. This example illustrates the reduction of the computation time by the utilization of heuristics to decrease the number of method calls.

| Method Name | Time Costs | Total Tick Count | Ticks / Call | Calling Method | Total Callings | Ratio: Callings |
|---|---|---|---|---|---|---|
| CompareProperties | 36,57 % | 66.617.105 | 97,77 | | | |
| | | | | Branch.CompareProperties | 38.839 | 5,70 % |
| | | | | Leaf.CompareProperties | 642.504 | 94,30 % |
| Property.CompareTo | 10,98 % | 20.007.863 | 1,47 | | | |
| | | | | CompareProperties | 13.626.860 | 100.0 % |
| CalculateSpecificWeight | 10,06 % | 18.372.437 | 29,19 | | | |
| | | | | CalculateSpecificWeight | 629.455 | 100.0 % |
| CalculateTotalForce | 8,39 % | 629.455 | 24,27 | | | |
| | | | | CalculateForce | 629.455 | 100.0 % |
| CalculateSingleForce | 4,83 % | 8.801.504 | 6,90 | | | |
| | | | | CalculateDistance | 1.275.743 | 100.0 % |
| .. | .. | .. | .. | .. | .. | .. |

**Table 2.7:** DT *Algorithm Example Profiling Log*;
*Data Model*: 1.000 elements with 20 features, distributed in 10 Gaussian fields of different sizes;
*Algorithm*: DT Algorithm;
Only the first five entries are shown; the left side represents the time costs of all involved methods, the right side shows the calling distribution of the according method.

## 2.3.2 The Dimensional Transposition Framework in Detail

Here the significant components of the DT Framework are presented. Section 2.3.2.1 starts with the property and element comparison techniques, forming the base of the *MDS*. The resulting EQ Algorithm in Section 2.3.2.2 represents the core concept and is classified as a precise, but slow variant of the DT Algorithm. The three following Sections describe the bases of the subsequent algorithm enhancements, i.e. the internal tree construction in Section 2.3.2.3, the branch radius calculation in Section 2.3.2.4 and also the data model representation in a such a hierarchy in Section 2.3.2.5. These data preprocessing methods enable an accelerated version of the EQ Algorithm in Section 2.3.2.6. The next three Sections 2.3.2.7 - 2.3.2.9 illustrate general heuristics to improve the tree-based EQ Algorithm even further. The final DT Algorithm in Section 2.3.2.10 represents a fast , but slightly less precise *MDS* techniques in comparison to the EQ Algorithm.

Figure 2.5 illustrates the dependencies of the introduced components within the DT Framework.

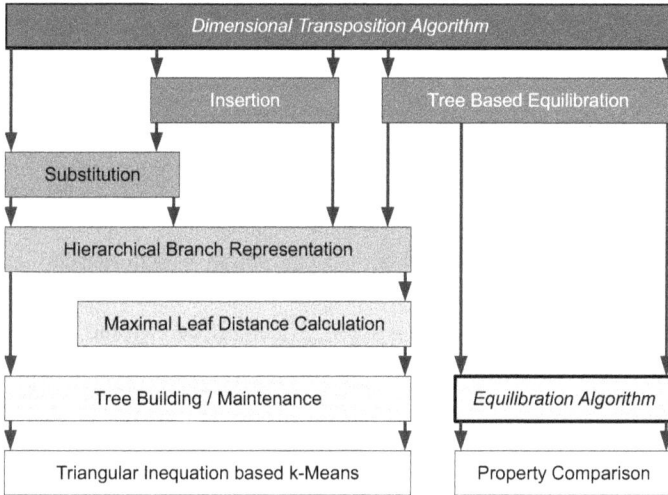

**Figure 2.5:** DT *Framework's Component Dependencies;* On the right side the *Property Comparison* (*PC*) component, which calculates affinities and representative distances between elements, represents the base of the EQ Algorithm. This subalgorithm alters the elements' position vectors iteratively regarding their affinities, until they are arranged in representative distances to each other (Sections 2.3.2.1 and 2.3.2.2).

On the left side an enhanced version of the *k-Means* algorithm, making use of the *triangle inequality* [Elkan, 2003; Inequality, 2014], forms the base of the *Tree Building* (*TB*) and *Tree Maintenance* (*TM*) mechanisms, which provide functionalities to create and maintain hierarchical data structures (Sections 2.3.2.3 and 2.3.2.7). The *Maximal Leaf Distance* (*MLD*) calculation module utilizes these structures to evaluate metrics of the tree's branches (Section 2.3.2.4). Each of the *TB*, *TM* and *MLD* component are required by the *Hierarchical Branch Representation* (*HBR*) concept, which encapsulates the elements' properties and positions hierarchically (Section 2.3.2.5).

Utilizing the *HBR* during the EQ process, the *Tree Based Equilibration* (*TB-EQ*) provides a faster, but more inaccurate technique to find equilibration states (Section 2.3.2.6). In addition to that the *Substitution* (*SUB*) and *Insertion* (*INS*) components improve the handling of already found affine elements or branches and the initial element positions before the equilibration process starts (Sections 2.3.2.8 and 2.3.2.9). The combination of the last three components forms the *Dimensional Transposition Algorithm* (DT Algorithm), which equilibrates given elements regarding their properties (Section 2.3.2.10). The EQ and the DT components are marked by thick frames, representing the algorithm variations, which were compared during the analysis (Section 2.3.3).

## 2.3.2.1 Property Comparison as Computing Base

The motivation of the *Property Comparison* (*PC*) concept is to provide an universal functionality to compare elements, which is required by the subsequent algorithm components. The concept assumes that $i$) each element consists of a set of *properties*, describing the element, $ii$) the sets are equal in length and order and $iii$) the user provides compare functions for each property type and a global optimal distance function. The compare function can be declared as:

$$aff^{prop_k^i, prop_k^j} = Func_{aff}^k(prop_k^i, prop_k^j) \qquad (2.6)$$

For a pair of properties $prop_k^i$ and $prop_k^j$ of the property type $k$ a user-defined compare function $Func_{aff}^k(..)$ is required, which returns an *affinity* value $aff^{prop_k^i, prop_k^j}$ between $0$ (maximal dissimilar) and $1$ (identical).

Hence the affinity $aff^{e^i,e^j}$ of two elements $e^i$ and $e^j$ is defined as:

$$aff^{e^i,e^j} = \frac{\sum_{k=0}^{k_{max}} \left( aff^{prop_k^{e^i}, prop_k^{e^j}} \cdot priority^k \right)}{\sum_{k=0}^{k_{max}} priority^k} \tag{2.7}$$

where $k_{max}$ is the length of the property sets, $aff^{prop_k^{e^i}, prop_k^{e^j}}$ is the affinity of the $k$-th property of $e^i$ and $e^j$ (Formula (2.6)) and $priority^k$ is a user-defined weighting factor for property $k$. In this way the compared properties of the elements $e^i$ and $e^j$ define the affinity of both elements, utilizing the user-defined compare functions and priorities.

Given the pair's affinity, the *optimal distance* $dist_{opt}^{e^i,e^j}$ between the elements $e^i$ and $e^j$, describing an Euclidean distance in the target space, is defined as:

$$dist_{opt}^{e^i,e^j} = Func_{dist_{opt}}(aff^{e^i,e^j}) \tag{2.8}$$

where $Func_{dist_{opt}}(..)$ stands for a global user-defined function, which takes the elements' affinity $aff^{e^i,e^j}$ (Formula (2.7)) as argument and returns a value between 0 (elements are identical regarding their properties) and $dist_{opt}^{max}$ (elements are maximal dissimilar). For cases where the property value pairs have less or no influence onto the affinity and hence the optimal Euclidean distance of the elements in the target space, the *relevance* is introduced, describing the properties' relative importance and thus providing a greater degree of freedom to the element arranging subalgorithms:

$$rel^{prop_k^i, prop_k^j} = Func_{rel}^k(prop_k^i, prop_k^j) \tag{2.9}$$

For a pair of properties $prop_k^i$ and $prop_k^j$ of the property type $k$, a user-defined relevance function $Func_{rel}^k(..)$ is required, which returns a relevance value $rel^{prop_k^i, prop_k^j}$ as a scalar factor ($>0$). The relevance $rel^{e^i,e^j}$ of two elements $e^i$ and $e^j$ is defined as:

$$rel^{e^i,e^j} = \frac{\sum_{k=0}^{k_{max}} \left( rel^{prop_k^{e^i}, prop_k^{e^j}} \cdot priority^k \right)}{\sum_{k=0}^{k_{max}} priority^k} \tag{2.10}$$

where $k_{max}$ is the length of the property sets and $rel^{prop_k^{e^i}, prop_k^{e^j}}$ is the affinity of the $k$-th property of $e^i$ and $e^j$ (Formula (2.9)). In this way the properties of the elements $e^i$ and $e^j$ define also their relevance for the resulting Euclidean distance in the target space.

The outcome of the *PC* concept is the reduction of an arbitrary amount of element properties of different value types and ranges to an universal representation. This representation is a space of low dimensionality with unit-less axes, in which the elements are positioned and where their distances to each other describe their property-based affinities. That means the single properties are encapsulated in the target system and the elements' affinities are computable as Euclidean distances.

To illustrate the influence of the elements' affinities and relevances (Formulas (2.7) and (2.10)) onto the resulting optimal distances (Formula (2.8)), Figure 2.6 shows an example constellation of several elements, utilizing an optimal distance function which returns small distances for high affinities. While affinity and relevance values close to 1 cause small element optimal distances (elements are affine), elements with affinities close to 0 are placed far distant (elements are dissimilar). For small relevances the optimal distances become irrelevant, i.e. the later introduced arrangement mechanism has a greater degree of freedom. In this way the figure shows a representation of the element relations, utilizing only spatial element positioning to describe the affinities of the elements.

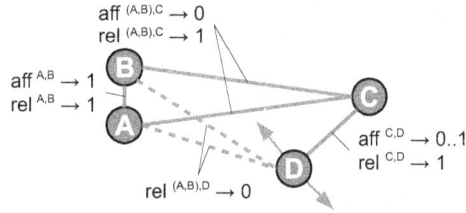

**Figure 2.6:** *PC Concept Example;* Shown are the elements **A**..**D** arranged regarding the affinities and relevances of their properties (not shown). While the affinity $aff^{i,j}$ (Formula (2.7)) of element $i$ and $j$ causes an optimal element distance (regarding the given user functions, Formula (2.8)), the relevance $rel^{i,j}$ indicates the priority according to which the optimal distance should be established (Formula (2.10)). This example utilizes an optimal distance function which returns small distances for high affinities. From the close neighborhood of **A** and **B** can be concluded, that the affinity of their properties tends to 1, while the large distances to element **C** indicates affinities of around 0 of **A** and **B** on one side and **C** on the other. The affinity of **C** and **D** is between 0 and 1, which results in a mid-range distance. In contrast to the relevances of 1 for all other relations, the relevance between **D** and **A** (or **B**) is 0, i.e. their optimal distances are irrelevant for the sought representation and **D** should be placed only in the desired distance to **C** (marked by the dotted circle).

To evaluate the meaningfulness of a given element arrangement, the *transposition quality* can be described as follows:

$$aff^{e^i,e^j}_{transp.} = \overline{Func_{aff}}(dist^{e^i,e^j})$$

$$quality^{e^i,e^j} = \frac{min(aff^{e^i,e^j}, aff^{e^i,e^j}_{transp.})}{max(aff^{e^i,e^j}, aff^{e^i,e^j}_{transp.})} \quad (2.11)$$

The *transposition affinity* $aff^{e^i,e^j}_{transp.}$ of two elements $e^i$ and $e^j$ is defined by an user-provided inverse affinity function $\overline{Func_{aff}}(..)$, taking the Euclidean distance $dist^{e^i,e^j}$ as argument and returning an affinity value as equivalent for the provided distance. The quality of the transposition for these two elements, $quality^{e^i,e^j}$, is defined by the quotient of the property based and the distance based affinity, $aff^{e^i,e^j}$ and $aff^{e^i,e^j}_{transp.}$, i.e. the closer the quotient is to 1, the better the element positions represents the elements' affinity. The terms $min(..)$ and $max(..)$ define functions which return the minimal and maximal values of the given arguments respectively.

To illustrate a typical form of the user provided functions, Figure 2.7 shows plots of functions defined as:

$$Func_{aff}(prop^i, prop^j) = 1 - (value_{prop^i} - value_{prop^j})^2 \quad (2.12)$$

$$Func_{dist_{opt}}(aff) = 1/aff - 1 \quad (2.13)$$

where $Func_{aff}(..)$ describes the affinity of two properties $prop^i$ and $prop^j$ (see also Formula (2.6)), and $Func_{dist_{opt}}(..)$ stands for a global optimal distance function, according to which the elements should be arranged (see also Formula (2.8)).

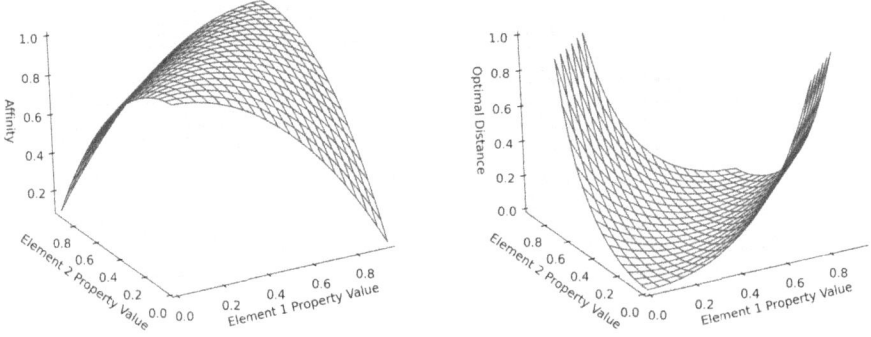

**Figure 2.7:** *Visualization of Example User Functions*; Shown are the surface plots of the example affinity function (Formula (2.12)) and the optimal distance function (Formula (2.13)). The left side depicts the affinity of two values, whose range is between 0 and 1. For nearly equal values the affinity approaches 1 (diagonal ridge), otherwise it decreases to 0. The right side shows a fraction of the optimal distance behavior resulting from the example definitions and the property values as input. The optimal distance tends to 0, if the values are equal (diagonal valley) and to $\infty$ ($dist_{opt}^{max}$), if the difference increases to its maximum (1).

### 2.3.2.2 The Equilibration Algorithm as Core of the Framework

To create an element representation, describing the elements' affinities by spatial element positions and resulting distances (Section 2.3.2.1), the *Equilibration Algorithm* (EQ Algorithm) is introduced. Its task is to find an arrangement, so that the average transposition quality ($quality^{e^i,e^j}$) of all element pairs $e^i$ and $e^j$ is maximal, i.e. close to 1 (Formula (2.11)).

The approach is to assign the elements initially to random start positions within the target space and then move the elements iteratively along their affinity gradients until an equilibrium,i.e. no further improvements are possible, is found. The affinity inflicted movement vector is referred to as *force*. The force of a single element to another is equivalent to their difference vector, whose magnitude is defined by the current (based on the current position vectors) and the optimal distance (based on the property affinity, Formula (2.8)). Hence the *force base scaling* for a single element $e^i$, inflicted by a target element $e^j$, is introduced as:

$$scaling_{base}^{e^i,e^j} = \frac{dist^{e^i,e^j}}{dist_{opt}^{e^i,e^j} + dist^{e^i,e^j}} - \frac{1}{2} \tag{2.14}$$

where $dist^{e^i,e^j}$ is the current Euclidean distance of the elements $e^i$ and $e^j$, derived from their position vectors $\overrightarrow{pos}^{e^i}$ and $\overrightarrow{pos}^{e^j}$, and $dist_{opt}^{e^i,e^j}$ is their optimal distance, defined by the affinity of their properties (Formula (2.8)). Figure 2.8 shows that the resulting value varies between $+0.5$ for a current distance smaller than the optimal distance, and $-0.5$ for the contrary. If the current distance becomes equal to the optimum, the force base scaling is 0, i.e. currently there exists no force between by this element pair. In this way $scaling_{base}^{e^i,e^j}$ reflects that the elements approach to each other ($>0$), move apart ($<0$) or are in an equilibrium (=0).

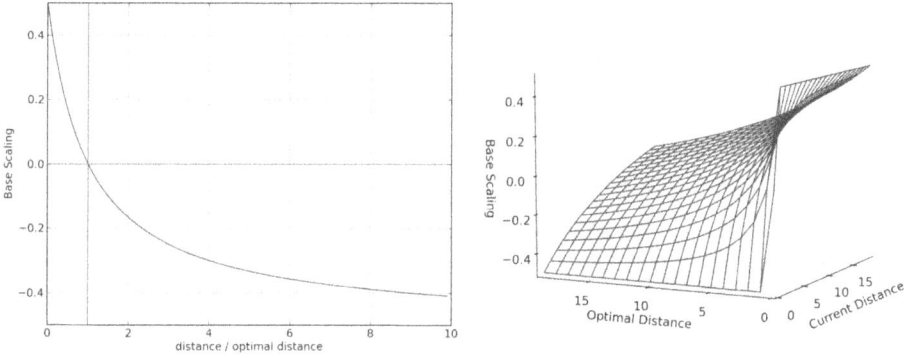

**Figure 2.8:** *Visualization of the Force Base Scaling Function*; Shown are plots of the base scaling function, defining the EQ Algorithm's behavior. The left side depicts the function dependent on the ratio of the current distance of two elements and their optimal distance (Formula (2.14)), which results from their affinity (Formula (2.8)). The scaling tends to 0, if the current distance is equal to the optimal distance. In the other cases it approaches to +0.5 or −0.5, reflecting the EQ Algorithm's tasks to move the elements towards each other or move them apart respectively. The right side shows the scaling influenced by both distances independently.

The maximal and minimal values of +0.5 and −0.5 are chosen with regard to the fact, that the forces are also calculated for the target element, hence the half of the difference vector as maximum for the force is sufficient. Taking into account the elements' relevance (Formula (2.10)) the resulting *force scaling* of an element pair $e^i$ and $e^j$ is:

$$scaling_{force}^{e^i,e^j} = scaling_{base}^{e^i,e^j} \cdot rel^{e^j,e^i} \tag{2.15}$$

Utilizing this scaling factor, the affinity and distance based *force vector* $\overrightarrow{force}^{e^i,e^j}$ of the element pair is the scaled difference vector, resulting from the position vectors $\overrightarrow{pos}^{e^i}$ and $\overrightarrow{pos}^{e^j}$:

$$\overrightarrow{force}^{e^i,e^j} = scaling_{force}^{e^i,e^j} \cdot \left( \overrightarrow{pos}^{e^j} - \overrightarrow{pos}^{e^i} \right) \tag{2.16}$$

To illustrate the data flow, starting from the elements' properties on the one side and the element position vectors on the other side, Figure 2.9 shows the calculation path towards the resulting force vector. Applying the Formulas (2.6) - (2.16), the single property affinities and relevances are merged to the elements' affinity and relevance. While the affinity leads to an optimal distance, the position vectors of both elements provides the current distance. Both values are combined with the relevance to a scalar factor. Finally the single force vector from one element to the other is calculated as the fraction (scalar factor) of their difference vector.

While Formula (2.16) defines the force vector of one single element to another one, for the *total force vector* $\overrightarrow{force}^e$ of an element $e^i$ applies:

$$\overrightarrow{force}^{e^i} = \frac{\sum_{e^j} \overrightarrow{force}^{e^i,j} \cdot weight^{e^j}}{\sum_{e^j} weight^{e^j}} \tag{2.17}$$

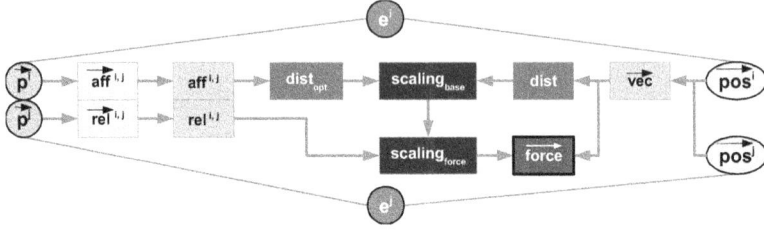

**Figure 2.9:** *Dataflow from the Elements' Properties and Positions to the Resulting Single Force Vector for Equili-*
*bration;* Given two elements $e^i$ and $e^j$ including their properties $\vec{p}^i$ and $\vec{p}^j$. The prioritized combination (property priorities are not shown) of all affinities and relevances leads to the element affinity
$aff^{i,j}$ and the element relevance $rel^{i,j}$ (Formula (2.7) and (2.10)). Applying the user-defined optimal
distance function $.dist_{opt}$ is calculated (Formula (2.8)). On the other side the position vectors $\overrightarrow{pos}^{i,j}$ are
utilized to determine the Euclidean distance $dist$, which leads in combination with the optimal distance
to the force's base scaling ($scaling_{base}$) (Formula (2.14)). Taking into account the total relevance, the
current force scaling ($scaling_{force}$) can be calculated (Formula (2.15)). The final force vector $\overrightarrow{force}$,
which element $e^j$ causes for element $e^i$, results from the difference vector $\vec{vec}$ of both positions and
the force scaling (Formula (2.16)).

where $\overrightarrow{force}^{e^{i,j}}$ stands for the single force vector from $e^i$ to $e^j$ (Formula (2.16)) and $weight^{e^j}$ describes
the weight of element $e^j$. In this way $\overrightarrow{force}^{e^i}$ describes the moving direction and magnitude of a single
element during a single iteration. This calculation is done for every element within the model to evaluate
their affinity gradients. Before the element positions are updated by the vectors, the following *global*
*scaling* applies during each iteration $n$:

$$scaling_{global}^n = \frac{dist_{max}^{elements} \cdot f_{SA}^{(n)}}{magnitude_{max}} \tag{2.18}$$

Every calculated $\overrightarrow{force}^e$ is scaled with the factor $scaling_{global}^n$, where $dist_{max}^{elements}$ and $magnitude_{max}$ are
the current maximal element distance and force vector magnitude respectively, and $f_{SA}^{(n)}$ stands for a *SA*
factor to the power of $n$. This scaling is chosen, so that the maximal force magnitude of an element
is a fraction ($f_{SA}^{(n)}$) of the current data model size, which is defined by the maximal element distance.
In this way a potential oscillating over multiple iterations is avoided and the force magnitudes decrease
in an exponential manner, which drives the elements to their equilibrium positions, where the average
transposition quality is maximal. Hence the new position vector of an element $e$ during each iteration is:

$$\overrightarrow{pos}^e = \overrightarrow{pos}^e + \overrightarrow{force}^e \cdot scaling_{global}^n \tag{2.19}$$

To ensure the elements are able to move far enough to reach their equilibration points, the iteration count
$n$ has to be chosen in dependence of the *SA* factor. Also it has to be ensured that $n$ is small enough to
avoid unnecessary iterations with very small *SA* factors, which have no influence onto the final result.
For that reason the sum of applied *SA* factors can be expressed as a geometrical series, describing the
possible maximum of a normalized single element's path length:

$$sum^n_{f_{SA}} = \sum_{i=0}^{n} f_{SA}^{(i)}$$
$$= \frac{f_{SA}^{(n-1)} - 1}{f_{SA} - 1} \tag{2.20}$$

Assuming an infinite count of iterations, this path has a maximal normalized length of:

$$sum^{max}_{f_{SA}} = \lim_{n \to \infty} \frac{f_{SA}^{(n-1)} - 1}{f_{SA} - 1}$$
$$= \frac{1}{1 - f_{SA}} \tag{2.21}$$

Assuming that a fraction of the maximal normalized path length is sufficient to reach an equilibrium, an *utilization factor* is introduced, describing the ratio of the normalized path length given an iteration count $n$ to the maximal normalized path length based on the *SA* factor:

$$utilization = \frac{sum^n_{f_{SA}}}{sum^{max}_{f_{SA}}}$$
$$= 1 - f_{SA}^n \tag{2.22}$$

To ensure a *minimal path utilization* with regard to a given *SA* factor, the required minimal iteration count $n_{min}$ can be isolated:

$$n_{min} = \frac{log(1 - utilization_{min})}{log(f_{SA})} \tag{2.23}$$

Example: a typical value of the *SA* factor is 0.9, i.e. the global scaling of the forces, starting with a value of 1, is decreased by this factor during each iteration. Applying a minimal *utilization*$_{min}$ of 0.95, i.e. the elements should be moved by $95\%$ of their maximal movement distance (determined by $f_{SA}$), results in a minimal iteration count of $n_{min} = 28.43$ (Formula (2.23)). Hence the algorithm should compute at least 29 iterations to ensure each element is able to reach its equilibration point.

Figure 2.10 illustrates the scheme of the EQ Algorithm. It consists of a main loop of $n_{min}$ steps (Formula (2.23)), which firstly calculates the total force vector for each element (Formulas (2.14) - (2.17)), secondly updates the global *SA* scaling (Formula (2.18)) and thirdly applies the scaled force vectors to the element positions (Formula (2.19)). This successive force (affinity gradient) calculation with decreasing magnitudes and element position updates drives the element to, or at least near their equilibration points.

To illustrate the function of the algorithm's scheme, Figure 2.11 shows a simplified data model of several elements, which share different affinities among each other. The equilibration mechanism transposes the elements from an unbalanced state near to one of their equilibriums, changing their positions with regard to their current affinity gradients iteratively. The final state represents the elements in positions to each other, which reflect their property based affinities, i.e. the closer two elements, the more affine they

**Initialization:** random element positions, $scaling_{global} = 1$

**Initialization:** $magnitude_{max} = 0$, $dist_{max}^{elements} = 0$

for i in $[0 .. n_{min}]$

- $n_{min}$ ≙ transposition step counts
- $scaling_{global}$ ≙ current global scaling factor
- $magnitude_{max}$ ≙ maximal force magnitude
- $dist_{max}^{elements}$ ≙ maximal element distance
- $weight_{sum}$ ≙ element weight sum
- $\overrightarrow{pos}^{source}$ ≙ position vector source
- $scaling_{global}$ ≙ global force factor

foreach source in $list_{elements}$

**Force Calculation for Element source**

**Reset:** $\overrightarrow{force}^{source} = \overrightarrow{0}$

foreach target in $list_{elements}$

**Calculation / Update:**
$aff^{source,target}$  $rel^{source,target}$
$dist^{source,target}$  $dist_{opt}^{source,target}$
$scaling_{base}$  $scaling_{force}$
$\overrightarrow{force}^{source,target}$  $\overrightarrow{force}^{source}$
$dist_{max}^{elements} = \max(dist_{max}^{elements}, dist^{source,target})$

- $list_{elements}$ ≙ list of elements
- $source$ ≙ element, from which the force is calculated
- $target$ ≙ element, to which the force is calculated
- $aff^{source,target}$ ≙ affinity of source and target
- $rel^{source,target}$ ≙ relevance of source and target
- $dist^{source,target}$ ≙ distance between source and target
- $dist_{opt}^{source,target}$ ≙ optimal distance between source and target
- $scaling_{force}$ ≙ scalar force factor
- $\overrightarrow{force}^{source,target}$ ≙ resulting force vector from source to target
- $\overrightarrow{force}^{source}$ ≙ total force vector of source

**Global Force Scaling**

**Update:** $scaling_{global}$

**Update:** $\overrightarrow{force}^{source} = \overrightarrow{force}^{source} / weight_{sum}$
$magnitude_{max} = \max(magnitude_{max}, |\overrightarrow{force}^{source}|)$

**Element Movement**

foreach element in $list_{elements}$

**Reposition:** $\overrightarrow{pos}^{element} = \overrightarrow{pos}^{element} + \overrightarrow{force}^{element} * scaling_{global}$

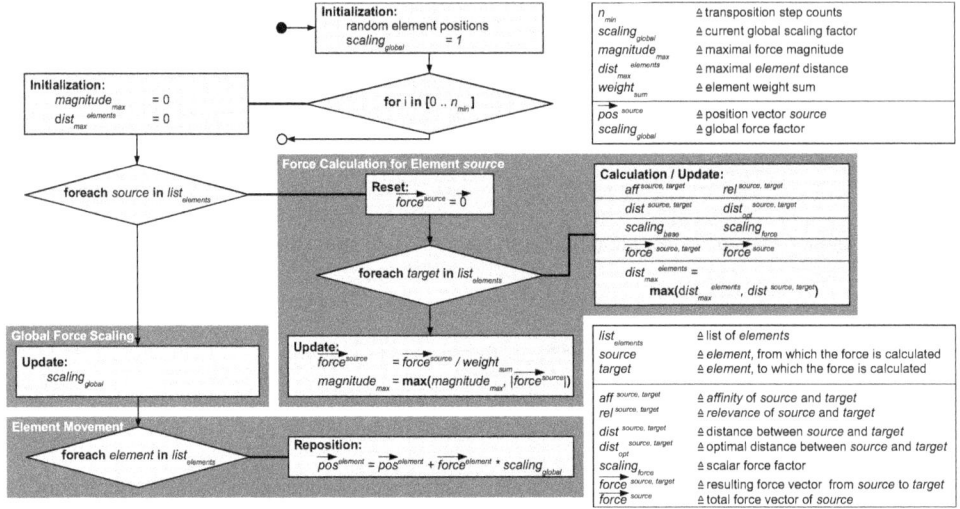

**Figure 2.10:** *Equilibration Scheme* to arrange elements according their optimal distances (affinities);
**Initialization:** Initially the elements are positioned randomly and the global scaling factor $scaling_{global}$ is initialized before the main loop is started to execute $n_{min}$ (Formula (2.23)) transposition steps. For each iteration the maximal force magnitude $magnitude_{max}$ and element distance $dist_{max}^{elements}$ is reset, before for each element *source* the force (movement vector) is calculated (Formula (2.17)).
**Force Calculation for Element *source*:** After reinitializing the current total force vector $\overrightarrow{force}^{source}$, the single force calculation (see also Figure 2.9) is performed for each counter element *target*. During this process $dist_{max}^{elements}$ is updated and the force vectors $\overrightarrow{force}^{source,target}$ of each element pair are accumulated in $\overrightarrow{force}^{source}$. Subsequent to this calculation the final total force vector $\overrightarrow{force}^{source}$ is weighted by the sum of the element weights ($weight_{sum}$) and the current maximal force magnitude ($magnitude_{max}$) is updated.
**Global Force Scaling:** According to Formula (2.18) the global scaling factor ($scaling_{global}$) is updated (*SA* concept).
**Element Movement:** After calculating the total forces and scaling factors, the force vectors are applied to their elements, updating their position vector $\overrightarrow{pos}^{element}$. This concludes a single transposition step. The transposition process is continued with the updated position vectors and the decreased global scaling factor to find the equilibrium state.

are.

In the following the transposition process is rendered for two different example data models. The user-defined affinity, relevance and optimal distance functions are:

$$Func_{aff}(prop^i, prop^j) = 1 - \frac{|value_{prop^i} - value_{prop^j}|}{255}$$

$$Func_{rel}(prop^i, prop^j) = 1 \tag{2.24}$$

$$Func_{dist_{opt}}(aff) = 1 - aff$$

The data models differ in the count and value distribution of their properties and are described in the Table 2.8. For the purpose of an intuitive visualization the properties are chosen as color portions. The combination of two or three color properties respectively results in visualizable element colors. Besides both data models have the same size, the same count of target dimensions and equal transposition

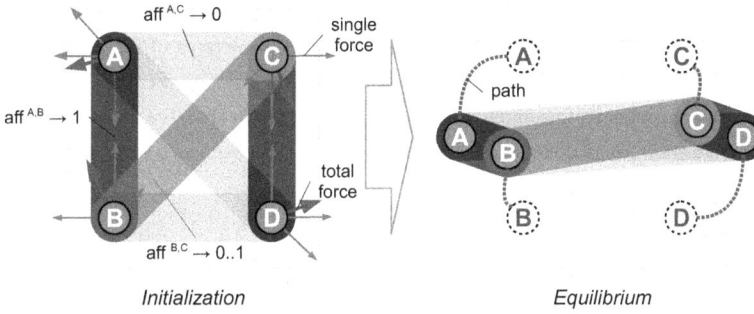

*Initialization*                                        *Equilibrium*

**Figure 2.11:** *Simplified Transposition Example* of the elements **A..D** in two dimensions, where the pairs (**A, B**) and (**C, D**) share a high affinity (visualized by dark gray bars) in contrast the pairs (**A, C**), (**A, D**) and (**B, D**), which are non-affine (light gray bars). Only the pair (**B, C**) has an mid-range affinity (medium gray bar). Starting at the initial positions (left side), the EQ Algorithm calculates the single forces for each element pair according to its affinity (red arrows), which result in the total forces (blue arrows), describing the direction and magnitude of the next movement step. While alternating the calculation and movement process several times, the elements are moved through the given metric system (paths as blue dashed lines) to their final positions, where the distances to each other are in relation to their affinities (right side). Note that the element pairs with high affinities are positioned close to each other, in contrast to distant pairs, whose distances represent low affinities.

parameters, the elements of the *CVD* data model exhibit an even value distribution of their two properties (Section 2.2.3). This results in a continuous distribution of the red and green portions (0..255), i.e. one group of even color gradients. In contrast, the *DVD* data model type provides only small (200..255) and disjointed value ranges, which results in three distinct color groups (red, green and blue).

| Name | Symbol | Value / Range | | |
|---|---|---|---|---|
| element count | $n_{element}$ | 1000 | | |
| dimensionality | $n_{dimension}$ | 2 | | |
| iteration count | $n_{iteration}$ | 29 | | |
| *SA* factor | $f_{SA}$ | 0.9 | | |
| *CVD* Data Model | | Group *Red-Green* | | |
| property value *red* | $p_{red}$ | 0..255 | | |
| property value *green* | $p_{green}$ | 0..255 | | |
| *DVD* Data Model | | Group *Red* | Group *Green* | Group *Blue* |
| property value *red* | $p_{red}$ | 200..255 | 0 | 0 |
| property value *green* | $p_{green}$ | 0 | 200..255 | 0 |
| property value *blue* | $p_{blue}$ | 0 | 0 | 200..255 |

**Table 2.8:** *Data Model Parameters of the rendered CVD and DVD Example*; see Figures 2.12 and 2.13

The following Figures 2.12 and 2.13 show the transposition process of the two defined example data models (*CVD* and *DVD* type, Table 2.8) in three steps. The first step shows the elements as colored dots randomly initialized in two dimensions, where the colors represent the property values. The algorithm's task is to rearrange these elements, so that their distances express the similarity of their colors, i.e. equal colored dots should be positioned closely and unequal colored ones distant from each other. The second step shows the models in a transient state, where the elements have already moved toward their equilibrium points, but have not yet reached them. The separation of equal colored dots becomes already apparent in addition to visualized the movement paths. In the third step the elements are in an equilibrium. In the *CVD* data model the elements with the four extrema of the property values span a rectangle containing the intermediate elements ordered by even color gradients. In contrast to that, the *DVD* data model is separated into three distinct groups with relatively large distances between. In this way the EQ Algorithm found data representations, describing the element affinities by the final element positions.

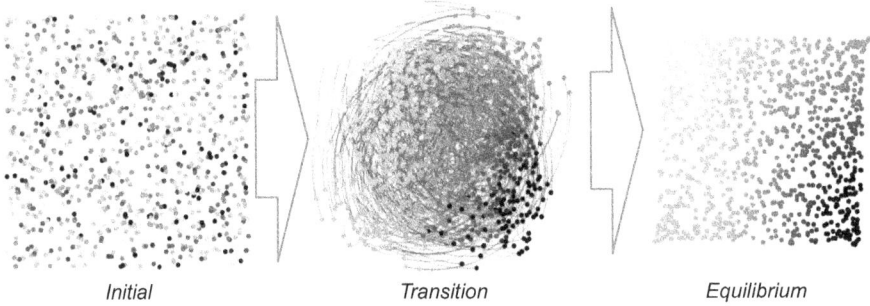

Initial                                      Transition                                   Equilibrium

**Figure 2.12:** *Rendered CVD EQ Transposition Example*; Shown is the transition from the initial state to the equilibrium state.

**Initial:** The elements of the example data model are initialized with random starting positions between $(0,0)$ and $(1,1)$ and two random continuous properties according to the *CVD* type (Section 2.2.3 and Table 2.8). The dots' color depicts the values of both properties (red and green).

**Transition:** The EQ Algorithm calculates forces iteratively for each element according to the affinities and distances to all other elements and moves them step-wisely (movement tracks as blue paths) to their equilibrium positions, i.e. the locations where all distances are maximally close to their optima.

**Equilibrium:** After the termination of the algorithm the elements are arranged according to their affinities, i.e. the space between the four extrema in the four corners (bottom-right: $p_{red/green} = 0$; bottom-left: $p_{red/green} = 0/255$; top-right: $p_{red/green} = 255/0$; top-left: $p_{red/green} = 255$) is filled continuously with elements.

Section 4.3 introduces an application example, which utilizes significantly more than three element properties, illustrating the approach's usefulness for large numbers of data item features.

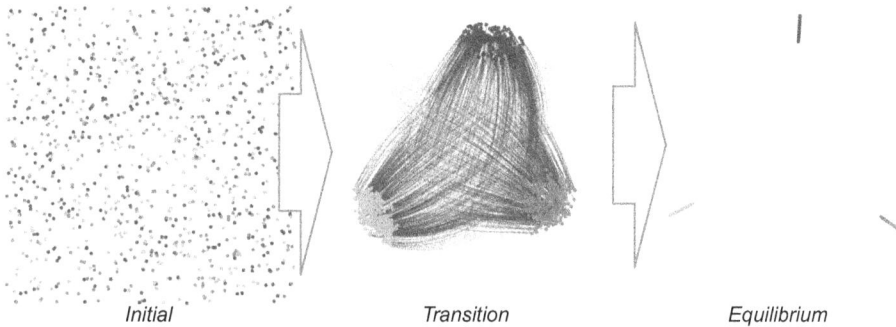

Initial                                    Transition                                    Equilibrium

**Figure 2.13:** *Rendered DVD* EQ *Transposition Example*; Shown is the transition from the initial state to the equilibrium state.

**Initial:** The elements of the example data model are initialized with random starting positions between $(0, 0)$ and $(1, 1)$ and with one out of three property presets according to the *DVD* type (Section 2.2.3 and Table 2.8). The dots' color depicts the values of the three properties (red, green, blue).

**Transition:** The EQ Algorithm calculates forces iteratively for each element according to the affinities and distances to all other elements and moves them step-wisely to their equilibrium positions (movement tracks as blue paths), i.e. the locations where all distances are maximally close to their optima.

**Equilibrium:** After the termination of the algorithm the elements are arranged according to their affinities, i.e. in three distinct groups, each containing the elements with affine properties (top: $p_{blue} = 200..255$; right: $p_{red} = 200..255$; left: $p_{green} = 200..255$).

### 2.3.2.3 Internal Tree Building

It is obvious in the EQ Algorithm's scheme (Figure 2.10) and the force definitions (Formula (2.17)), that every element has to be compared with every other element within the model. This implies a quadratic runtime behavior. To reduce the computation effort, the following heuristic is considered: From a single element's point of view, the more distant target elements are, the more they can be combined into a element groups, with average position vectors, average property values and the sums of their weights. These element groups replace the target elements during the force calculation, thus reducing the count of calculations.

However, this approach requires a preliminary executed grouping or precluster step, which embeds the elements as *leafs* in an *element tree* with regard to their positions (this Section). Furthermore, a mechanism to determine the size of branches is necessary (Section 2.3.2.4), as well as subalgorithms, which provide average properties to the branches (Section 2.3.2.5). Subsequently these precluster subalgorithms are utilized during the a tree-based equilibration process (Section 2.3.2.6).

To accomplish the tree building task, an accelerated version of *k-Means* algorithm, the *Triangular Inequality based k-Means Algorithm* (TRIN K-MEANS Algorithm) by Elkan [2003], which takes advantage of the triangular inequality, is applied recursively to find $k$ subbranches $b_{sub}$ beneath an already existing branch $b$ which contains $n$ leafs. The TRIN K-MEANS Algorithm distributes all $n$ leafs to one of the $k$ subbranches. For any subbranch, whose leaf count exceeds a given limit $size_{max}$, the TRIN K-MEANS Algorithm is executed again to split the branch even further. Finally the center coordinates of branch $b$ are calculated as the geometric center:

$$\overrightarrow{pos}^b = \begin{pmatrix} \left[ max(pos_0^{b^0_{sub}}, .., pos_0^{b^k_{sub}}) + min(pos_0^{b^0_{sub}}, .., pos_0^{b^k_{sub}}) \right] /2 \\ .. \\ \left[ max(pos_d^{b^0_{sub}}, .., pos_d^{b^k_{sub}}) + min(pos_d^{b^0_{sub}}, .., pos_d^{b^k_{sub}}) \right] /2 \end{pmatrix} \tag{2.25}$$

where $\overrightarrow{pos}^b$ is the position vector of branch $b$, $pos_j^{b^i_{sub}}$ the single position of subbranch $i$ in dimension $j$, and $d$ the count of dimensions. This center calculation is performed to reduce the radius of $b$, i.e. the maximal branch – subbranch distance, in contrast to the traditional *k-Means* algorithm, which utilizes the arithmetic center.

The *Tree Building* (*TB*) subalgorithm is configured by three parameters: $i)$ $size_{opt}$, which defines the optimal count of subbranches per branch, or the optimal branch size, $ii)$ $size_{max}$, which stands for the maximal count of subbranches per branch, or the maximal branch size, without triggering a further split and $iii)$ $size_{min}$, describing the minimal count of subbranches, or the minimal branch size, before a branch collapse is performed. Hence, for $size^b$ of each branch $b$ within a completed tree applies:

$$size_{min} \leq size^b \leq size_{max} \tag{2.26}$$

Figure 2.14 depicts the recursive scheme of the *TB* process for the given elements as leafs. The process is initialized with a new root branch and all given leafs beneath it. Subsequently a recursive check and modification process is applied to each branch in a recursive manner: it checks, whether the subbranch count is within the allowed range (see Formula (2.26)). If it is outside the range, one of the following operations is executed:

*Split:* If the branch is too large, i.e. $size^b > size_{max}$, a split is performed by applying the TRIN K-MEANS Algorithm to remaining leafs with $size_{opt}$ as given target cluster count. This step creates up to $size_{opt}$ new subbranches, which contain the leafs now. The *TB* process is continued for every newly created subbranch.

*Collapse:* In contrast to that, if the branch is too small, i.e. $size^b < size_{min}$, the branch is collapsed by reassigning its leafs to the super branch and removing the branch. In this case the *TB* process doesn't descend deeper into the tree from this position on.

Finally, in a bottom-up manner, the branch positions are calculated within the $n$-dimensional space as the geometric center (see Formula (2.25)), basing on the positions of the current subbranches.

Due to its great time performance while handling large element and small cluster counts [Elkan, 2003], the TRIN K-MEANS Algorithm determines more compact branches in acceptable times, in contrast to *kd-tree*-based mechanisms, which divide leaf groups along dimensional axes [Bentley, 1990; Pelleg and Moore, 1999; Kanungo et al., 2002]. This approach results in branches, differing greatly in size and density, but being more compact and distinct from each other. Hence the tree is not balanced. Furthermore, trees built with the TRIN K-MEANS Algorithm are independent from the selection and order of split dimensions and therefore suffer less from the *curse of dimensionality* [Indyk and Motwani, 1998].

Additionally, but not shown in the figure, the super distances $dist_{super}^{b^0_{sub}}..dist_{super}^{b^k_{sub}}$ from the branch to their subbranches are calculated, describing the distances between $b$ and all its subbranches $b^0_{sub}..b^k_{sub}$. These metrics are required for the branch radius calculation in Section 2.3.2.4, see Formula (2.29).

To clarify the recursive process of *TB*, Figure 2.15 illustrates the subalgorithm's flow using a sim-

$size_{opt}$ ≙ optimal count of leafs / subbr.
$size_{max}$ ≙ maximal count of leafs / subbr.
$size_{min}$ ≙ minimal count of leafs / subbr.

$b$ ≙ current branch
$size^b$ ≙ subbranch count of b
$\overrightarrow{pos}^b$ ≙ position vector of b
$b_{super}$ ≙ super branch of b
$b_{sub}$ ≙ subbranch of b

**Initialization:**
  $b$ = new branch as root
  Assign all leafs to $b$

**Recursion**

Recursive Call: $b = b_{sub}$

foreach $b_{sub}$ beneath $b$

**Calculation:**
  $\overrightarrow{pos}^b$ = geometric center of all subbranches

**Branch Check**

$size^b > size_{max}$ ?  yes

$size^b < size_{min}$ ?  yes

no

**Branch Modifcation**

**Split:**
  Create $size_{opt}$ subbranches out of all *leafs* beneath b

**Collapse:**
  move *subbranches* of b to $b_{super}$
  remove b

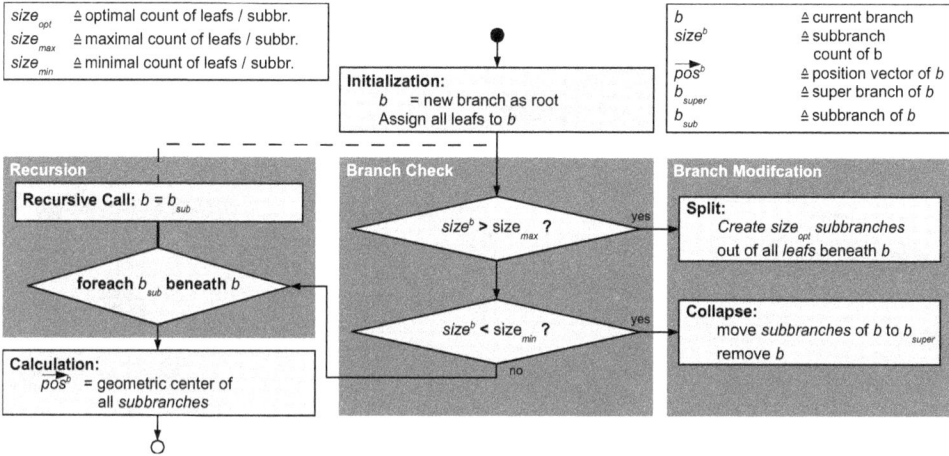

**Figure 2.14:** *TB scheme for the Element Tree Construction;*
**Initialization:** The current branch $b$ is set to the tree and all leaf are assigned to it.
**Branch Check:** The subbranch count $size^b$ of the current branch $b$ is checked: if it exceeds the upper limit $size_{max}$, $b$ will be split, if it falls below the lower limit $size_{min}$, $b$ is collapsed, otherwise.
**Branch Modification:** A split executes the TRIN K-MEANS Algorithm and divides the leafs into $size_{opt}$ groups. For each group a new subbranch $b_{sub}$ is created, to which the according leafs are subordinated. During a collapse, the subbranches or leafs of $b$ are relocated to $b$'s super branch $b_{super}$ and $b$ is removed from the tree.
**Recursive Call:** If the current branch isn't collapsed, the subbranches $b_{sub}$ are processed in the same manner as $b$, by calling the *TB* process recursively, handling $b_{sub}$ as $b$.
**Calculation:** Finally the position $\overrightarrow{pos}^b$ of the current branch $b$ is calculated as the geometric center of all subbranches.

plified example. A count of given leafs should be embedded in a tree (vertical), here spreading across only one dimensions (horizontal). The building process is initialized with a new root branch (see also Figure 2.14, Initialization). Advancing to every newly created branch (Figure 2.14, Branch Checking and Branch Modification) the checks reveal the branches as too large and split them up until the leaf's count falls below the given limit ($size_{max}$). In the case that a branch has too few subbranches after a split step, i.e. the subbranch count falls below $size_{min}$, a collapse is performed, reassigning the branch's subbranches to its super branch and removing the branch. This technique results in an unbalanced tree, which depth increases in regions of high leaf density and avoids malformed branches, i.e. too many or too few subbranches.

Regarding the requirement that the subbranches of a branch should be distributed evenly across the given ranges to achieve balanced branches, the following optimization can be applied: During the creation of the subbranches while performing a split, the new branches can be positioned evenly within the radius of the super branch to avoid unnecessary cluster updates by the TRIN K-MEANS Algorithm. In the case that $size_{opt} < 2^{n_{dimension}}$, this can be done by assigning the subbranches to the centers of the resulting orthants[9] [Roman, 2007] spanned by the dimension axes. This optimization is not shown in scheme or the *TB* examples.

Finally the rendering of a tree, created by the *TB* subalgorithm is presented. Table 2.9 shows the

---

[9] generalization of a quadrant in the plane or an octant in three dimensions for a $n$-dimensional Euclidean space

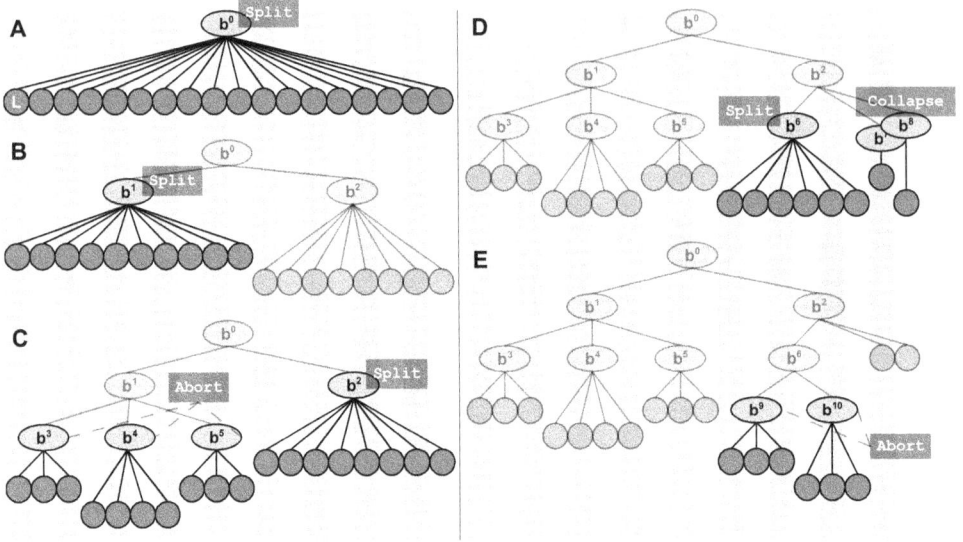

**Figure 2.15:** *TB Subalgorithm Flow Example* with $size_{min} = 2$, $size_{opt} = 3$ and $size_{max} = 5$;

    **A:** A new tree is initialized with a root branch $b^0$ (light green ellipse) and 18 leafs ($L$, dark green circles); the subbranch count of $b^0$ exceeds $size_{max}$ (18), therefor it will be split up (utilizing the TRIN K-MEANS Algorithm);

    **B:** $b^0$ is divided into two subbranches $b^1$ and $b^2$, the *TB* steps down to $b^1$ and detects that its subbranch count exceeds $size_{max}$ (10), so $b^1$ will be split up;

    **C:** $b^3$, $b^4$ and $b^5$ are the new subbranches of $b^1$ and their subbranch counts are in the allowed range between $size_{min}$ and $size_{max}$, so the *TB* returns to branch $b^2$, which will be also split up, due to its subbranch count (8).

    **D:** The split reveals a new subbranch $b^6$, whose subbranch count (6) is greater than $size_{max}$, which triggers a split, and two subbranches $b^7$ and $b^8$, whose have too few leafs in comparison to $size_{min}$ and will be collapsed.

    **E:** $b^7$ and $b^8$ are removed and now their subbranch (leafs) are assigned to the super branch $b^2$; $b^6$ is split further into $b^9$ and $b^{10}$, which completes the *TB* process.

data model's parameters, as well as the algorithm's configuration, which is chosen for the purpose of an appropriate tree representation.

| Name | Symbol | Value |
|---|---|---|
| element count | $n_{element}$ | 1000 |
| dimensionality | $n_{dimension}$ | 2 |
| optimal branch size | $size_{opt}$ | 4 |
| minimal branch size | $size_{min}$ | 3 |
| maximal branch size | $size_{max}$ | 8 |

**Table 2.9:** *Data Model / Algorithm Parameters of the TB Example*

    The tree example is shown in Figure 2.16. It visualizes the branches of different hierarchy levels as well as the elements as leafs. Due to the data model's element count ($n_{element}$) and the branch split size ($size_{opt}$) the tree has an average depth of $\frac{log(n_{element})}{log(size_{opt})} = 4.98$.

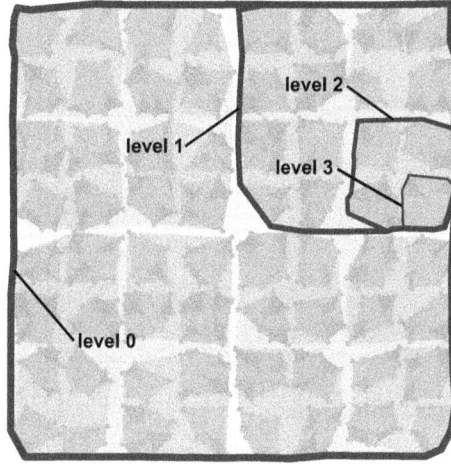

**Figure 2.16:** *Rendered TB Example;* shown is a tree of $1k$ evenly distributed elements (green dots) build with the *TB* subalgorithm. The branches of each level (gray polygons) consists of up to eight subbranches until the leaf level is reached. Example branches of each level are visualized by red frames. See also Table 2.9.

### 2.3.2.4 Maximal Leaf Distance Calculation

Based on the static trees in Section 2.3.2.3, the *Maximal Leaf Distance* (*MLD*) component encapsulates the mechanism to find the maximal *branch center – leaf* distance, which defines the radius of the branch. The mechanism utilizes the hierarchical tree structure to improve the search speed and yields an essential metric used by further subalgorithms as explained in the following Sections 2.3.2.6 - 2.3.2.9.

First the calculation scheme is explained in Figure 2.17. The intention is to find a distance $dist^b_{maxleaf}$ for every branch $b$, for which applies:

$$dist^b_{maxleaf} = max \left( |\overrightarrow{pos}^b - \overrightarrow{pos}^{l^0}|, .., |\overrightarrow{pos}^b - \overrightarrow{pos}^{l^m}| \right) \tag{2.27}$$

where $l^0, .., l^m$ are all leafs beneath $b$, directly or indirectly, $\overrightarrow{pos}^b$ and $\overrightarrow{pos}^{l^i}$ are the position vectors of the branch $b$ and the leaf $l^i$ and $|\overrightarrow{pos}^i - \overrightarrow{pos}^j|$ is the spatial distance between the tree items (branch or leaf) $i$ and $j$.

After initializing the *MLD* subalgorithm with the root branch of the given tree, the process for the most distant leaf is done in a bottom-up manner, thus accelerating the search by starting with the deepest tree items, which are not leafs. This is done by a first recursion step and ensures that for each subbranch of a branch the *MLD* is already known and can be utilized to accomplish an abort estimation, see Formula 2.31.

After processing all subbranches of the current branch, the search for its most distant leaf starts: First the search values are initialized, such as the current *MLD* and also the *current path length*, which is required for the abort estimation, as well as the *current search position* with the current branch as the search's origin. Subsequently the following recursive process is performed: every subbranch of the current search position is checked for being a leaf. If a leaf is found, the distance to the search's origin is calculated and compared to the former found *MLD*. In this way the *MLD* is updated if a larger distance is found. If the subbranch is not a leaf, a check is performed to decide whether to continue the search from this position.

For that purpose the following upper bound is utilized: For the distance between the search's origin $b$ and a subbranch $b_{sub}$ applies:

$$|\overrightarrow{pos}^b - \overrightarrow{pos}^{b_{sub}}| \leq length_{path} + dist_{maxleaf}^{b_{sub}} \tag{2.28}$$

with

$$length_{path} = \sum_{i=0}^{k-1} |\overrightarrow{pos}^i - \overrightarrow{pos}^{i+1}| = \sum_{i=1}^{k} dist_{super}^{b^i} \tag{2.29}$$

where $length_{path}$ describes the path length from tree item 0 to item $k$, following the hierarchical tree edges downwardly, and $dist_{maxleaf}^{b_{sub}}$ stands for the *MLD* of subbranch $b_{sub}$ as well as $dist_{super}^{b^i}$ for the distance between a branch $b^i$ and its super branch. Therefor the expression

$$dist_{maxleaf}^b < |\overrightarrow{pos}^b - \overrightarrow{pos}^{b_{sub}}| \tag{2.30}$$

only applies if

$$dist_{maxleaf}^b \leq length_{path} + dist_{maxleaf}^{b_{sub}}. \tag{2.31}$$

Utilizing this as an abort estimation, Formula 2.31 applies only if $b_{sub}$ contains a leaf, which is more distant from $b$ than the former found leaf. Otherwise the search can be aborted at this point, reducing computation effort, due to maintaining the current search path length is cheaper in comparison to the expensive distance calculation $|\overrightarrow{pos}^i - \overrightarrow{pos}^j|$ and all other utilized tree metrics ($dist_{maxleaf}^{b^i}$, $dist_{super}^{b^i}$) are static.

After passing the check, the search is continued recursively with the subbranch as the current search position and an updated path length. In this way the most distant leaf beneath a given branch can be found, checking every leaf beneath the search's start branch or aborting the search at a certain positions, if former *MLD* calculations reveal, that no more distant leaf can be found.

To further visualize the mechanism of the essential abort check, Figure 2.18 shows the three cases with regard the estimation. Illustrated is a (partial) example tree with a center branch and several sub-branches. Known from the *TB* process are the super distances, i.e. the distances between a branch and its super branch (Section 2.3.2.3). These data enable the upper bound estimation for the branch distances following Formula (2.29), summarizing the path lengths. In the figure the path is visualized by connected small gray circles, towards to the considered subbranches. The estimated distances are represented as virtual branches at the most possible distant positions. Furthermore, the *MLD*s of all subbranches are known, due to the bottom-up scheme, see Figure 2.17, first recursion. By applying Formula (2.31), the sum of the path lengths to a subbranch and its *MLD* allows an upper bound prediction about the potential *MLD* from the center to the considered subbranch. The example involves three cases:

- *case* $b^1$, *search abort*: the sum of the path lengths and the known *MLD* is smaller than the current *MLD*, i.e. even if the path runs straight toward the subbranch and its most distant leaf (which has

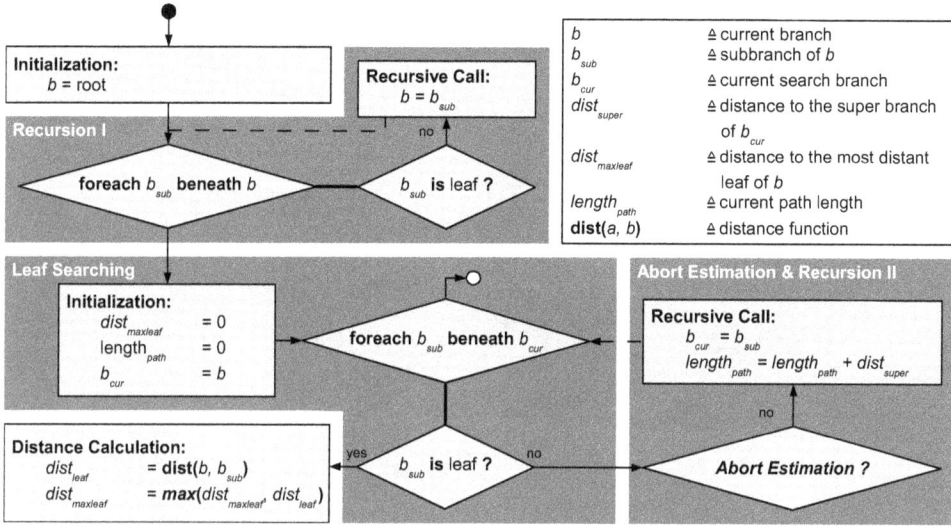

**Figure 2.17:** *MLD Calculation Scheme* for a given element tree (Section 2.3.2.3);

    **Initialization:** The calculation is done bottom-up, this means the search assumes the actuality of $i$) the distance to the current super branch $dist^b_{super}$ and $ii$) the *MLD* $dist^{b_{sub}}_{maxleaf}$ of all subbranches $b_{sub}$. Initially the current branch $b$ is set to the tree root, and the

    **Recursion (I):** is repeated for every subbranch $b_{sub}$ until it is marked as a leaf.

    **Leaf Searching:** First, the *MLD* $dist_{maxleaf}$ and the current path length $length_{path}$ are reset. Furthermore, the current search position $b_{cur}$ is set to the current branch $b$ before every subbranch $b_{sub}$ beneath the current search position $b_{cur}$ is checked for being a leaf. If $b_{sub}$ is a leaf, the

    **Distance Calculation** is processed: the distance between the origin of the search ($b$) and the current subbranch ($b_{sub}$) is calculated and replaces the current *MLD* $dist_{maxleaf}$, if it exceeds the current value. If $b_{sub}$ is not a leaf, an

    **Abort Estimation** is calculated: only if $dist_{maxleaf}$ falls below the length of the current search path, it is possible to find a more distant leaf (upper bound, see also Formula (2.31) and Figures 2.18 and 2.19). In this case a second

    **Recursion (II)** is performed: the current search position $b_{cur}$ is set to the current subbranch $b_{sub}$, the current path length $length_{path}$ is updated by the distance to the super branch and the search continues one level deeper.

to be in opposition to the center branch in the best case), the distance couldn't exceed the current one, allowing the search abort at this point.

- *case* $b^2$, *unsuccessful search continue*: the abort check is passed with a larger sum of path lengths and *MLD*, but the later distance calculation reveals the potential leafs as closer to the center than the current one.

- *case* $b^3$, *successful search continue*: after a passed abort check a more distant leaf is found.

The abort check is done to avoid expensive distance calculations from $b_0$ to potential leafs or branches, using static metrics out of the former analyzed sub structure.

To explain the *MLD* subalgorithm's flow in more detail, Figure 2.19 depicts the *MLD* calculation for an example tree with several branches and leafs. The intention is to find the distance to the most distant leaf from the center branch. Due to the *TB* mechanism and the bottom-up scheme, both the super − sub

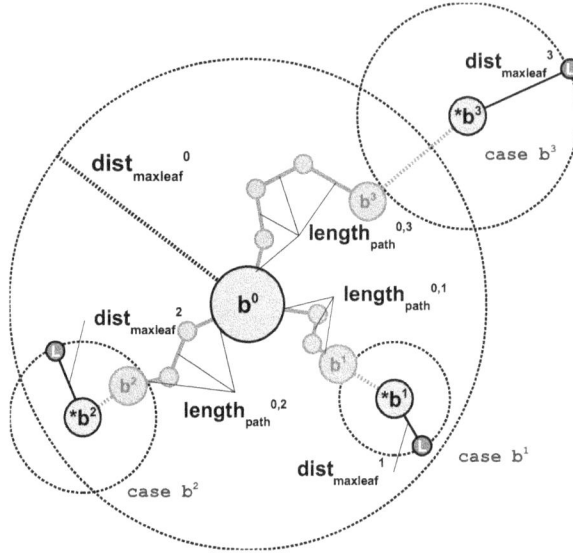

**Figure 2.18:** *Case Analysis of the Abort Estimation during the MLD Calculation;*
    **All cases:** Given is an example tree with the branch $b^0$ and its subbranches $b^1 - b^3$ (light green cir-
cles). The current *MLD* of $b^0$ is $dist^0_{maxleaf}$ (dotted line and circle) and the subalgorithm tries to find a
leaf, which is more distant than the current one (which is not shown here). Gray lines and small dark
green circles represent hierarchical relations and further subbranches, dotted circles stand for known
*MLD*s. Following Formula (2.28), the distance $dist^{b^0,b^i}$ is limited by the path length $length^{0,i}_{path}$, shown
here as virtual branches $*b^i$, distant equal to path length from $b^0$. All branches $b^0 - b^3$ may have more
subbranches or leafs (which are not shown in this example).
    **Case $b^1$:** the distance between $b^0$ and $b^1$ can be estimated by $dist^{b^0,b^1} \leq length^{0,1}_{path}$ as path length
downward the tree hierarchy. Hence, the potential *MLD* from $b^0$ involving $b^1$ is $dist^{0,1}_{maxleaf} \leq$
$length^{0,1}_{path} + dist^1_{maxleaf}$ (see Formula (2.31)). In this case the estimation $length^{0,1}_{path} + dist^1_{maxleaf}$ falls
below the current *MLD* $dist^0_{maxleaf}$ (small dotted circle is completely inside the large dotted circle) and
the search can be aborted at this position, as $b^1$ can contain no more distant leaf.
    **Case $b^2$:** For the potential *MLD* applies $*dist^{0,2}_{maxleaf} = length^{0,2}_{path} + dist^2_{maxleaf}$ as maximal possible
value. In this case $*dist^{0,2}_{maxleaf}$ exceeds the current value $dist^0_{maxleaf}$ (small dotted circle is partly outside the large dotted circle), but the later distance calculation between $b^0$ and the leaf $L$ beneath $b^2$
reveals that it is not more distant than the current one.
    **Case $b^3$:** The estimation shows, that the most distant leaf of $b^3$ has a maximal distance $*dist^{0,3}_{maxleaf} =$
$length^{0,3}_{path} + dist^0_{maxleaf}$ from $b^0$, which exceeds the current *MLD* $dist^0_{maxleaf}$ (small dotted circle is partly
outside the large dotted circle). The later distance calculation yields that the leaf $L$ beneath $b^3$ is more
distant from $b^0$ and updates $dist^0_{maxleaf}$.

branch distances are known, as well as the *MLD*s of all subbranches (Section 2.3.2.3 and Figure 2.17,
*Recursion (I)*). The shown tree is parsed recursively starting from the center, which matches recursion
(II) in Figure 2.17. The three analyzed cases in Figure 2.18 also occur here: $b_1$ (*search abort*) in region
C, $b_2$ (*unsuccessful search continue*) in region D, and $b_3$ (*successful search continue*) in region B and
D. The example shows that only a fraction of all subordinated leafs have to be considered, if utilizing
the static metrics of the *TB* process and the subordinated *MLD* calculations in the abort check, explained
above.
    Finally the rendering of a tree, including the *MLD* for each branch, created by the *MLD* subalgorithm

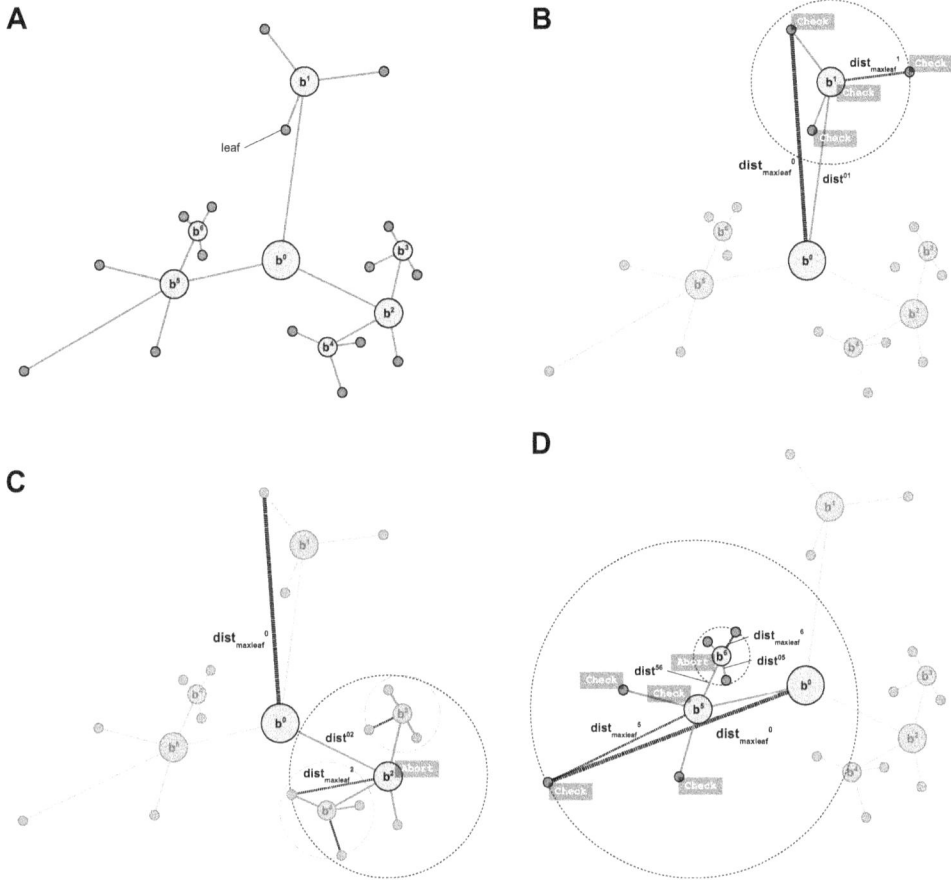

**Figure 2.19:** *Example of MLD Calculation Subalgorithm Flow;*
A: Given is an example tree with seven branches $b^0..b^6$ (light green circles) and 16 leafs (dark green circles) in total, gray lines symbolize hierarchical relations. The search tries to find the most distant leaf from $b^0$ and assumes that the subordinated *MLDs* $dist^1_{maxleaf}, .., dist^6_{maxleaf}$ and the super distances $dist^{0,1}, .., dist^{5,6}$ of all subbranches $b^1, .., b^6$ are known.
B: The search starts at $b^1$ and no abort estimation applies, because the current *MLD* $dist^0_{maxleaf}$ is 0. After calculating the distance to every leaf beneath $b^1$, a new $dist^0_{maxleaf}$ is found (thick dotted line). The gray dotted circle describes the local *MLD* of $b^1$.
C: The search checks $b^2$, but because of the abort estimation ($dist^0_{maxleaf} > dist^{0,2} + dist^2_{maxleaf}$, see Formula (2.31)) reveals that no leaf beneath $b^2$ can be more distant than the current one, the search aborts at this position, skipping the entire branch $b^2$. The dotted circles represent the *MLDs* of $b^2..b^4$.
D: The search reaches $b^5$ and passes the abort check ($dist^0_{maxleaf} \leq dist^{0,5} + dist^5_{maxleaf}$). Before checking the direct leafs of $b^5$, branch $b^6$ fails to pass the estimation ($dist^0_{maxleaf} > dist^{0,5} + dist^{5,6} + dist^6_{maxleaf}$) and is skipped. After that, the most distant leaf is found (thick dotted line). The *MLDs* of both $b^5$ and $b^6$ are shown as dotted circles.

is presented in Figure 2.20. Table 2.10 shows the data model's parameters, as well as the algorithm's configuration, which are chosen for the purpose of an appropriate representation of the *MLDs* of different hierarchy levels.

| Name | Symbol | Value |
|------|--------|-------|
| element count | $n_{element}$ | 300 |
| dimensionality | $n_{dimension}$ | 2 |
| optimal branch size | $size_{opt}$ | 4 |
| minimal branch size | $size_{min}$ | 3 |
| maximal branch size | $size_{max}$ | 8 |

**Table 2.10:** *Data Model / Algorithm Parameters of MLD Example*

The branches themselves in the Figure are not shown, but are represented by the *MLD*s as the branch radii. The figure shows that these distances are appropriate expressions for the branches' extent.

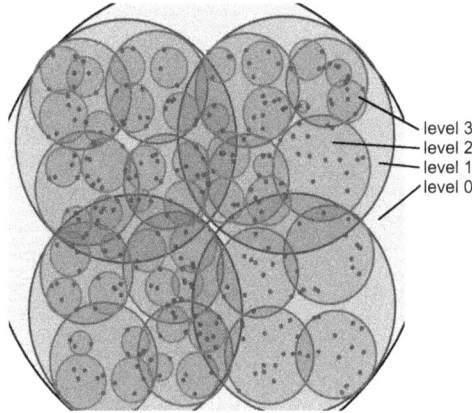

**Figure 2.20:** *Rendered MLD Example;* Shown are leafs (blue dots) embedded in a tree (for parameters see Table 2.10), whose branches are visualized by their *MLD*s (red circles). The higher the branch is in the tree's hierarchy, the larger its *MLD* circle.

### 2.3.2.5 Hierarchical Branch Representation as Meta Heuristic

Besides the *MLD* as radius of a branch (Section 2.3.2.4) the following data have to be evaluated for every branch to enable the subsequent equilibration algorithm, which utilizes the precalculated *Hierarchical Branch Representation (HBR)* (Section 2.3.2.6).

**Property Update:**  While the position vector of a branch represents the average positions of all sub-branches and leafs (elements), a mechanism is required to create branch properties to cover the property values of the subordinated tree items. For that reason additional user-defined functions are necessary to calculate representative values for each property type based on the corresponding property values of the underlying items:

$$p^b_{type} = Func_{type}(b^1_{sub}, .., b^n_{sub}) \tag{2.32}$$

Here the typed branch property $p^b_{type}$ of the branch $b$ is the result of the user-defined function $Func_{type}(..)$, which takes the $n$ subbranches or leafs $b^1_{sub}..b^n_{sub}$ of $b$ as arguments. In this way the properties of each non leaf, i.e. each branch can be created, by applying the provided function in a bottom-up manner to the tree.

As an example for branch property calculation, the property update function $Func_{color}^b$, based on the color properties defined in Formula (2.24) in Section 2.3.2.2 can be described as weighted average function:

$$Func_{color}^b = \frac{\sum_{b_{sub}} value_{p_{color}}^{b_{sub}} \cdot weight^{b_{sub}}}{\sum_{b_{sub}} weight^{b_{sub}}} \tag{2.33}$$

where $b_{sub}$ are all direct subbranches (branches or leafs) of $b$ and $value_{p_{color}}^{b_{sub}}$ is the value of the according color property $p_{color}$ of one of the $b_{sub}$. The properties' influence is weighted by the weights ($weight^{b_{sub}}$) of the subbranches. This function can be applied several times to calculate the branch property values for each color property.

To visualize the application of the branch property update subalgorithm, Figure 2.21 shows the rendered *DVD* data model as described in Table 2.8 in Section 2.3.2.2. The elements are embedded in a tree, build by the *TB* subalgorithm (Section 2.3.2.3) and its branches are supplemented by color properties, whose value are calculated by the function defined in Formula (2.33).

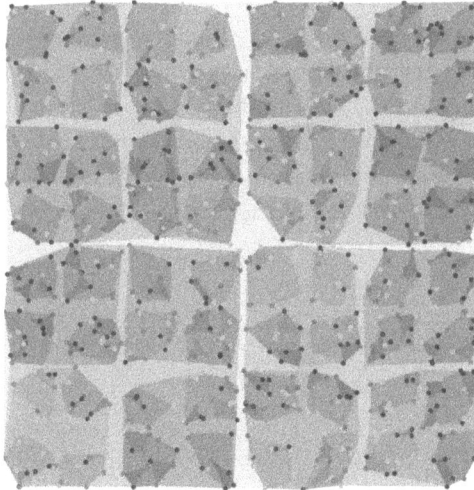

**Figure 2.21:** *Rendered HBR*; Shown is the *DVD* data model of Table 2.8 embedded in an extended tree. The elements are visualized as colored dots, where the colored represents the properties of the element (red, green or blue group). The branches, depicted as semi-transparent color polygons, now contain properties average values of their subbranches' properties (see example function in Formula (2.33)). In this way the branches represent their subelements or -branches regarding their positions and properties.

**Inferior Affinity:**  In addition to the branch properties, a metric is required to express how representative a branch is with respect to the properties of the subordinated elements. For that reason the *Inferior Affinity (IA)* $aff_{inferior}^b$ of a branch $b$ is introduced:

$$aff^b_{inferior} = \frac{\sum_{b_{sub}} aff^{b,b_{sub}} \cdot aff^{b_{sub}}_{inferior} \cdot weight^{b_{sub}}}{\sum_{b_{sub}} weight^{b_{sub}}} \quad (2.34)$$

The term $aff^{b,b_{sub}}$ is referred to as the element affinity (Formula (2.7) in Section 2.3.2.1), which can be applied to the branches $b$ and its subbranch $b_{sub}$, as they now contain comparable properties after the branch property update (see above). The affinity of a branch to its subbranch is scaled by the subbranch's IA $aff^{b_{sub}}_{inferior}$, as $b$ can only be as representative for the encapsulated elements as the subbranches $b_{sub}$. Finally $weight^{b_{sub}}$ stands for the weight of the subbranch $b_{sub}$.

Similar to the branch property update, the IAs are calculated in a bottom-up manner, beginning with the leafs (elements, for which apply: $aff^{leaf}_{inferior} = 1$) and ending at the tree's root. Hence, the IAs varies between 0 (not representative) and 1 (identical property values).

The application of the three supplements $i$) MLDs (Section 2.3.2.4), $ii$) branch properties and $iii$) IAs for each branch of an element tree (Section 2.3.2.3), results in the hierarchical model description, which is utilized by the subsequently extended equilibration process (Section 2.3.2.5).

### 2.3.2.6 Tree Based Equilibration Algorithm

To shorten the runtime of the EQ Algorithm (Section 2.3.2.2), this section discusses the *Tree Based Equilibration (TB-EQ)* subalgorithm, which is based on the *HBR* as introduced previously. The mechanism functions in the same manner as the EQ Algorithm, but utilizes the precalculated tree (Section 2.3.2.3) and branch data (Sections 2.3.2.4 and 2.3.2.5) to calculate forces to branches (element groups) of higher level with increasing distances instead of forces to all single elements.

To take into account the branches' representation capability of subordinated elements, expressed by the IA (Formula (2.34)) the position vector of a target branch is artificially distorted in the following manner:

$$noise_{max} = 1 - aff^b_{inferior}$$

$$random_{min, max} = Func_{random}(min, max)$$

$$\overrightarrow{pos}^{b,e}_{noised} = \overrightarrow{pos}^b + \begin{pmatrix} dist^b_{maxleaf} \cdot noise_{max} \cdot random^0_{-1,1} \\ .. \\ dist^b_{maxleaf} \cdot noise_{max} \cdot random^d_{-1,1} \end{pmatrix} \quad (2.35)$$

The value $noise_{max}$ expresses the maximal relative position noise and $Func_{random}(min, max)$ a random generator, which provides pseudo-random numbers between $min$ and $max$. The noisy position $\overrightarrow{pos}^{b,e}_{noised}$ of a branch $b$ in relation to an element $e$, whose forces are currently calculated, is equal to the branch's original position vector $\overrightarrow{pos}^b$, varied by a random fraction ($noise_{max} \cdot random^i_{-1,1}$) of the branch's radius $dist^b_{maxleaf}$ for each dimension $i$. The result is a position, that lays inside a (hyper) rectangle centered around the original position. The smaller the IA and the larger the MLD of the branch, the larger is the rectangle. This branch's distortion is determined for every single force calculation and represents an artificial fuzziness, which increases at decreasing IA, avoiding the approach of all elements towards the branch's center.

Consequently the single force calculation of an element $e$ to a branch $b$ changes from Formula (2.16) to:

$$\overrightarrow{force}^{e,b} = scaling_{force}^{e,b} \cdot \left( \overrightarrow{pos}_{noised}^{b,e} - \overrightarrow{pos}^{e} \right) \tag{2.36}$$

where $\overrightarrow{pos}_{noised}^{b,e}$ is permanently reevaluated and $scaling_{force}^{e,b}$ is equivalent to Formula (2.15). Furthermore, a branch with a low *IA* has less influence on the total force vector, hence the adapted weight of $b$ is introduced as:

$$weight_{adapted}^{b} = weight^{b} \cdot aff_{inferior}^{b} \tag{2.37}$$

This adapted weight is applied to the total element force calculation in Formula (2.17), where $b^i$ represents all considered branches, thus covering all other elements within the model:

$$\overrightarrow{force}^{e} = \frac{\sum_{b^i} \overrightarrow{force}^{b^i} \cdot weight_{adapted}^{b^i}}{\sum_{b^i} weight_{adapted}^{b^i}} \tag{2.38}$$

This updated calculation scheme is utilized during the *TB-EQ* process as described in Figure 2.22. The displayed scheme replaces the force calculation for single target elements in Figure 2.10 in Section 2.3.2.2.

The branch based calculation starts at the super branch of the current element, whose total force vector should be calculated. For each tree item (branch or leaf) beneath this branch the single forces are calculated and accumulated according to the Formulas (2.15) - (2.23) and (2.35) - (2.38). This covers the force calculation of the direct neighborhood of the current element. Subsequently the current branch is set to its super branch, extending the scope of calculation to the next hierarchy level. The force calculation is continued by considering alternately the sub items of the current branch and ascending in the hierarchy level further until the tree's root. This is done for every element within the model in each iteration. In this way the count of force calculations is reduced to a fraction of $n_{element}$, considering only branches of increasing sizes with growing distances at the costs of less accuracy. After each iteration, i.e. after each element position update the tree has to be rebuild or maintained to keep it up to date.

Figure 2.23 visualizes the different scopes of branch considerations from a single element's point of view, taken from the example data model of Table 2.9 in Section 2.3.2.3. Starting at the lowest hierarchical level (the level of the element), the forces to its direct neighbors are calculated. Subsequently the scope is raised step-wisely, calculating the forces to branches of increased size with every recursion step. The figure shows, that all elements are covered by few calculations steps and the size of the processed branches corresponds to the distance to the requesting element, as far as the tree structure allows it.

To visualize the subalgorithm's processing, the Figures 2.24 and 2.25 show the same data models as described in Table 2.8 in Section 2.3.2.2 and their transpositions as the Figures 2.12 and 2.13 in this chapter. Again, the data models differ only regarding the property value distribution type (*CVD* and *DVD*) and hence in the transposition results. The *TB-EQ* transposition utilizes the same user-defined functions (Formula (2.24)) as the previous examples. In addition to the EQ Algorithm's flow, these Figures show also the underlying element trees, which serve to accelerate the algorithm's execution. Note that the transposition results are visually similar to the results in Section 2.3.2.2 in spite of the reduced calculation effort.

| | |
|---|---|
| $source$ | ≙ *element*, for which the force is calculated |
| $branch$ | ≙ *branch*, to its subbranches the force is calculated |
| $branch_{super}^{b}$ | ≙ super branch of $b$ |
| $target$ | ≙ *branch / element*, to which the force are calculated |
| $aff^{source,\,target}$ | ≙ *affinity* of *source* and *target* |
| $rel^{source,\,target}$ | ≙ *relevance* of *source* and *target* |
| $\overrightarrow{pos}_{noised}^{\,target}$ | ≙ *noised position vector* of *target* |
| $dist^{source,\,target}$ | ≙ *distance* between *source* and *target* |
| $dist_{opt}^{source,\,target}$ | ≙ *optimal distance* between *source* and *target* |
| $scaling_{force}$ | ≙ *scaling force factor* |
| $\overrightarrow{force}^{\,source,\,target}$ | ≙ *resulting force vector* from *source* to *target* |
| $\overrightarrow{force}^{\,source}$ | ≙ *total force vector* of *source* |
| $magnitude_{max}$ | ≙ *maximal force magnitude* |
| $weight_{adapted}^{sum}$ | ≙ *adapted weight sum* |

**Force Calculation for Element *source***

Reset / Initialization:
$$\overrightarrow{force}^{\,source} = \vec{0}$$
$$branch = branch_{super}^{source}$$

**foreach *target* beneath *branch***

Calculation / Update:
$$aff^{source,\,target} \qquad rel^{source,\,target}$$
$$\overrightarrow{pos}_{noised}^{\,target} \qquad weight_{adapted}^{sum}$$
$$dist^{source,\,target} \qquad dist_{opt}^{source,\,target}$$
$$scaling_{base} \qquad scaling_{force}$$
$$\overrightarrow{force}^{\,source,\,target} \qquad \overrightarrow{force}^{\,source}$$

**Recursion**

Recursive Call:
$$branch = branch_{super}^{branch}$$

yes

$branch_{super}^{branch}$ **exists?**

no

Update:
$$magnitude_{max} = \textbf{max}(magnitude_{max}, |\overrightarrow{force}^{\,source}|)$$
$$\overrightarrow{force}^{\,source} = \overrightarrow{force}^{\,source} / weight_{adapted}^{sum}$$

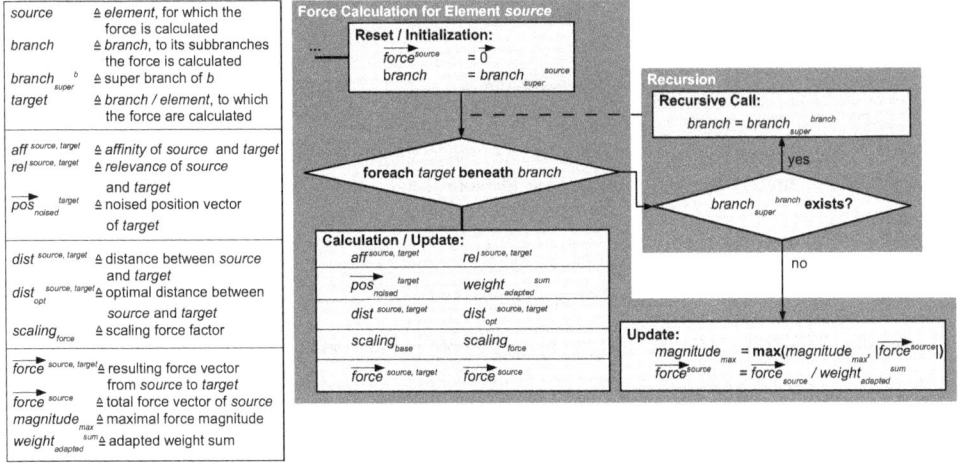

**Figure 2.22:** *TB-EQ Scheme*; Shown is the replacement of the *Force Calculation for Element source* during the EQ process, explained in Section 2.3.2.2, Figure 2.10. In contrast to this concept, which handles elements, the *TB-EQ* considers branches of higher levels with growing distances, i.e. the subalgorithm requires an updated tree as *HBR*.

After reseting the current movement vector $\overrightarrow{force}^{source}$, the current scope is set to the super branch $branch_{super}^{source}$ of the processed element *source*. That means, first, the elements or branches (*target*), which are beneath the same branch as *source*, are considered. The force calculation follows the same scheme as in the EQ Algorithm (Formulas (2.15) - (2.23)). In addition, a noised position $\overrightarrow{pos}_{noised}^{target}$ is calculated and $weight_{adapted}^{sum}$ is updated, which modify some calculation instructions (Formulas (2.35) - (2.38)).

After the force to each neighbor was calculated, the scope is raised by one level, setting the *branch* to its super branch $branch_{super}^{branch}$ and the force calculation to current neighbors is resumed. During this recursion the processing of the neighbor branches is repeated and the scope is raised further, until it reaches the root of the tree. In this way *TB-EQ* calculates forces with decreasing granularity, the more distant the branches are to the *source* element.

Assuming a data model of $n_{element}$ elements, the conventional way of affinity evaluation and force calculation would indicate a total comparison amount of

$$n_{calculation} = \frac{n_{element} \cdot (n_{element} - 1)}{2} \tag{2.39}$$

Applying the introduced precalculated tree structure with $size_{opt}$ subbranches per branch, an element $e$ has an average depth of

$$depth^{e} = \frac{ln(n_{element})}{ln(size_{opt})} \tag{2.40}$$

Following the described scheme of the bottom-up *TB-EQ* subalgorithm, the amount of single force calculations per element $e$ is reduced to an average of

$$n_{calculation}^{e} = depth^{e} \cdot (size_{opt} - 1) \tag{2.41}$$

Extending this to all elements within the data model, the number of all calculations per iteration on average can be described as

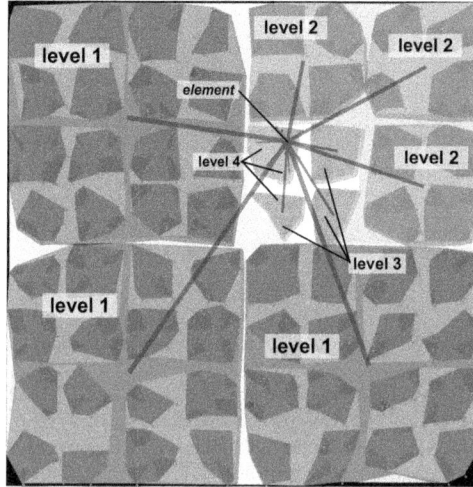

**Figure 2.23:** *Branch Consideration Scheme during Force Calculation for a Single Element*; Shown is the rendered *HBR* of the model described in Table 2.9. An example *element* (center of the blue lines) is chosen to visualize the branch consideration scheme, as described in Figure 2.22. During each transposition step the force calculation of the example *element* is based on the according sub items beneath the *element*'s super branches of the level 4 to 1 (blue lines), including the tree's root. In this way the count of required force calculations are reduced to a fraction of all element – element relations at the cost of less of accuracy.

$$n_{calculation}^{TB\text{-}EQ} = n_{element} \cdot depth^e \cdot (size_{opt} - 1) \tag{2.42}$$

when utilizing the tree structure. The resulting average calculation reduction is

$$r_{calculation} = \frac{n_{calculation}^{TB\text{-}EQ}}{n_{calculations}}$$

$$r_{calculation} = 2 \cdot \frac{ln(n_{element}) \cdot (size_{opt} - 1)}{ln(size_{opt}) \cdot n_{element}} \tag{2.43}$$

Example: The calculation reduction factor for $n_{element} = 10^6$ with a branch size of $size_{opt} = 16$ is $r_{calculation} = 0.00015$, i.e. the amount of calculations is reduced to $0.015\%$ of its original value. According to Formula (2.43) the reduction increases for larger data models.

On the other hand establishing the tree structure, including the averaged features, increases the memory consumption. A tree contains the following number of branches on average:

$$n_{branch} = \sum_{i=0}^{depth^e - 1} size_{opt}^{(i)} \tag{2.44}$$

where $depth^e - 1$ is the first branch level above the leafs within the tree (Formula (2.40)). Neglecting the memory consumption of the tree structure itself and contained operating variables, such as the *MLD*, the minimal memory increase factor $r_{memory}^{min}$ is:

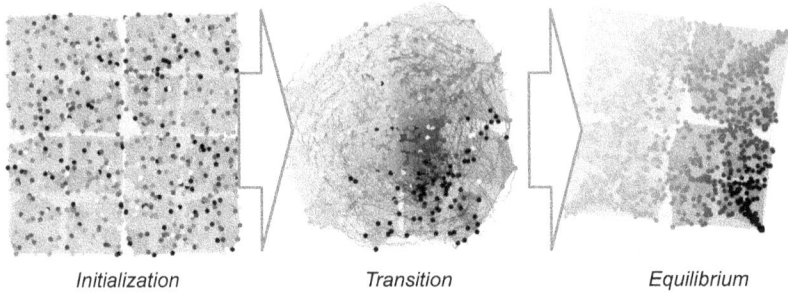

*Initialization*                    *Transition*                      *Equilibrium*

**Figure 2.24:** *Rendered CVD TB-EQ Transposition Example*; Shown is the model transition from the initial to the equilibrium state.

**Initialization:** The elements of the example data model are initialized at random starting positions between $(0,0)$ and $(1,1)$ and two random continuous property values according to the *CVD* type (Section 2.2.3 and Table 2.8). The dots' color depicts the values of both properties (red and green). The colored semi-transparent polygons represent the branches of the underlying element tree.

**Transition:** The *TB-EQ* subalgorithm calculates forces iteratively for each element according to the affinities and distances to all neighbor items, following the scheme described previously, and moves them step-wisely to their equilibrium positions (gray paths), i.e. the locations where all distances are maximally close to their optima.

**Equilibrium:** After the termination of the *TB-EQ* subalgorithm the elements are arranged according their affinities, i.e. the space between the four extrema in the four corners (bottom-right: $p_{red/green} = 0$; bottom-left: $p_{red/green} = 0/255$; top-right: $p_{red/green} = 255/0$; top-left: $p_{red/green} = 255$) is filled with elements continuously. The visualization shows, that the equilibration state of the *TB-EQ* subalgorithm is comparable to the result of the EQ Algorithm (compare Figure 2.12).

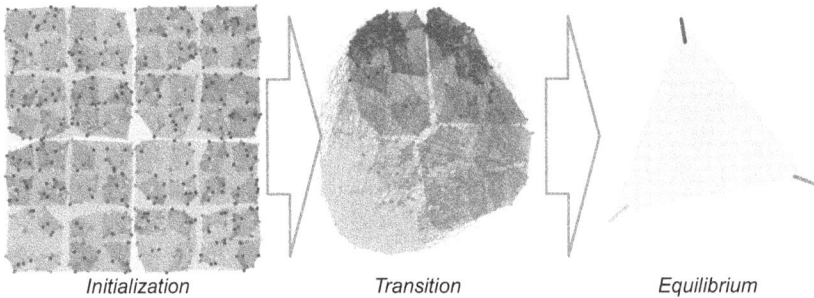

*Initialization*                    *Transition*                      *Equilibrium*

**Figure 2.25:** *Rendered DVD TB-EQ Transposition Example*; Shown is the modeltransition from the initial to the equilibrium state.

**Initialization:** The elements of the example data model are initialized at random starting positions between $(0,1)$ and $(1,1)$ and with one out of three property presets according to the *DVD* type (Section 2.2.3 and Table 2.8). The dots' color depicts the values of the three properties (red, green, blue). The colored semi-transparent polygons represent the branches of the underlying element tree.

**Transition:** The *TB-EQ* subalgorithm calculates forces iteratively for each element according to the affinities and distances to all neighbor items, following the scheme described previously, and moves them step-wisely to their equilibrium positions (gray paths), i.e. the locations where all distances are maximally close to their optima.

**Equilibrium:** After the termination of the *TB-EQ* subalgorithm the elements are arranged according their affinities, i.e. in three distinct and distant groups, each containing the elements with affine properties (top: $p_{blue} \rightarrow 255$; right: $p_{red} \rightarrow 255$; left: $p_{green} \rightarrow 255$). The visualization shows, that the equilibration state of the *TB-EQ* subalgorithm is comparable to the result of the EQ Algorithm (compare Figure 2.13).

$$r_{memory}^{min} = 1 + \frac{n_{branch}}{n_{element}}$$

$$r_{memory}^{min} = 1 + \frac{\sum_{i=0}^{depth^e-1} size_{opt}^{(i)}}{n_{element}} \tag{2.45}$$

Example: The minimal memory increase factor for $n_{element} = 10^6$ with a branch size of $size_{opt} = 16$ is $r_{memory}^{min} = 1.07$, i.e. 7% of the original elements' memory consumption is additionally required to store the branches. With regard to the high dimensionality of the original data, the neglected consumption caused by the tree structure and operating variables is comparable small and $r_{memory}^{min}$ serves as a sufficient approximation.

### 2.3.2.7 Algorithm Enhancement by Tree Maintenance

As shown in the previous section, the *TB-EQ* subalgorithm requires an element tree with updated hierarchical relations, branch positions and branch metrics (Section 2.3.2.3 - 2.3.2.5). Counteracting this, the frequent element position updates invalidate the tree. The position vectors' modification may shift element away from their original branches, towards other branches in other hierarchy levels. This is equivalent to a growing effect of the branches, so that the branches overlap strongly and the tree looses the capability to represent the data model in a hierarchical and compact way. Hence the previous approach relies on periodic tree rebuilds, which occupies comparable much runtime.

Therefore the subalgorithm *Tree Maintenance* (TM) is introduced to replace the costly tree rebuilds with tree maintenance steps, which reassign the shifted elements to their closest branches, i.e. the main task of the *TM* process is to find the closest branch in the lowest hierarchy level for each element. First, the distance between an element $e$ and a branch $b$ is introduced as:

$$dist^{e,b} = |\vec{pos}^b - \vec{pos}^e| \tag{2.46}$$

To reduce the number of distance calculations the following estimation, based on the triangular inequality [Inequality, 2014], can be applied: If the distance $dist^{b_{super},e}$ of an element $e$ to a super branch $b_{super}$ is known, the minimal distance $dist_{min}^{e,b}$ from $e$ to one of its subbranches $b$ can be estimated, utilizing the also known super branch distance $dist^{b,b_{super}}$:

$$dist_{min}^{e,b} = |dist^{b_{super},e} - dist^{b,b_{super}}|$$

$$dist_{min}^{e,b} \leq dist^{e,b} \tag{2.47}$$

Based on this inequality an abort criteria can be formulated as:

$$dist_{min}^{e,b} > dist^{e,b_{best}} \tag{2.48}$$

which means that if the minimal distance $dist_{min}^{e,b}$ exceeds the current best distance, the distance calculation can be skipped and the search can be aborted at this point, as $b$ or one of its subbranches can not be closer to $e$ than the current best branch.

As shown in Figure 2.26, the *TM* subalgorithm firstly detects the closest branch for each element.

This is done in a top-down manner, beginning at the tree's root and advances down the tree's hierarchy, following a path of the closest branches of each hierarchy level. At each level the closest branch of the current super branch is determined, which serves as the next super branch during the subsequent recursion step. Utilizing the minimal distance estimation (Formula (2.48)) reduces the number of distance calculations during each step. After the hierarchically lowest and closest branch is found, the current element is assigned to it as its new leaf.

Subsequent to the reassignment of all elements to their new branches, a tree update step is performed in a similar manner as described in the *TB* process (Section 2.3.2.3). A branch is denoted as changed, if it or any of its direct or indirect subbranches lost or gain elements during the reassignment step. That means from the tree's root to the leafs (elements) every changed branch is checked regarding its number of sub items, which may cause a split or a collapse. Before the *MLD*s are reevaluated (Section 2.3.2.4), the invalid branch position vectors are updated. The maintenance step is finalized by the recreation of the branch properties and the calculation of the *IA*s of changed branches (Section 2.3.2.5).

To illustrate the reassignment process further, Figure 2.27 shows the branch switch of a single element, caused by its current force. After changing its position vector, the element is no longer within the domain of its current branch, hence it is sorted in at the tree's root and sinks down the tree, following the respective closest branches until it reaches the tree's bottom. Due to the leaf number changes of the former and the current branch, tree operations (split or collapse) may be necessary.

In addition to schematic algorithm flow, Table 2.11 describes a *CVD* model, which is processed by the *TB-EQ* subalgorithm (Section 2.3.2.6), utilizing the *TM* mechanism described above. A transient state of the transposition process is shown in Figure 2.28. Starting from the elements' embedding in a tree after a recent position update, it is apparent, that some elements left the domain of their branches, which leads to branch overlappings (especially in the center of the tree), causing distorted subsequent *TB-EQ* iterations.

The standard approach to handle this issue is to discard the tree completely and rebuild it (*TB* subalgorithm, Section 2.3.2.3). The faster way to deal with the tree distortions is to apply the *TM* mechanism. Both results are shown for comparison. Note that the results of both approaches are visually comparable, but not identical. However, tests showed that maintained trees provide the same functionality to the *TB-EQ* subalgorithms, as rebuilt ones. The speed gain grows with growing iteration counts, as if the positions of only few elements change only slightly, the tree maintaining effort is reduced.

| Name | Symbol | Value / Range |
|---|---|---|
| element count | $n_{element}$ | 300 |
| dimensionality | $n_{dimension}$ | 2 |
| | | Group *Red-Green* |
| property value *red* | $p_{red}$ | 0..255 |
| property value *green* | $p_{green}$ | 0..255 |
| property value *blue* | $p_{blue}$ | 0 |

**Table 2.11:** *Data Model Parameters of TM Example*

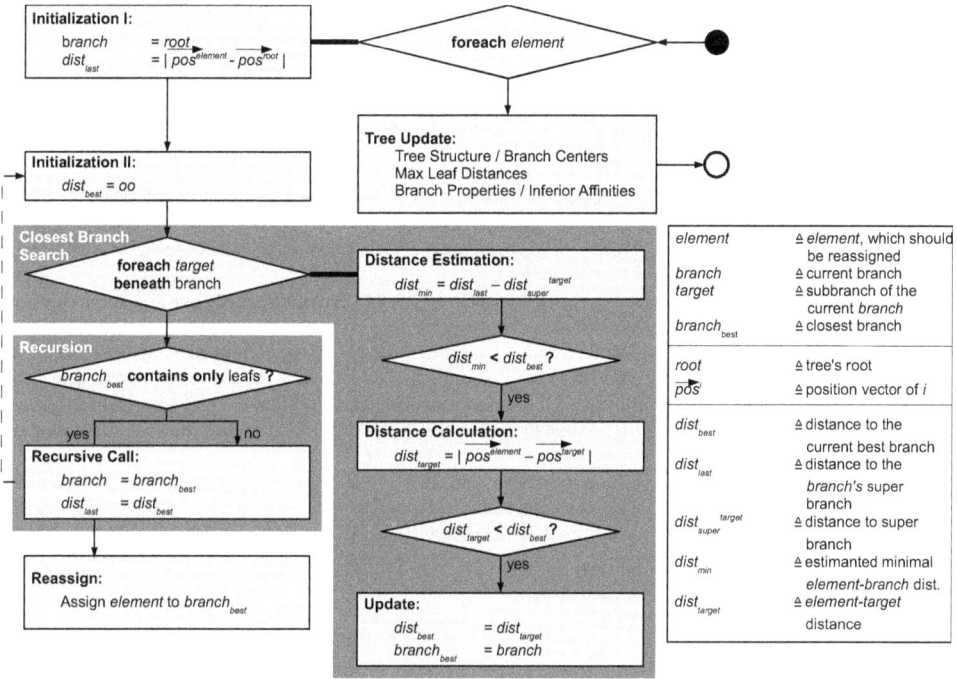

**Figure 2.26:** *TM Scheme*; Shown is the tree maintaining subalgorithm, which reassigns elements to their closest and deepest branches.

**Initialization:** First, the subalgorithm initializes the main loop, which processes every element. The search starts for each element at the tree's root. Furthermore, the distance of the last hierarchy level $dist_{last}$, which is required by the abort estimation, and the current best (minimal) branch distance $dist_{best}$ are initialized.

**Closest Branch Search:** At this point the subalgorithm searches the closest branch to the given *element* at the current hierarchy position (the *branch*). First, for each subbranch *target* beneath the current *branch* the abort estimation is performed (Formula (2.48)), checking whether the target could be a candidate as the closest subbranch. If the check passes, the element – target distance is calculated and the closest branch is stored.

**Recursion:** After the determination of the closest branch of the current hierarchy level, the search is continued recursively by setting the current branch to the best found subbranch $branch_{best}$ and updating the best distance of the last hierarchy level $dist_{last}$.

**Reassign:** After the detection of the closest branch of the lowest hierarchy level, the *element* is assigned to this branch.

**Tree Updates:** After each element is reassigned to its closest branch, the tree is updated in the same manner as described in the *TB*, the *MLD* and the *HBR* subalgorithms. First, the tree structure is checked (split and collapse) and the branches' centers are updated (Section 2.3.2.3). Secondly the *MLD*s are recalculated (Section 2.3.2.4), followed by the update of the branch properties and the *IA*s (Section 2.3.2.5).

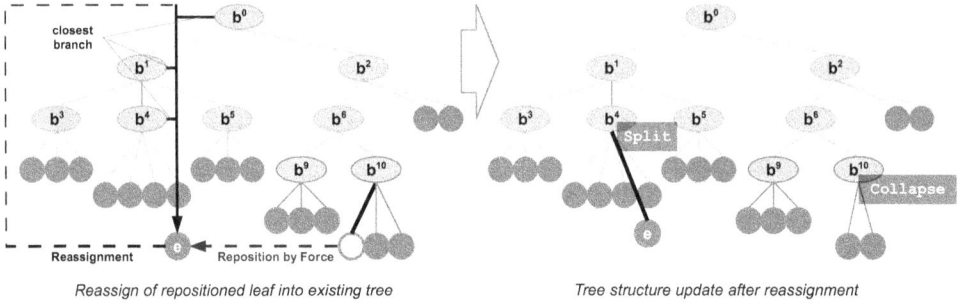

Reassign of repositioned leaf into existing tree                Tree structure update after reassignment

**Figure 2.27:** *TM Example of a Single Element Repositioning*; Shown is the reassignment processing of a single element.
    **Left:** Given is an example tree of 10 branches $b^0..b^{10}$, which spreads in only one dimension (horizontal), and a processed element $e$. After applying the force vector of the last iteration (Section 2.3.2.6) to the position vector of $e$, its location within the tree has changed, thus making an update of the element's branch association necessary. Starting at the tree's root, the *TM* subalgorithm detects the closest branch for each hierarchy level in a recursive manner (Figure 2.26, *Closest Branch Search*). In this case the branches $b^0$, $b^1$ and $b^4$ are identified, defining a "search path" down to the closest and deepest branch.
    **Right:** After assigning $e$ from its previous branch $b^{10}$ to the new tree location $b^4$, the subsequent update mechanisms recognize both branches as too large or small, respectively, and splits and collapses them (not shown, see also Figure 2.26, *Tree Update*).

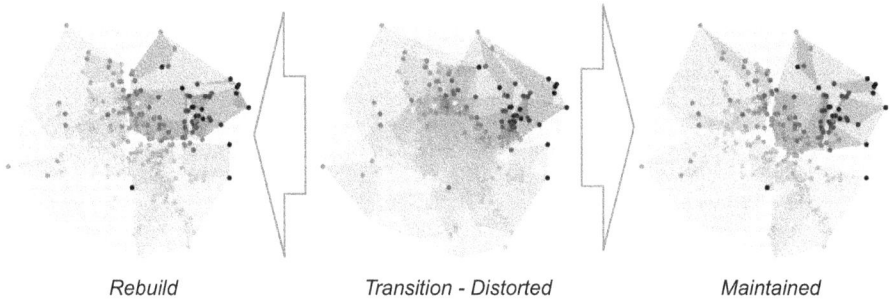

Rebuild                          Transition - Distorted                          Maintained

**Figure 2.28:** *Rendered TM / TB Example;* shown is a visualized comparison of a distorted tree and its rebuild or maintained version, respectively.
    **Middle:** According to Table 2.11 a *DVD* data model is shown in transition during the *TB-EQ* equilibration process. The last elements' position update distorted the tree, so that the branches are misshaped and overlap. These distortions hinders the continue of the *TB-EQ* process, making a rebuild or a maintenance necessary.
    **Left:** As described in Section 2.3.2.3 the tree can be completely rebuild, enabling the *TB-EQ* subalgorithm to continue with a valid tree.
    **Right:** As alternative to the rebuild, the tree can be maintained as described above (Figure 2.26). After the reassignment process and the update of the tree structure and metrics, it is revealed, that the maintained tree is comparable to the rebuilt tree regarding the branches' shape and the degree of overlapping. In this way the computation effort can be saved by the maintenance of the distorted tree section, avoiding the complete and costly rebuild of the tree.

### 2.3.2.8 Algorithm Enhancement by Substitution

After the consideration of the tree maintenance in the previous section, the following optimization approach is discussed: As the number of elements has at least linear influence onto the runtime, the execution speed can be improved by replacing very affine and very close elements by a substitute. To accomplish this the existing tree structure is utilized, so that all elements beneath a branch $b$ can be substituted, if the following applies:

$$aff^b_{inferior} \geq aff^{limit}_{inferior} \tag{2.49}$$

where $aff^b_{inferior}$ is the $IA$ of the branch (Section 2.3.2.5) and $aff^{limit}_{inferior}$ is a global threshold. Typically this threshold has a value of $0.99$, which stands for almost identical elements. According to this requirement the *Substitution* (*SUB*) process tests every branch to be inferior affine enough, so that the underlying elements can be removed from the model temporarily. The branch is altered to a substitute element with an accumulated weight and an average position as well as with the properties of the former branch. Consequently the substitute elements can be substituted again by assigning their encapsulated original elements to the new substitute.

At the end of the transposition process all substitutions are reverted. Due to the loss of the elements' original position vectors $\overrightarrow{pos}^e$, the restored vectors have to be compliant with:

$$|\overrightarrow{pos}^{b_{subst}} - \overrightarrow{pos}^e| \leq Func_{dist_{opt}}(aff^{b_{subst}}_{inferior}) \tag{2.50}$$

where $\overrightarrow{pos}^{b_{subst}}$ is the position vector of the substitute and $Func_{dist_{opt}}(..)$ is the user-defined optimal distance function (Formula (2.8), Section 2.3.2.1). As argument serves the $IA$ $aff^{b_{subst}}_{inferior}$ of the substitute (which is equal to $aff^{limit}_{inferior}$ or higher). The result is an $IA$ dependent circular area around the substitute, in which the former encapsulated elements can be placed.

### 2.3.2.9 Algorithm Enhancement by Insertion

The last subalgorithm *Insertion* (*INS*), completing the subsequent DT Algorithm (Section 2.3.2.10), handles the issue that unfavorable initial element positions (compare Section 2.3.2.2) may cause additional iterations to find the equilibrium, thus reducing the efficiency or the final transposition quality (Formula (2.11)). Hence a mechanism is required to improve the initial element positions as much as possible prior to the actual equilibration process.

The *INS* subalgorithm utilizes the inherent irregularity of the branch properties, i.e. the fact that the average property values of each branch tend to one direction each. It follows that for each element $e$ a branch $b$ exists in each hierarchy level, which is most affine to $e$ with regard to their properties. The adapted affinity takes into account the $IA$ $aff^b_{inferior}$ of the branch (Formula (2.34), Section 2.3.2.5):

$$aff^{e,b}_{adapted} = aff^{e,b} \cdot aff^b_{inferior} \tag{2.51}$$

where $aff^{e,b}$ is equivalent to the total affinity of $e$ and $b$ (Formula (2.7)). Thus, following a track down the hierarchical structure, along the branches of the highest adapted affinity of each hierarchy level, each element can be assigned to a branch with no further subbranches and a high adapted affinity. While assigning an element $e$ to that branch $b$, the element's position vector $\overrightarrow{pos}^e$ changes, so that the following applies:

$$|\overrightarrow{pos}^e - \overrightarrow{pos}^b| < dist^b_{maxleaf} \qquad (2.52)$$

which means, that the distance to the branch's center $\overrightarrow{pos}^b$ is equal to or lesser than its *MLD* $dist^b_{maxleaf}$, i.e. the element is located within the radius of the branch.

The assignment mechanism of all elements to one of their most affine and most deepest branches is illustrated in Figure 2.29. For each iterated element, the scheme initializes the best affinity value and the start branch at the tree's root. Subsequently the subalgorithms tests each subbranch whether it is more affine than the current branch and advances to the best branch in a recursive manner.

For the purpose of acceleration the current branch's precalculated *IA* is compared to the current best adapted affinity first, since the result of Formula (2.51) can not be larger than that value. Thus, in this case the calculation is skipped. Otherwise the adapted affinity is calculated, compared and stored if necessary. If the most affine and deepest branch is found, the element is assigned to it as a leaf and the element position is altered, so that the inequality of Formula (2.52) is fulfilled.

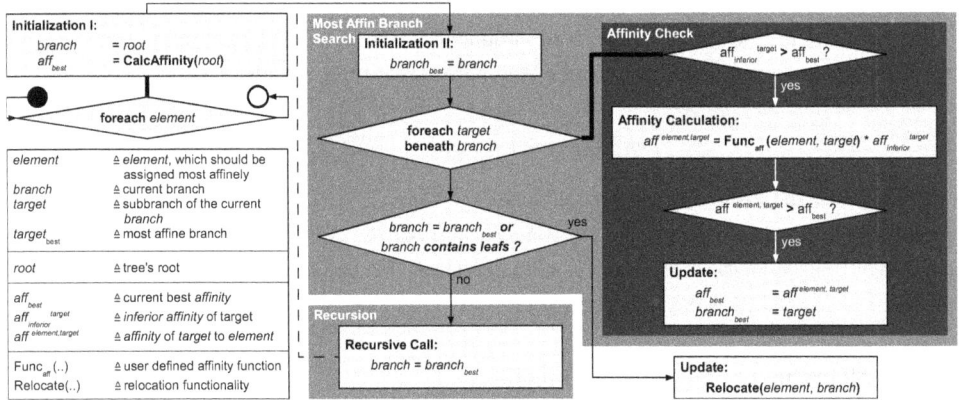

**Figure 2.29:** *Most Affine Assignment subalgorithm Scheme*; Shown is the sub mechanism of the *INS* subalgorithm, which assigns all elements to the most affine branch within the tree.
**Initialization:** The main loop starts for each *element* at the tree's root and initializes the current best adapted affinity $aff_{best}$ with the affinity to the root according to Formula (2.51).
**Most Affine Branch Search + Recursion:** First, it is assumed, that the current *branch* is the most affine one. Subsequently all subbranches *targets* are checked for being more affine than the current *branch*. In the case a more affine subbranch is found, the search continues recursively one hierarchy level deeper, setting the current *branch* to the more affine subbranch *branch$_{best}$*.
**Update:** In the case no more affine subbranch is found or the current *branch* contains only leafs (elements), the Relocate sub functionality is utilized, assigning the *element* to the *branch* and changing the element's position vector, so that Formula (2.52) applies.
**Affinity Check:** First, the test for the *element – target* affinity checks whether the *IA* of the *target* branch $aff^{target}_{inferior}$ exceeds the current best affinity $aff_{best}$, otherwise the check can be aborted, for that the final affinity can not be larger. Subsequent to this check, the *element – target* affinity $aff^{element, target}$ is calculated (Formula (2.51)). If it exceeds the current best affinity, the value is stored together with the updated most affine branch *target*.

The recursive search for the tracks along the most affine branches down the hierarchy is visualized in

Figure 2.30. Here a set of elements with different property values is assigned to the most affine and most deepest branches, i.e. branches with similar colors and no subbranches. Due to the elements' random initial positions, their super-ordinated branches differ in the average property values. In this case suitable, i.e. affine branches exist for all element groups. Beginning at the tree's root, each element follows a track down, which leads to an affine branch, in whose radius the element is placed (Formula (2.52)). After the reassignment of all elements some branches might be abandoned and thus can be removed.

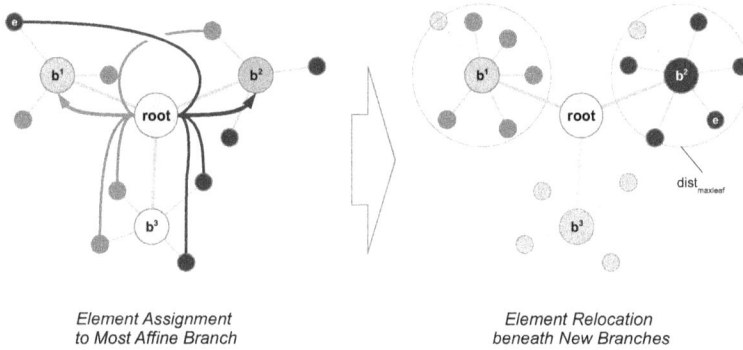

Element Assignment
to Most Affine Branch

Element Relocation
beneath New Branches

**Figure 2.30:** *Most Affine Assignment Example*; Shown is an example tree of three branches $b^1$, $b^2$ and $b^3$ and 10 elements ($e$), which are assigned to the branches (gray lines). The elements have different properties (visualized as red and green), whereas the branches represent the average properties of their subordinated elements (light green, orange and yellow).
**Left:** During the *Most Affine Assignment* Process (Figure 2.29) the subalgorithm detects the most affine branch for each element, starting at the tree's root and recursively progresses to lower levels of the hierarchy. Hence, following their most affine path, all red and green elements pass the yellow root before approaching to the green dominated branch $b^1$ (green elements) or to the red dominated one $b^2$ (red elements). The assignment tracks are depicted as colored arrows.
**Right:** After reaching their most affine branch, the elements are randomly placed within the radius (*MLD*) of the branch, visualized as red dotted circles. In this way the indifferent branch $b^3$ becomes obsolete, can be discarded and the elements are coarsely sorted and placed regarding their properties (in this case: their colors).

Although after the most affine assignment step the elements are located close to affine neighbors, they are probably fragmented in several distant subgroups and the distances to each other don't correspond to their affinities (Formula (2.8)). Hence the *INS* subalgorithm extends the most affine assignment step to a scheme, which is shown in Figure 2.31:

After the elements' relocation to appropriate branches, the tree is maintained and updated, utilizing the *TM* and the *HBR* subalgorithms (Sections 2.3.2.7 and 2.3.2.6). Taking advantage of the fact that affine elements are now located close to each other, a substitution step similar to the *SUB* concept (Section 2.3.2.8) is performed, which replaces branches with only leafs beneath by substitutes, thus reducing the total amount of elements. Subsequently these substitute elements are processed by the EQ Algorithm (Section 2.3.2.2), to arrange them in a manner, that their distances correspond to their affinities. This step may be regarded as a preliminary coarse and flat equilibration step, handling only groups of strongly affine elements. Finally the substitution is reversed, replacing the substitute elements by their encapsulated originals (compare Section 2.3.2.8). After this process all elements are positioned coarsely regarding their affinities to each other.

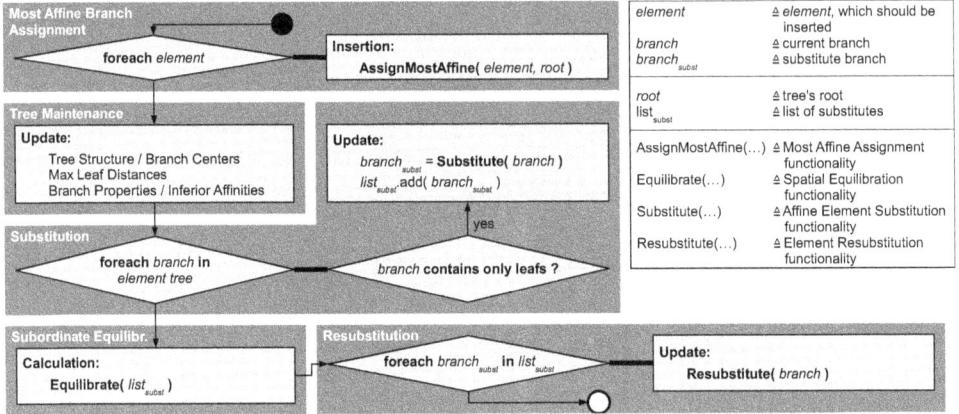

**Most Affine Branch Assignment**
foreach *element*

**Insertion:**
AssignMostAffine( *element*, *root* )

**Tree Maintenance**
Update:
  Tree Structure / Branch Centers
  Max Leaf Distances
  Branch Properties / Inferior Affinities

Update:
  $branch_{subst}$ = **Substitute**( *branch* )
  $list_{subst}$.add( $branch_{subst}$ )

**Substitution**
foreach *branch* in element tree

*branch* contains only leafs ?    yes

**Subordinate Equilibr.**
Calculation:
  Equilibrate( $list_{subst}$ )

**Resubstitution**
foreach $branch_{subst}$ in $list_{subst}$

Update:
  Resubstitute( *branch* )

| | |
|---|---|
| *element* | ≙ *element*, which should be inserted |
| *branch* | ≙ current branch |
| $branch_{subst}$ | ≙ substitute branch |
| *root* | ≙ tree's root |
| $list_{subst}$ | ≙ list of substitutes |
| AssignMostAffine(...) | ≙ Most Affine Assignment functionality |
| Equilibrate(...) | ≙ Spatial Equilibration functionality |
| Substitute(...) | ≙ Affine Element Substitution functionality |
| Resubstitute(...) | ≙ Element Resubstitution functionality |

**Figure 2.31:** *INS Subalgorithm Scheme*; Preliminary process to initialize the elements' positions coarsely.

**Most Affine Branch Assignment:** The *INS* subalgorithm starts with the assignment of every element to its most affine branch, as introduced above (Figure 2.29). After this process the tree is distorted and the elements are sorted coarsely, but not in the desired distance to each other. Hence the subsequent steps handle this issue.

**Tree Maintenance:** Reassigning elements to other branches and changing their position vectors requires a tree maintenance step. This task is accomplished by the *TM* subalgorithm (Section 2.3.2.7).

**Substitution:** Subsequent to the tree update, it can be assumed that branches hold comparable affine elements. Hence, branches, containing only leafs (elements), are substituted by container elements with according weights and properties to accelerate the subsequent step (similar to Section 2.3.2.8).

**Subordinate Equilibration:** In a subordinated EQ step (Section 2.3.2.2) the substitutes are rearranged to optimize their distances regarding their affinities.

**Resubstitution:** Finally the substitutes are replaced by their elements, locating them within a radius, which is defined by the inferior affinity of the former substitute (Section 2.3.2.8, Formula (2.50)).

An illustration of the *INS* subalgorithm flow is shown in Figure 2.32, processing the *CVD* data model described in Table 2.8 in Section 2.3.2.2. First, the elements are positioned randomly and evenly across a two-dimensional plane. The previously explained most affine assignment step (Figures 2.29 and 2.30) detects affine and deep subbranches and relocates the elements within the *MLD* radius of these new super branches. The result of this step is depicted as several element groups of affine colors and different sizes and densities. Subsequently these groups are substituted by corresponding container elements, utilizing the existing branches within the tree and reducing the total computation effort. This is visualized by larger substitution elements. Because the elements are affine to their neighborhood, but not in affine-equivalent distances, a subsequent EQ process rearranges the substitution elements instead of handling all elements. The equilibrium state is shown as repositioned substitution elements, whose color similarities are expressed by their distances. That means the distances between these elements correspond to their affinities, and thus to the average affinities of the encapsulated original elements. Finally the substitution is reversed, in the Figure the restored original elements replace their gray substitutes.

The *INS* result shows a coarse arrangement of all elements, which is done with only a fraction of the *TB-EQ* algorithm's computation effort and which is already similar to the final result (Figure 2.24, Section 2.3.2.6).

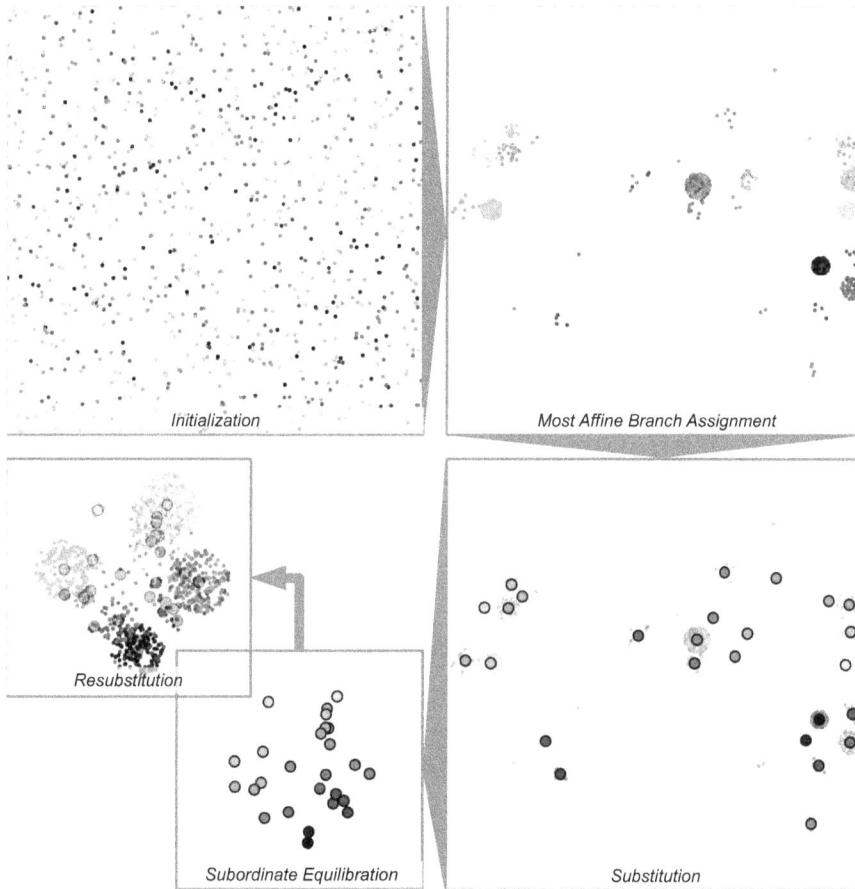

**Figure 2.32:** *Rendered Insertion Example*; Shown is the visualization of the *INS* subalgorithm (Figure 2.31) processing the data model described in Table 2.8. The subalgorithm's task is to sort and arrange the elements coarsely prior to the actual equilibration.

**Initialization:** The elements are shown as colored dots, representing their properties, and are placed initially randomly in two dimensions within the range $(0, 1)$ before the construction of the tree (Section 2.3.2.3) is performed (tree is not shown).

**Most Affine Assignment:** This step (Figure 2.29) assigns all elements to according branches, which are most affine regarding their properties, and changes the elements' position to place them near these branches. The figure shows that affine elements (similar colors) are already placed close to each other. This is followed by a tree maintenance step (Section 2.3.2.7) to remove distortions from the tree and update the branches' properties (not shown).

**Substitution:** Due to the fact that after the previous steps all branches contain similar elements, they can be substituted to reduce the amount of elements to process. The substitutes are visualized by large color dots, which are placed on top the encapsulated elements (gray).

**Subordinate Equilibration:** Following the same scheme as described in Section 2.3.2.2 the substitutes are rearranged force-driven regarding their properties until they reach an equilibrium state.

**Resubstitution:** Finally the substitution is reversed, positioning the elements (small colored dots) around the locations of their substitutes (gray circles). Note that after this process the arrangement of the elements is already similar to the final state, as shown in the Figure 2.24 in Section 2.3.2.6.

## 2.3.2.10 The Final Dimensional Transposition Algorithm

Concluding the presentation of the single algorithmic components, the *Dimensional Transposition Algorithm* (DT Algorithm) illustrates the global scheme, integrating all mechanisms introduced previously. As described in Section 2.3.2.1 the overall task is to arrange the elements in a low-dimensional metric space, so that their Euclidean distances express their property based affinities. The EQ approach in Section 2.3.2.2 provides an algorithm, which already fulfills this task, but exhibits an at least quadratic runtime behavior. The issue is discussed in Sections 2.3.2.3 - 2.3.2.5, representing mechanisms to create a *HBR* of the elements, and in Section 2.3.2.6, which describes the *TB-EQ* subalgorithm. This approach utilizes the *HBR* to accomplish the task similar to the EQ Algorithm, but in less time with acceptable quality losses. The subsequent Sections 2.3.2.7 - 2.3.2.9 cover the issues to reduce the *HBR* maintenance time, handle partial results efficiently and improve the element position initialization, i.e. three optimizations are represented, which are based on the *TB-EQ* subalgorithm.

Figure 2.33 shows the overall scheme, which combines all subalgorithms described in Section 2.3.2. After the element initialization the underlying tree is constructed for the first time, utilizing the *TB, MLD* and *HBR* subalgorithms (Sections 2.3.2.3 - 2.3.2.5).

It follows the *INS* step (Section 2.3.2.8), which improves the initial element positions: After the element assignments to their most affine branches, the *TM* subalgorithm (Section 2.3.2.7) is utilized in combination with the branch update mechanisms to ensure the tree's validity. Subsequently the resulting element groups are substituted and a subordinate EQ process (Section 2.3.2.2) rearranges these groups according to their average affinities. The *INS* step is concluded by the resubstitution as described in the *SUB* approach.

Subsequently to the coarse element prearrangement, the fine-granular equilibration follows, handling single elements. First the element tree is constructed a second time, utilizing the same components as described above. Based on this tree, the *TB-EQ* subalgorithm (Section 2.3.2.6) is performed to position the elements more precisely: During each iteration the *SUB* mechanism replaces very affine element groups by substitutes, reducing the total element count. Then the next position vectors of all elements are calculated utilizing a bottom-up strategy within the element tree. An iteration is concluded by a tree maintenance and a branch update step.

Finally the substitutes, created during the *TB-EQ* process are reversed according to the *SUB* subalgorithm (Section 2.3.2.8).

The illustration of the transposition process of a larger data model finalizes this section. Table 2.12 describes a *CVD* data model and the parameters of the utilized DT Algorithm, while Figure 2.34 shows the major steps of this transposition. The same user-defined functions (Formula (2.24)) as in the previous examples are utilized.

It starts with the initial random element placement within the target space, depicted as a square of colored dots. The construction of an underlying tree (not shown) enables the *INS* subalgorithm to arrange the elements coarsely regarding their affinities, i.e. to reposition them in (circular) groups of affine properties (colors) in distances to each other, which represent groups' average affinities. Note that both extrema of the property sets ($p_{red/green} = 255/0$ and $p_{red/green} = 0/255$) are located maximal distant from each other. Subsequently the element tree is constructed a second time, i.e. the Figure shows the *INS* result embedded in the tree, whose branches consist of four subbranches each. This data model representation is utilized by the *TB-EQ* subalgorithm, substituting very affine element groups and calculating

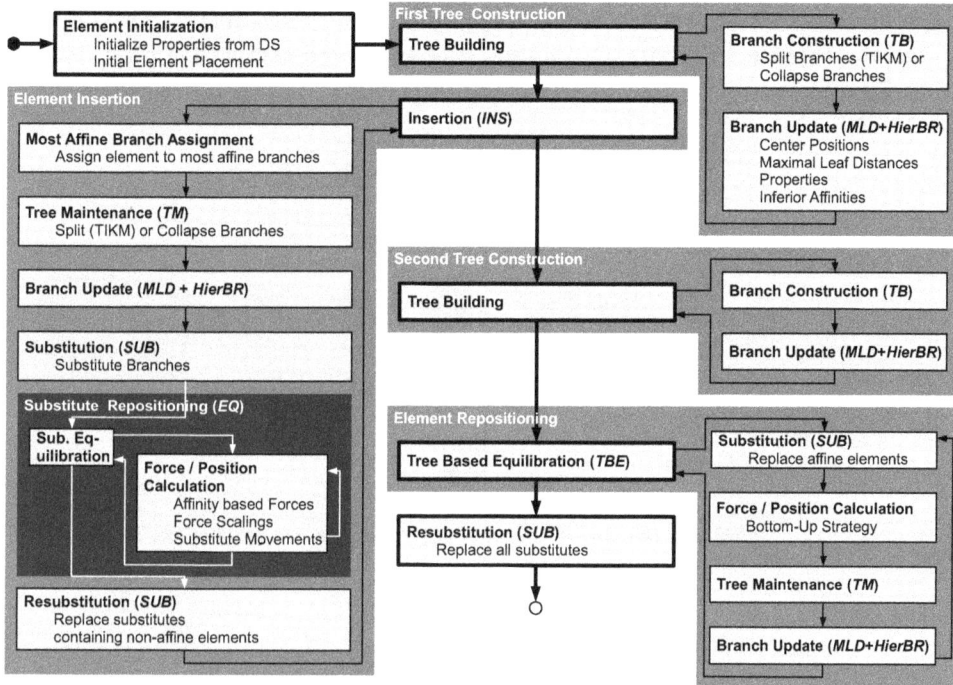

**Figure 2.33:** DT *Algorithm Scheme*; Shown is the algorithm's main sequence. The task is to arrange elements regarding their properties in a low-dimensional space for subsequent *CA*s.

**Element Initialization:** First, the elements, carriers of properties, are initialized from a given data source (DS) and placed randomly within the target space.

**First / Second Tree Construction:** During the algorithm process the element tree, an essential data structure for many subalgorithms, is constructed two times (before and after the *INS* subalgorithm step). The construction is done by the *TB* subalgorithm (Section 2.3.2.3) and is followed by the calculation of several branch metrics, i.e. by the *MLD* and the *HBR* subalgorithms (Sections 2.3.2.4 and 2.3.2.5).

**Element Insertion:** After the first tree construction the *INS* subalgorithm is performed (Section 2.3.2.9) to sort and arrange the elements coarsely before the actual equilibration starts. This process assigns the elements to their most affine branches, maintains and updates the tree to handle the occurred distortions and keep the branches valid (Sections 2.3.2.7, 2.3.2.4 and 2.3.2.5) and substitutes the resulting branches similar to the *SUB* component (Section 2.3.2.8). These substitutes are finally equilibrated by the EQ Algorithm (Section 2.3.2.2) to optimize their Euclidean distances to each other before the substitution is reversed.

**Element Positioning:** Subsequent to the second tree construction the final equilibration is performed by the *TB-EQ* subalgorithm (Section 2.3.2.6). It executes the *SUB* component iteratively to reduce the amount of processed elements and utilizes a bottom-up strategy within the tree to calculate the elements' forces and position vectors efficiently. These steps are followed be tree update and branch update steps to keep the tree valid.

**Resubstitution:** The main algorithm is concluded by the resubstitution of the encapsulated elements. Now the elements are positioned, so that the distances to each other are as close as possible to their optimal values, which is defined by the elements' property affinities.

the next element positions alternately. The result of this step is presented with the substitutes' positions (the more affine neighbor items are, the smaller the distance) and their paths to this positions. Finally the substitution is reversed, presenting the elements' equilibrium as the final result of the DT process.

Note that the arrangement is visually comparable to the smaller versions of this data model, processed by the EQ Algorithm and the stand-alone *TB-EQ* algorithm (Figures 2.12 and 2.24 in Sections 2.3.2.2 and 2.3.2.6).

| Name | Symbol | Value |
|------|--------|-------|
| element count | $n_{element}$ | $10k$ |
| dimensionality | $n_{dimension}$ | 2 |
| iteration count | $n_{iteration}$ | 29 |
| SA factor | $f_{SA}$ | 0.9 |
| minimal substitution affinity | $aff_{inferior}^{limit}$ | 0.99 |
| minimal branch size | $size_{min}$ | 3 |
| optimal branch size | $size_{opt}$ | 4 |
| maximal branch size | $size_{max}$ | 8 |
|  |  | Group *Red-Green* |
| property value *red* | $p_{red}$ | 0..255 |
| property value *green* | $p_{green}$ | 0..255 |
| property value *blue* | $p_{blue}$ | 0 |

**Table 2.12:** *Data Model Parameters of* DT *Algorithm Example Run*

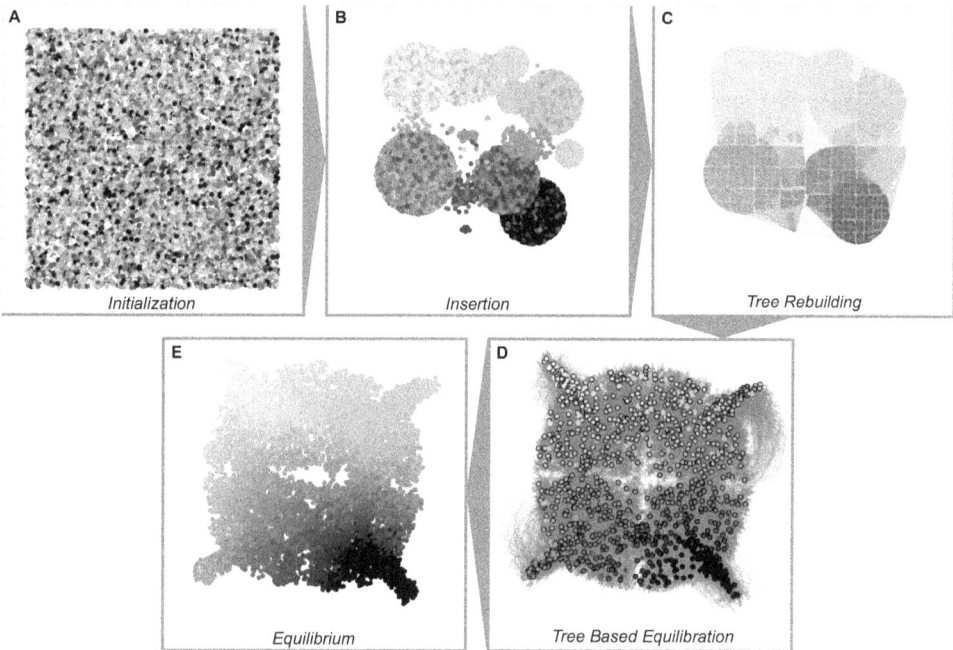

**Figure 2.34:** *Rendered* DT *Algorithm Result Example* based on a data model described in Table 2.12;

    **A:** Initially the elements (colored dots) are positioned randomly in two dimensions in a range of (0, 1) for each dimension. The elements' properties are represented by the elements' colors (fractions of red and green).

    **B:** The *INS* step (Section 2.3.2.9) builds a tree (*TB* subalgorithm), updates the branches and metrics (*MLD* and *HBR* subalgorithms), see Sections 2.3.2.3 - 2.3.2.5, and reassigns each element to its most affine branch. Subsequent to this, a subordinate equilibration step (EQ subalgorithm, Section 2.3.2.2) is performed, arranging the element groups (*SUB* subalgorithm, Section 2.3.2.8) regarding their average properties. The result is a coarsely sorted and arranged preliminary equilibration state. As visualized in the figure, affine (similar colored) elements are grouped together and positioned in distances to each other, which correspond with the colors' similarities.

    **C:** After executing the coarse sorting steps, the second tree is constructed and prepared to start the fine granular arrangement process. The figure shows the *INS* result embedded in a tree.

    **D:** Utilizing the second *HBR*, the *TB-EQ* (Section 2.3.2.6) sub process is performed to transpose the model to an equilibrium, i.e. arrange the elements iteratively regarding their properties (gray paths). Close and affine element groups are substituted (*SUB* subalgorithm), which are visualized by circles with dark contour lines. The figure depicts the equilibrium state, where all elements or their substitutes are positioned so that the distances to all other elements correspond to their affinities (similar colored neighborhood).

    **E:** The substitution of affine elements is reversed and the original elements are inserted, which finalizes the DT process. Now the elements are positioned, so that the distances to each other are as close as possible to their optimal values, which is defined by the elements' property affinities (*PC*, Section 2.3.2.1).

## 2.3.3 Result and Performance Measurements

In the following the measurement results of the tests as described in Section 2.2 are illustrated. The first three Sections compare the EQ and DT Algorithm's quality of result regarding *i*) the dependency of the algorithm granularity, i.e. its specified iteration count (Section 2.3.3.1), $ii$) the influence of the underlying hierarchical structure shape (Section 2.3.3.2), and $iii$) the target dimension count, limiting the algorithms' degree of freedom (Section 2.3.3.3).

The other crucial aspect, the algorithms' runtimes and runtime complexities, are considered in the Sections 2.3.3.4 and 2.3.3.5, investigating the runtime while processing large and very large data models.

### 2.3.3.1 Transposition Quality Dependency of the Simulated Annealing Factor

In the following representative transposition quality measurements of *test scenario* 1 are illustrated (Section 2.2.4, Table 2.4).

The Figures 2.35 and 2.36 show surface plots of the average transposition quality (Formula (2.4)) of processed *DVD* and *CVD* data models. To cover different sizes of the data models, the element count as well as the property count were altered by constant steps ($x$- and $y$-axes). The utilized transposition approaches were the slow, but precise EQ Algorithm (left side) and the fast, but less precise DT Algorithm (right side). Both algorithms were executed with different *SA* factors (Formula (2.18)), visualized as surfaces of varying colors. The results of the EQ Algorithm are regarded as references for transposition results of best possible quality.

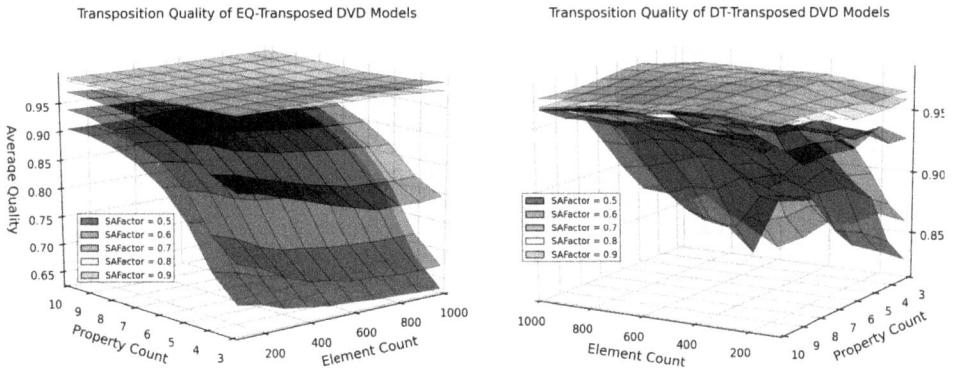

| Data Model Class ($type_{distribution}$) | Algorithms ($type_{algorithm}$) | Element Count ($n_{element}$) | Property Count ($n_{property}$) | Dimensions Count ($n_{dimension}$) | SA Factor ($f_{SA}$) | Optimal Branch Size ($size_{opt}$) |
|---|---|---|---|---|---|---|
| DVD | EQ (left) / DT (right) | $100..1k$ | $3..10$ | $n_{property}$ | $0.5..0.95$ | 8 |

**Figure 2.35:** *Average Quality of Result ($quality_{average}$) of the* EQ *and* DT *Algorithms applied to DVD Data Models, dependent on Element and Property Count, as well as SA factor.*

First, the following observation can be made: if the *SA* factor is high enough ($f_{SA} \geq 0.9$) the average transposition quality becomes independent of the data model's element count and their property counts, except if the DT Algorithm processes *CVD* data models. Here the most interesting region is that of high element counts ($n_{element} = 1k$) and high property counts ($n_{property} = 10$), where at least a local minimum exists, but its environment indicates a quality saturation in this range, i.e. the transposition quality becomes invariant even for larger data models at a comparable high values of about $0.85$.

The second observation is, that the majority of all transpositions exhibits a high quality ($quality_{average} \geq$

| Data      Model | **Algorithms** | Element | Property | Dimensions | SA      Factor | Optimal |
| Class | $(type_{algorithm})$ | Count | Count | Count | $(f_{SA})$ | Branch    Size |
| $(type_{distribution})$ | | $(n_{element})$ | $(n_{property})$ | $(n_{dimension})$ | | $(size_{opt})$ |
|---|---|---|---|---|---|---|
| CVD | EQ (left) / DT (right) | 100..1k | 3..10 | $n_{property}$ | 0.5..0.95 | 8 |

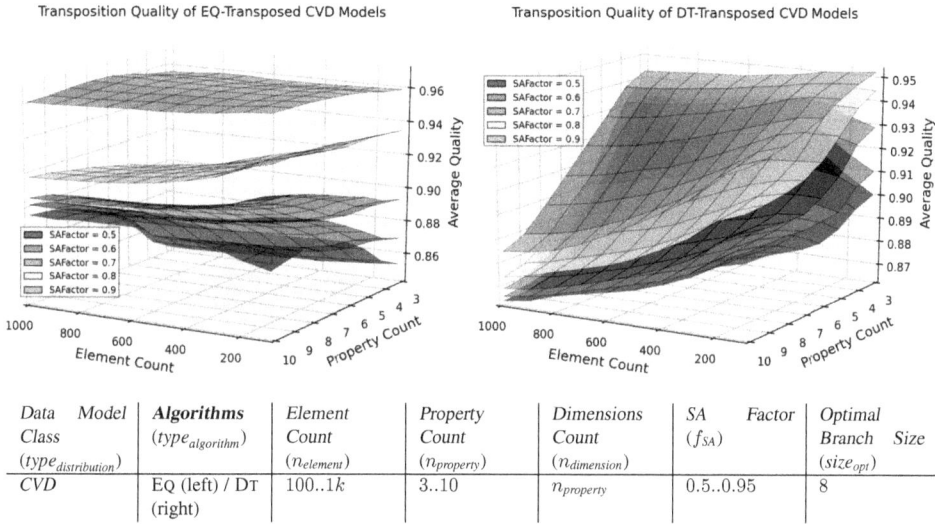

**Figure 2.36:** *Average Quality of Result ($quality_{average}$) of the* EQ *and* DT *Algorithms applied to CVD Data Models, dependent on Element and Property Count, as well as SA Factor.*

0.95), which is close to ideal transpositions.

A comparison of the DT Algorithm results with the EQ references shows that the faster DT Algorithm algorithm is able to deliver transposition results of almost the same quality, except for the most difficult case, that is the continuous property value distribution (*CVD*) of many properties encapsulated by many elements, but with still acceptable quality. Considering the second criteria, that is the worst single transposition qualities (Formula (2.4)) and for which the results can be found in the Appendix Section 6.1.2.1, a similar pattern is observable.

This leads to the conclusion, that a *SA* factor of $f_{SA}^{best} = 0.9$ is sufficient to enable the fast DT Algorithm to produce transposition results of a quality, which is sufficient and comparable to the best possible results delivered by the EQ method. As the *SA* factor determines the count of iterations and hence the runtime, it is desirable to set the factor to this lowest acceptable value. The chosen $f_{SA}^{best}$ is equivalent to an iteration count of 30 (Formula (2.23)).

### 2.3.3.2 Transposition Quality Dependency of the Branch Size

In the following representative transposition quality measurements of *test scenario 2* are illustrated (Section 2.2.4, Table 2.4).

Figure 2.37 shows surface plots of the average transposition quality (Formula (2.4)) of data models, which were altered in the same manner as described previously in Section 2.3.3.1. The transposition was done only by the DT Algorithm, processing element trees of different shapes to examine the influence of the tree parameters and detect their optimal values (see *TB* mechanism in Section 2.3.2.3). The count of subbranches per branch $size_{opt}$ was incremented in equidistant steps, visualized as surfaces of varying colors.

To illustrate the parameter's influence, Figure 2.38 depicts the qualities mapped onto the average quality – element count plane, where each point represents the average for all according property count dependent values.

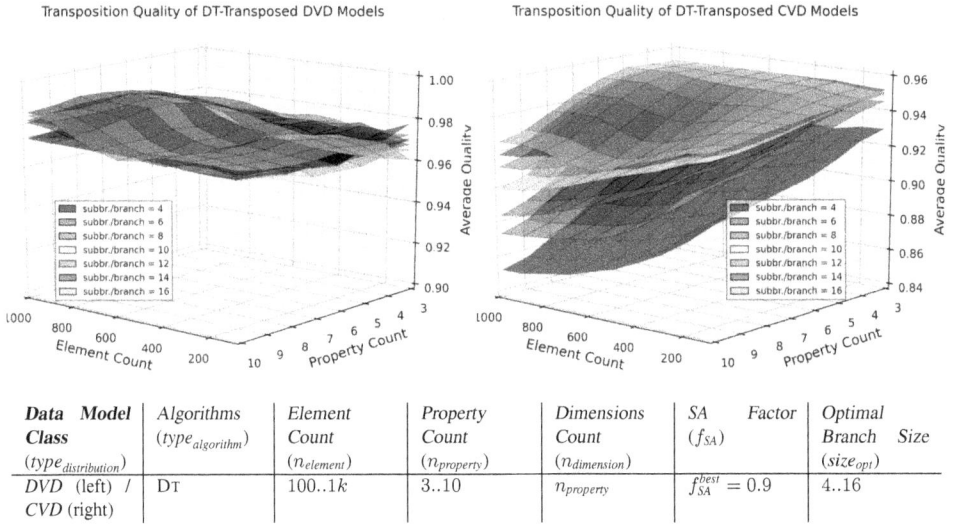

Transposition Quality of DT-Transposed DVD Models      Transposition Quality of DT-Transposed CVD Models

| Data Model Class (type_{distribution}) | Algorithms (type_{algorithm}) | Element Count (n_{element}) | Property Count (n_{property}) | Dimensions Count (n_{dimension}) | SA Factor (f_{SA}) | Optimal Branch Size (size_{opt}) |
|---|---|---|---|---|---|---|
| DVD (left) / CVD (right) | DT | 100..1k | 3..10 | $n_{property}$ | $f_{SA}^{best} = 0.9$ | 4..16 |

**Figure 2.37:** *Average Quality of Result (quality_{average}) of the DT Algorithm applied to DVD and CVD Data Models dependent on Element and Property Count, as well as Branch Size.*

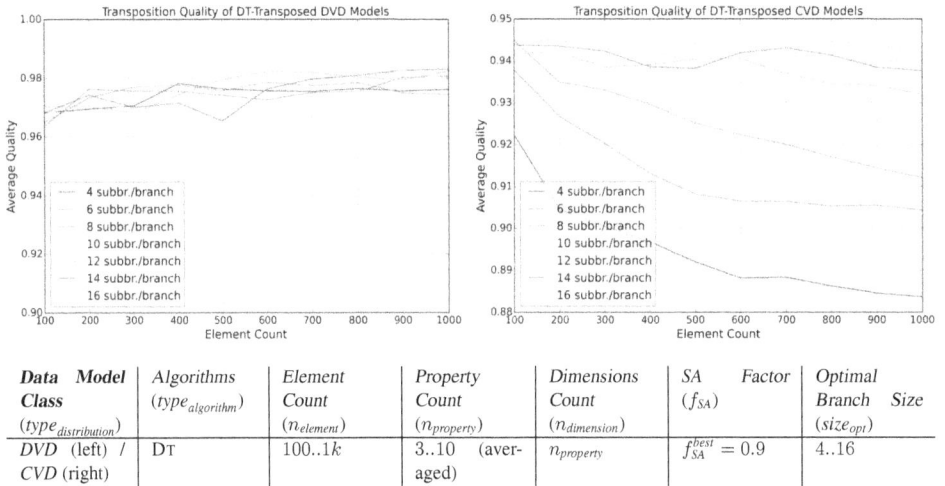

Transposition Quality of DT-Transposed DVD Models      Transposition Quality of DT-Transposed CVD Models

4 subbr./branch
6 subbr./branch
8 subbr./branch
10 subbr./branch
12 subbr./branch
14 subbr./branch
16 subbr./branch

| Data Model Class (type_{distribution}) | Algorithms (type_{algorithm}) | Element Count (n_{element}) | Property Count (n_{property}) | Dimensions Count (n_{dimension}) | SA Factor (f_{SA}) | Optimal Branch Size (size_{opt}) |
|---|---|---|---|---|---|---|
| DVD (left) / CVD (right) | DT | 100..1k | 3..10 (averaged) | $n_{property}$ | $f_{SA}^{best} = 0.9$ | 4..16 |

**Figure 2.38:** *Average Quality of Result (quality_{average}) of the DT Algorithm applied to DVD and CVD Data Models dependent on Element Count and Branch Size.*

The results show that the count of subbranches per branch has almost no influence at *DVD* data models, but changes the average transposition quality at *CVD* data models significantly: the larger $size_{opt}$, the higher the average transposition quality is, especially for larger data models ($n_{element} \rightarrow 1k$ and $n_{property} \rightarrow 10$). For values of $size_{opt} = 16$ the quality saturates. Considering the worst single transposition qualities (Formula (2.4), see Appendix Section 6.1.2.2), the same behavior is revealed.

As enlarging the branch size slows down the algorithm in a quadratic manner (Section 2.3.2.6), it is desirable to determine the lowest possible value of $size_{opt}$. In consequence it can be said, that a subbranch count of $size_{opt}^{best} = 16$ is sufficient to ensure that the transposition quality is within an acceptable range.

### 2.3.3.3 Transposition Quality Dependency of the Dimension Count

In the following representative transposition quality measurements of *test scenario* 3 are illustrated (Section 2.2.4, Table 2.4).

After the determination of the *SA* factor and branch sizes for the DT Algorithm to achieve sufficient transposition qualities (see previous Sections), the data model transpositions to low-dimensional spaces is considered. As described in Section 2.1, the less the axes, the easier the data are to be analyzed.

Figure 2.39 shows transposition quality surface plots for *CVD* data models transposed to spaces of different dimensionalities. The data models consist of $1k$ elements each, which contain different numbers of properties. The higher the property count, the more different are the sought distances between the elements and thus the less fit the models into spaces of lower dimensionality, and the lower are the expected transposition quality. In the upper row the average transposition qualities are shown for each property – dimension count combination, in the lower row are the according standard deviations. Additionally the $z$-planes of each chart provides a heatmap, which illustrates the quality distributions in a different way. The profiles of the surfaces are mapped onto the $x$- and $y$-planes to clarify the quality dependencies of the property and dimension count.

One the one hand the results show that the average transposition qualities are settled in a comparable high range, i.e. $0.81 \leq quality_{average} \leq 0.97$, which is close to ideal transpositions. One the other hand the quality drops for data models with many properties ($n_{property} \rightarrow 10$) in spaces of very low dimensionality ($n_{dimension} \rightarrow 2$) to a local minimum. However, with respect to the element count the quality's slope is already saturated at this point (the dimension count can not be reduced any more in a sensible way), which suggests that it will remain invariant for further growing data models. The most interesting and difficult case (transposing elements with 10 properties to 2 dimensions) reveals an only slight quality difference of 0.81 for the DT Algorithm compared to 0.86 for the EQ Algorithm.

As shown in Section 2.3.3.1, the precise EQ method delivers transposition results close to the ideal. This suggests that the current EQ Algorithm's qualities (left column) are also close to the theoretical maximum, which may be significant below 1, due to the reduction of the spatial dimensions in comparison to the property counts. Considering the similarity of the EQ and DT results, it can be said, that the transposition qualities of the DT Algorithm (right column) are also comparatively close to their maximum and thus acceptable.

The evaluation of the qualities' standard deviation yields a significant increase for low-dimensional transpositions, i.e. the final quality varies more between different data model seeds. This means, with decreasing dimensionality the theoretical quality maximum becomes more seed dependent or the algorithms' quality of result becomes more start condition dependent. The graphs show, that the standard deviations of both methods are nearly identical, i.e. the DT approach is affected in the same manner from this issue than the precise EQ Algorithm.

In addition to the results for transposed *DVD* data models and the consideration of the single worst transposition qualities in the Appendix 6.1.2.3, it can be concluded, that the transposition of data models

Transposition Quality of EQ-Transposed CVD Models　　　Transposition Quality of DT-Transposed CVD Models

Transposition Quality of EQ-Transposed CVD Models　　　Transposition Quality of DT-Transposed CVD Models

| Data Model Class ($type_{distribution}$) | **Algorithms** ($type_{algorithm}$) | Element Count ($n_{element}$) | Property Count ($n_{property}$) | Dimensions Count ($n_{dimension}$) | SA Factor ($f_{SA}$) | Optimal Branch Size ($size_{opt}$) |
|---|---|---|---|---|---|---|
| CVD | EQ (left) / DT (right) | $1k$ | $3..10$ | $2..10$ | $f_{SA}^{best} = 0.9$ | $size_{opt}^{best} = 16$ |

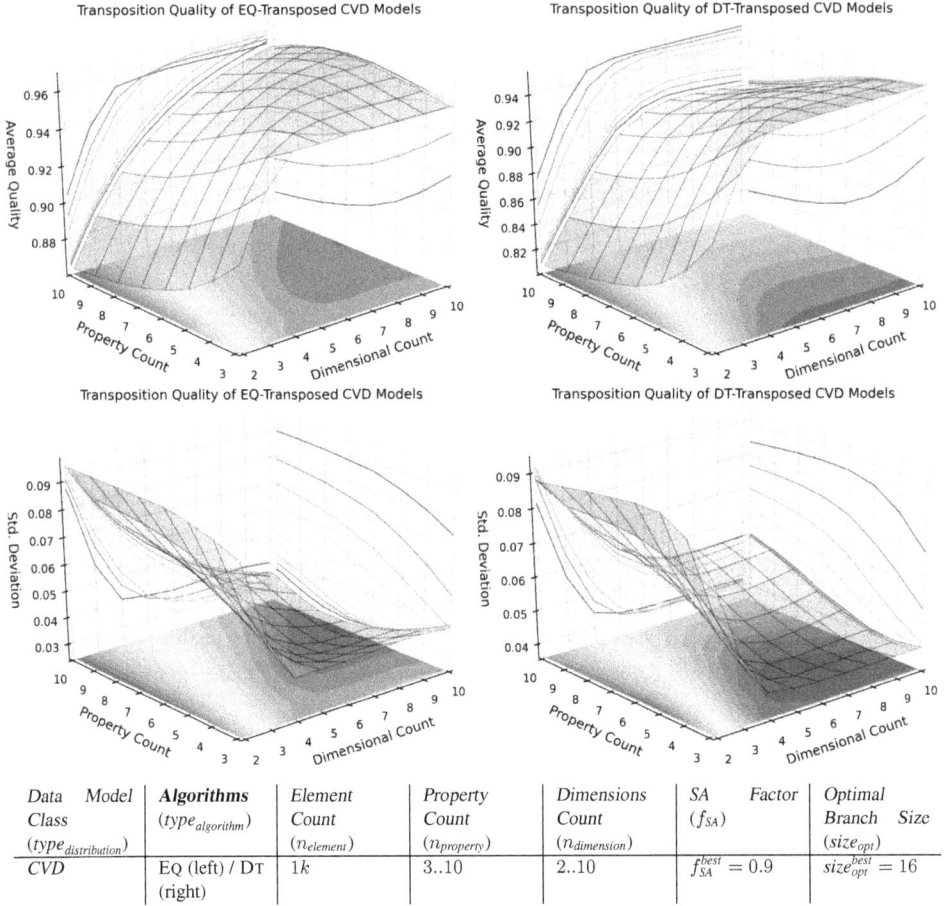

**Figure 2.39:** *Average Quality of Result ($quality_{average}$, top) and its Standard Deviation ($quality_{SD}$, bottom) of the EQ and DT Algorithms applied to CVD Data Models, dependent on Dimension and Property Count.*

with high property counts to low-dimensional target spaces can be realized with acceptable quality losses, especially with the fast DT Algorithm.

### 2.3.3.4 Runtime Dependency of the Data Model Size

In the following representative runtime measurements of *test scenario* 4 are illustrated (Section 2.2.4, Table 2.4).

Subsequently to the DT Algorithm's transposition quality considerations and evaluations in the previous Sections 2.3.3.1 - 2.3.3.3, the current test scenario investigated the runtime dependencies of the data model size, which is defined by the element and property count. In addition to valid transpositions, i.e. acceptable transposition qualities, the runtime and its progression are crucial criteria of the algorithm's application to large data sets.

In Figure 2.40 runtime measurements and their according runtime slopes are shown as surface plots.

The data models were transposed to an uniform space of three dimensions and have different element and property counts, whereby the property values either follow the *DVD* (left column) or the *CVD* (right column) distribution scheme. The upper row illustrates the runtimes dependency of the data model size for both distribution types, the lower row shows the according runtime slopes (see Formula (2.5)). The charts' $z$-planes provide heatmaps of the runtimes and their slopes respectively, the $x$- and $y$-planes shows the mapped profiles of the measured values to clarify the size parameters' influence.

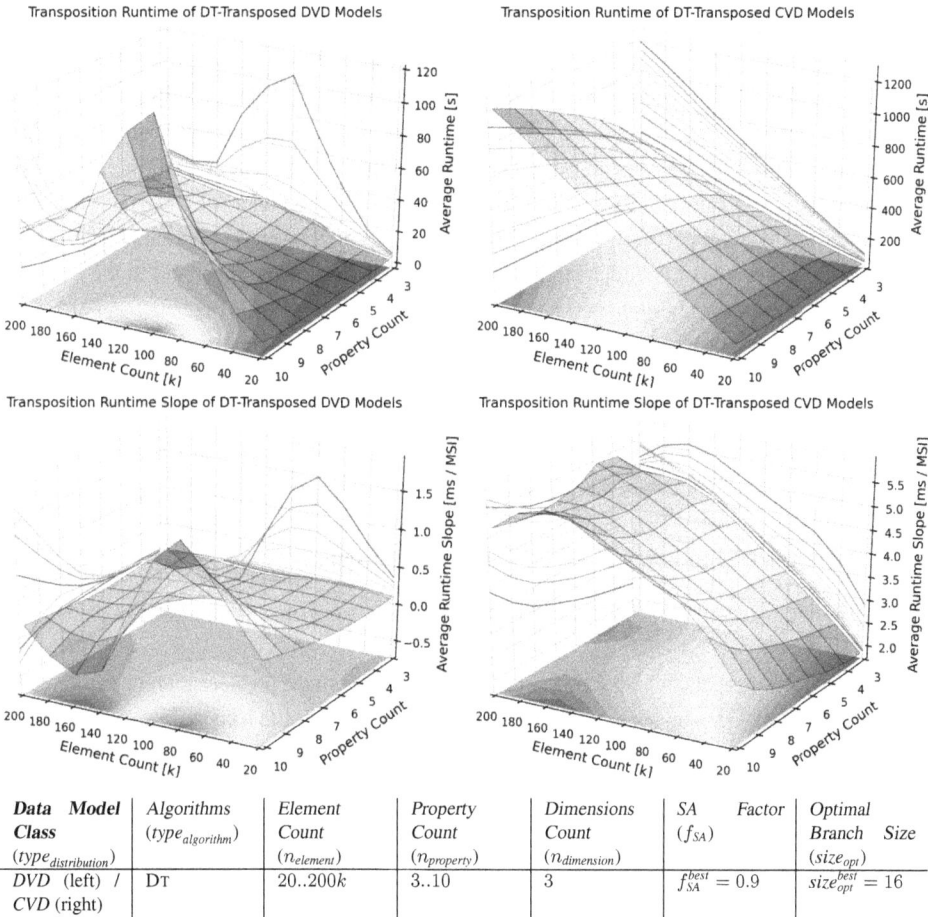

| Data Model Class | Algorithms | Element Count | Property Count | Dimensions Count | SA Factor | Optimal Branch Size |
|---|---|---|---|---|---|---|
| $(type_{distribution})$ | $(type_{algorithm})$ | $(n_{element})$ | $(n_{property})$ | $(n_{dimension})$ | $(f_{SA})$ | $(size_{opt})$ |
| DVD (left) / CVD (right) | DT | $20..200k$ | $3..10$ | $3$ | $f_{SA}^{best} = 0.9$ | $size_{opt}^{best} = 16$ |

**Figure 2.40:** *Average Runtime ($t_{run}$, top) and its Slope ($slope_{runtime}$, bottom) of the DT Algorithm applied to DVD and CVD Data Models, dependent on Element and Property Count.*

First, the runtime's magnitude and the difference between the value distribution types are reflected. While data models of the *DVD* type (left column) are transposed in $5..42$ seconds (with exception of the irregular peak), *CVD* data models (right column) require a runtime of $100$ up to $1200$ seconds. With regard to the slightly lower transposition quality of *CVD* data models (Section 2.3.3.3), this observation reflects the more complex task to arrange elements with continuously distributed property values, than elements with unambiguously grouped values. In contrast to the mainly even runtime course of the

*DVD* data model transpositions, an irregular peak at $n_{element} = 110k$ and $n_{property} = 10$ is visible. This behavior is probably caused by the fixed branch sizes of the utilized precalculated tree (Section 2.3.3.2), which lead to disadvantageous tree structures at certain data model sizes.

Regardless of the runtime magnitudes, both runtime charts indicate an approximately linear runtime increment for linear growing data models, which suggests practical runtimes even for larger models. The plots of the runtime slopes (lower row) allow a closer look at this issue and show that indeed the runtimes for transposed *DVD* data models increase linearly (with an approximately constant value of $slope_{runtime} < 0.1\frac{ms}{MSI}$, except at the irregular peak. In contrast, the runtime slope for *CVD* data model transpositions is not a constant value, but varies in the range of $slope_{runtime} \in [1.8, 5.7]\frac{ms}{MSI}$. In this cases the *MSI* is set to the test scenario's model step sizes, i.e. the maximal slope is equivalent to an average runtime increment of 5.7 milliseconds per additional element. Considering the saturation dependency of the element count and the comparable flat runtime slope caused by the property count suggests an only slight non-linear runtime increment in an low, and thus acceptable magnitude for *CVD* models.

Summarizing the runtime measurements, it can be said, that firstly the DT Algorithm yields practical runtimes for data models of both distribution types. Secondly, and more important, in the worst case the runtime increment is only slight non-linear at linear growing models (for larger element or property counts), which implies that the DT Algorithm is capable of handling very large data sets, in spite of the at least quadratically scaling problem. More runtime measurements, including smaller ranges of the model sizes, can be found in the Appendix in Section 6.1.2.4.

Section 4.3 illustrates the DT Framework's application to example model of much larger property counts. Their runtime measurements imply also linear runtime complexities.

### 2.3.3.5 Runtime Dependency of Large Scale Data Models

In the following representative runtime measurements of *test scenario* 5 are illustrated (Section 2.2.4, Table 2.4).

To further investigate the runtime and its progression, a subset of the previous runtime measurements (Section 2.3.3.4) was chosen and extended: the property count was set to a constant value and the data model size was evenly increased to 0.5 million elements. The DT Algorithm transposed *DVD* and *CVD* data models of these sizes to an uniform target space of 3 dimensions, while the runtimes were measured.

Figure 2.41 shows the runtimes, their standard deviations and their slopes for both distribution types. The results are presented in dependency of the data models' element counts, utilizing two $y$-axes: the average runtime (left axis) and the average runtime slope (right axis). Note that the axes are scaled to different magnitudes to simultaneously visualize the characteristics of all graphs. To clarify the slopes' global course, smoothed version are added.

The results show that the DT Algorithm requires up to 400 seconds on average to process *DVD* data models of 0.5 million elements. The larger the data model, the higher is the runtime's standard deviation, which suggests an increased start condition dependency at constant transposition qualities (Section 2.3.3.3). Additionally periodic peaks at certain data model sizes are observable, which can be attributed to disadvantageous branch size – element count combinations. The runtime slope is approximately constant except for data models larger than $300k$ elements, where the slope is increased slightly, but remains at low magnitudes of $slope_{runtime} < 3\frac{ms}{MSI}$ on average.

In contrast, transposing *CVD* data models occupies significantly more runtime, up to 4.200 seconds

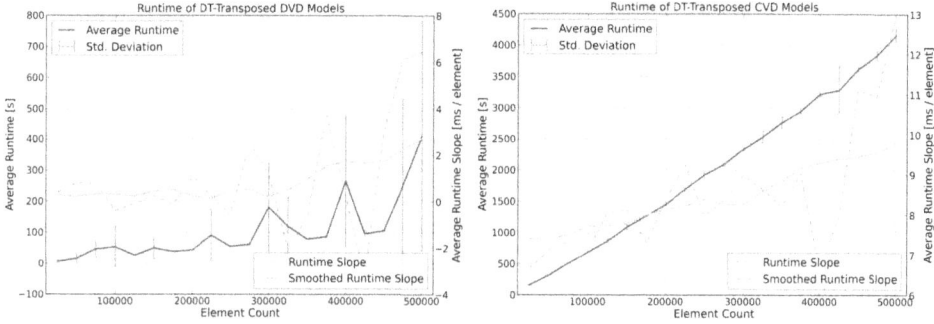

| Data Model Class ($type_{distribution}$) | Algorithms ($type_{algorithm}$) | Element Count ($n_{element}$) | Property Count ($n_{property}$) | Dimensions Count ($n_{dimension}$) | SA Factor ($f_{SA}$) | Optimal Branch Size ($size_{opt}$) |
|---|---|---|---|---|---|---|
| DVD (left) / CVD (right) | DT | 25..500k | 10 | 3 | $f_{SA}^{best} = 0.9$ | $size_{opt}^{best} = 16$ |

**Figure 2.41:** *Average Runtime ($t_{run}$) and its Slope ($slope_{runtime}$) of the DT Algorithm applied to DVD and CVD Data Models, dependent on Element Count.*

on average, but with more even runtimes, as the standard deviations indicate. Also the graph shows no remarkable peaks and appears approximately linear. The runtime slope grows approximately linearly, but with a comparable low magnitude, which is not visible in the runtime's graph. The slope reaches a maximum of $slope_{runtime} < 10 \frac{ms}{MSI}$ on average, i.e. a runtime increment of 10 milliseconds for each additional element for data models of 0.5 million elements.

In conclusion the graphs suggest that the DT Algorithm's runtimes grow virtually linearly for linearly increased data model sizes. The property value distribution type defines the runtime's slope mainly. For the most complex case, the CVD type shows an acceptable average time complexity of $slope_{runtime} = 8.5 \frac{ms}{MSI}$ when processing $500k$ elements.

## 2.3.4 Summary of the Insights regarding the Dimensional Transposition Framework

The following Section summarizes the results as overview of the developed algorithms in Section 2.3.4.1, their validations and performance evaluations in Section 2.3.4.2 and with respect to their potential applications and relations to other *MDS* solutions in Section 2.3.4.3.

### 2.3.4.1 The Dimensional Transposition Framework as a Novel MDS Technique

The DT Framework as an *Iterative Metric MDS* technique was presented. Its purpose is to transform large, complex and high-dimensional data models to a normalized low-dimensional form, enabling the access of common *CA* techniques or interpreters to the model. The data interface requires user-defined affinity and optimal distance functions as metrics in combination with averaging functions for each feature type to establish the internal data structures (Section 2.3.2.1). The underlying algorithms utilize these definitions to find an equivalent expression of the data items in a low-dimensional metric space, so that the Euclidean distances correspond to the items' similarities. The transposition is done by a force-driven process, called equilibration or EQ Algorithm, shifting the data items iteratively from random start positions to their final positions, where the affinity inflicted forces are in equilibrium. The framework interface provides these final position vectors as transposition result to subsequent *CA*s (Section 2.3.2.2).

To meet the requirement to achieve a time complexity significantly below $O(n^2)$, precalculated data structures are introduced to evaluate the forces more efficiently (Sections 2.3.2.3 - 2.3.2.7). These structures are also exportable to subsequent *CA* processes for later utilization. To further increase the performance regarding execution time as well as transposition quality and stability, several enhancements are added (Sections 2.3.2.8 and 2.3.2.9), which results in the final *MDS* technique referred to as DT Algorithm (Section 2.3.2.10).

The process of development was carried out by analyzing profiling runs (Section 2.3.1) of well defined development scenarios (Section 2.2.5, Table 2.5) to identify the current algorithm's issues and bottlenecks systematically.

### 2.3.4.2 Result and Performance Measurements

As analysis strategy test scenarios were designed, combining parameter setups and changes with selected measurements to gain different insights (Section 2.2.4, Table 2.4). To introduce a generalized testing approach, two corner cases of artificial data model types (*CVD* and *DVD*) of arbitrary size were defined. Utilizing these data models allows for the comparisons the DT Framework's results and performance with other algorithms, which process these data model types too. Furthermore, this approach offers more flexibility to analyze new *MDS* techniques than fixed size data models of an external sources, which may be unavailable at all.

The first two test scenarios fulfill the task to find appropriate parameter values for the *SA* factor ($f_{SA}$), which controls the gradual reduction of the data item movement as well as the number of iterations, and for the optimal branch size ($size_{opt}$) of the precalculated data structure, setting the underlying tree's width and depth. To measure the quality of the affinity mapping to Euclidean distances, these are retransformed by inverse affinity functions, calculating the according affinity. The average relation of these achieved distances-expressed affinities to their originals serves as meaningful measurement for the transposition quality.

It could be shown, that a *SA* factor of $f_{SA} = 0.9$ (which correlates with an iteration count of 30) is sufficient to achieve transposition qualities of $quality_{average} \geq 0.9$ (Section 2.3.3.1), i.e. an average deviation of 10% of the elements' distance based affinities to their property based original affinities, regardless of the element or property count. It can be said that in general the higher $f_{SA}$, the closer $quality_{average}$ is to 1, i.e. the ideal transposition. In comparison to the more accurate EQ Algorithm, the DT Algorithm reaches nearly the same transposition qualities.

For the second fixed parameter $size_{opt}$ it could be shown, that above a value of 16 no further quality increase occurs, thus the branch size was set to this value for the remaining tests (Section 2.3.3.2). For *DVD* data model transpositions $size_{opt}$ has no observable influence, *CVD* data model transpositions show significant $quality_{average}$ gains from 0.86 to 0.93. Similar to the behavior depending on the $f_{SA}$ parameter, higher branch sizes cause higher transposition qualities in general until it saturates with the said branch sizes.

Regarding the question of the DT Algorithm's ability to transpose high-dimensional data models to low-dimensional equivalents, the more accurate EQ Algorithm was utilized as best-case reference. For all data model types it could by shown, that the average transposition quality ($quality_{average}$) is within acceptable ranges of 0.86..0.99 (EQ) and 0.81..0.98 (DT), i.e. the results of the fast DT Algorithm are only slightly below the accurate but slow EQ Algorithm (Section 2.3.3.3). The minima of these results occur during the most challenging tests, i.e. the transposition of data items with many different features ($n_{property} = 10$) to low-dimensional target systems ($n_{dimension} = 2$). Nevertheless, the evaluated behaviors in many different tests show, that the transposition quality saturates with further growing feature counts, which implies a stable transposition quality for even more complex data items. The tests further revealed, that the standard deviation $quality_{SD}$ is significantly higher at low-dimensional target systems, implying a greater diversity among the pairwise affinity mappings for both algorithms. In general it can be stated, that the DT Algorithm is capable to process high-dimensional data models of arbitrary sizes and both distribution types with only slight concessions to the transposition quality in comparison to the EQ reference.

After the determination of the algorithm's fixed parameters and the analysis of the transposition quality, the runtime behavior was investigated as one of the major criterion. The execution time evaluation for transposing large data models ($n_{element} \leq 500k$, $n_{property} \leq 10$) to low-dimensional equivalents ($n_{dimension} = 3$) shows a virtually linear runtime complexity for both data model types and all sizes (Sections 2.3.3.4 and 2.3.3.5). Only the runtime slope analysis ($slope_{runtime}$) reveals a slight nonlinear behavior. In comparison to the runtime's order of magnitude (up to $10^3$ seconds), its worst increment from $7.5 \frac{ms}{MSI}$ to $9.75 \frac{ms}{MSI}$ is comparable small and hence acceptable. For *CVD* data models the runtime increases evenly with growing data item and feature counts, but with a larger overall slope, *DVD* data models are processed significantly faster, but the runtime exhibits some irregular peaks, which may be caused by the underlying tree structure.

### 2.3.4.3 Framework Application and Comparison

With regard to the framework objectives (Section 2.1.3), it can be stated, that the DT Framework achieves the stated requirements. Via its metric based input interface and position vector based export, it supports a large spectrum of data models and feature types. Furthermore, the low time complexity enables the framework's utilization in combination with current *CA* techniques (Section 2.1.2) to process large data sets. The results can be stored easily and allow the transpositions' restoration as well as the adaptation to changes of the original data or the user preferences. Besides the necessary feature metrics, averaging functions and target system definitions, the DT Framework requires no additional parameters, thus making it comparably user-friendly. Measurements with representative data models showed that the algorithms are capable of to transpose the given data models to meaningful low-dimensional equivalents (Section 2.3.3).

The work most closely related is the *PC* framework by Agarwal et al. [2010], which also implements a more general approach to handle *MDS* problems. But whereas this framework is based on user-defined cost functions, which should be minimized, the DT Algorithm requires metric functions to compare single types of features in a standardized way, an optimal distance function, which defines the aimed character of the transposition result, and an averaging function per feature type. The underlying algorithms handle the determination of the final data item positions in a comparable manner, starting with random initialization and improving the position vectors iteratively. Therefore both techniques are capable of to handle data and preference updates without discarding the current results. In contrast to the

*PC*'s complete comparison approach, the DT Framework utilizes precalculated data structures to compare items efficiently in a group-wise manner. Furthermore, the *PC* framework is specialized in spherical *MDS*, where the DT Framework aims at more general orthogonal-metric target systems.

As far as available there are no comparable runtime measurements of non-graph based data sets, which exceed the relative small size of $10^3$ items in combination with many features [Agarwal et al., 2010]. Most publications utilize prestructured data, such as graphs, rely on sampling methods, and investigate specialized data models of fixed size [de Silva and Tenenbaum, 2003; Brandes and Pich, 2007; Yehuda et al., 2002]. With respect to unstructured original data sets, this does not allow for statements regarding the transposition quality and the runtime dependency of item counts, feature counts or inherent data distributions. As such the presented analysis strategy herein can be regarded as a novel approach of systematic *MDS* technique investigation.

## 2.4 Discussion

With the DT Framework, embedding the DT Algorithm, a novel *MDS* technique was presented, which aims to provide a generalized data and user interface, as well as a sufficient efficiency, enabling the transposition and thus the later *CA* of a wide field of data sets.

As stated in Section 2.1 preprocessing methods as the *MDS*, are necessary preceding steps to ensure the analysability of large, high-dimensional and diverse data sets (Section 2.1.1). Recent approaches (Section 2.1.2) showed, that the transposition of such data is a complex task of non-linear time complexity, thus excluding data sets, which exceed a certain size or complexity, from the analysis. Hence the DT Framework's focus is on the transposition's abstraction and equivalence preservation, while minimizing the runtime slope (Section 2.1.3).

To validate the data sets' equivalence preservation and the low runtime complexity, artificial data models of arbitrary sizes were defined, covering the corner cases of possible data structure types. Utilizing these data models, a set of test scenarios were created, firstly to determine suitable algorithm parameters, and secondly to measure the equivalence preservation and the according runtimes (Section 2.2).

The objectives are achieved by defining an interface of property metrics and implementing the DT Algorithm to transform data models provided by this interface (Section 2.3.2). The predefined models were applied to the interface, enabling the algorithm's configuration, as well as the quality and runtime measurement of their transpositions (Section 2.3.3).

The measurements showed, that the DT Framework is able to transpose data of no and strong internal structure, while preserving the models' equivalence at a almost linear runtime increment with linearly growing model sizes. Hence this technique allows the processing and analysis of data sets, which were previously restricted due to their sizes, complexity and/or diversity.

In the following details of the advantages and disadvantages of the approach are discussed (Section 2.4.1), potential application listed (2.4.2), and an outlook on the further development presented (Section 2.3.4.3).

## 2.4.1 Advantages and Disadvantages

**Advantages:** The DT Algorithm provides a generalized data interface, based on metrics and optimal distance functions (compare the definitions in Section 2.3.2.1) as opposed to other available methods, which require cost functions and/or operate on the data items themselves (introduced in Sections 2.1.2, compared in Section 2.3.4). The approach allows the user to implement individual interpretations of feature differences and adapt the character of the transposition result to suit the chosen subsequent *CA* algorithms. In addition to that it is possible to include features of all domains, given that an appropriate compare metric is made available.

The almost linear runtime complexity for processing unstructured high-dimensional data outperforms other approaches so far and enables the framework to prepare large complex data models for *CAs*. As shown in Section 2.3.4.2 the runtime slope is virtually constant while processing linear growing models, i.e. enlarging the model sizes by ten times, implies runtime increases by the factor of only ten (instead of 100) in spite of the quadratic problem complexity.

In comparison to other preprocessing techniques the DT Framework is a virtually parameter-less and thus a comparable robust method. Other approaches (Section 2.1.2.2) require sets of scaling cost functions and weights in combination to algorithm parameters, which influence the transposition result. Section 2.3.4.2 shows, that in addition to the target dimensions count, only the *SA* factor as the degree of accuracy has to be set, determining the runtime and thus the result's transposition quality, but not its character.

Due to the strategy of the step-wise transposition during a fixed number of iterations as introduced in Section 2.3.2.2, the algorithm's convergence is guaranteed, optional parameter adjustments can refine the final result at the cost of higher runtimes.

**Disadvantages / Challenges:** On the other hand, the DT Framework is rather large and complex, making reimplementations expensive and suggests to use it as provided (Section 2.3.2). In contrast to other approaches as introduced in Section 2.1.2 a larger set of user-defined functions for data comparison is necessary (compare Section 2.3.2.1 for the comparison interface and Section 2.3.2.5 for the property averaging interface), which provides more degrees of freedom and flexibility, but makes the setup more extensive.

Furthermore, the accelerating underlying tree structures, including also the averaged features, increase the overall memory consumption, preventing large data models to fit in the memory and thus to be processed efficiently. The real memory consumption is not investigated so far, so that only approximations as in Section 2.3.2.6 are available.

Due to the metric comparison approach, non-metric feature comparisons are excluded from consideration, restricting the variety of processable data model types. Even with regard to generalized *MDS* concepts (Section 2.1.2.2) it is difficult to integrate non-metric comparisons into the framework. Although not analyzed yet, it is probable that high-dimensional data models with many noisy or irrelevant axes could lead to unsatisfactory results regarding the subsequent *CAs*, i.e. results of high transposition quality, but with cluster structures covered by noise.

As drawback of the flexibility provided by the interface, the user might distort the algorithms unintentionally by implementing inappropriate metric or averaging functions or increase the runtime by including performance bottlenecks in these functions. The risk potential is described in Section 2.1.2.3.

## 2.4.2 Potential Applications

As stated in Section 2.1.1, the automated investigation of unstructured data is a wide field, i.e. the pre-processing of large and complex high-dimensional data models to enable subsequent *CAs* is applicable to many data-intensive problems. The presented *MDS* technique can be deployed to the fields of web analysis, customer relationship management, marketing, medical diagnostics, computational biology and computer or social science [Berkhin, 2006; Soni and Ganatra, 2012], transposing the data to accomplish tasks like exploratory pattern Analysis, grouping, decision making, data mining, document retrieval, image segmentation or pattern classification [Jain et al., 1999; Jain, 2008]. Where common methods fail due to the size, their dimensionality and/or complexity of the data sets [McCallum et al., 2000; Can and Ozkarahan, 1990; Agrawal et al., 2005; Jiang et al., 2004], the presented framework is suitable for these problems, suggesting practical runtimes even for larger data models.

For example the mentioned tasks in Section 1, as the national or international finance controlling, taking into account many different aspects [Deb, 2001], the optimization of engineering processes like the creation of efficient methods and schedules, while saving resources [Marler and Arora, 2004], or the treatment of computer science challenges, such as the (heuristic) solving of abstract issues ($np^{10}$-hard problems) or the development support of novel hardware approaches [Brüderle et al., 2011], represent potential fields of application of this *MDS* technique.

## 2.4.3 Further Enhancements and Improvements

In the following potential improvements for the DT Framework are listed briefly, ordered coarsely from probably easy to difficult to implement. Besides, it could be worthwhile to consider the include of additional property value distribution types to extend and strengthen the analysis strategy.

To improve the algorithm's transposition quality or runtime, a subset of recent approaches is applicable to the DT Framework as introduced in Section 2.1.2: At the expense of granularity, e.g. sampling methods might reduce the runtime significantly. Only a (random) sample could be processed by the DT Algorithm, the remaining data is assigned later to the tree structure in a manner as proposed in Section 2.3.2.9. However, these subsequent assignments would import noise to the transposition result, which might be problematic during later *CAs*, as the noise could cover smaller structures. An other comparable straight forward way to improve the runtime is to parallelize the algorithm, as large parts involve independent long-running iterations, which their structures, introduced in the Sections 2.3.2.2 - 2.3.2.10, imply.

Also the branches of the internal tree structure are created during a single pass (Section 2.3.2.3), i.e. a single initialization of the branch position vectors and their iterative refinement by the TRIN K-MEANS Algorithm. As the branches should preferably represent compact and affine subelements, it is worthy to analyze, whether or not branch reinitialization steps to improve the starting positions serve the transposition quality and/or the runtime.

It was already stated in Section 2.4.1 that the existence of many noisy or irrelevant axes could imply difficulties for subsequent *CA* steps, in spite of high transposition qualities. As it is costly to handle a large number of axes, a promising approach might be to establish mechanisms to pre-process the axes first. By following the concepts of feature transformation and feature selection (Section 2.1.2.2) corre-

---

[10]non-deterministic in polynomial time

lated axes could by merged to single ones or irrelevant axes could be excluded, thus reducing runtime and memory consumption.

Furthermore, the algorithm investigation could be extended, e.g. by the analysis of the actual memory consumption or the influence of noisy axes regarding the covering of underlying structures. If the original data already contain structure information, e.g. graphs or precalculated similarities, these information could be utilized to improve the initialization of the position vectors, thus increasing the achieved transposition quality or reduce the runtime respectively.

# 3 Tree Fusion k-Means Algorithm

The following Section introduces a potential second component of the TRACS-MOO approach (Section 1.4) to analyze an uniformed model by methods of the *CA*, after its transposition by the DT Framework (Section 2). Section 3.1 considers the importance of the *CA* in general, particularly the well known *k-Means* algorithm along with the base concept of this second abstracted TRACS-MOO component. To evaluate and explore its fields of application and performance Section 3.2 describes the applied testing and analysis strategy. Subsequently the obtained insights, i.e. the developed algorithm components and the measurement results are presented in Section 3.3. Finally the results' meaning with respect to the proposed TRACS-MOO approach and to the *CA* in general are discussed in Section 3.4.

## 3.1 Introduction to the k-Means based Cluster Analysis and its Enhancements

First, the tasks of the *CA* as a sub field of the *MOO* are reflected in Section 3.1.1 with regard to the different existing clustering approaches. Subsequently the often utilized and adapted *k-Means* cluster algorithm, along with its fields of application, advantages and disadvantages as well as available enhancements is introduced in Section 3.1.2. Finally the motivation for and the base concept of the developed *CA* method, based on the *k-Means* algorithm, is presented.

### 3.1.1 The Meaning of Cluster Analysis

The *CA* as a group of methods to analyze unstructured data can be classified as a subset of the *MOO* [Ehrgott, 2005; Miettinen, 1999]: elements of a given data model should be summarized to groups (clusters) in a manner, so that [Berkhin, 2006; Soni and Ganatra, 2012]:

$i)$ the similarity of all elements within a single cluster is maximal (*Affinity*),

$ii)$ the degree of similarity within a single clusters is even (*Homogeneity*),

$iii)$ and the number of clusters is minimal (*Meaningfulness*).

These aspects can be formulated as several, competing optimization objectives, thus expressing the *CA*'s task as a *MOO* problem [Pulido and Coello Coello, 2004; Taboada and Coit, 2007].

Analogue to the *EVA*, methods of the *CA* in general cover the entire solution and target space, which favors the search for global (pareto) optima. Moreover, current techniques are capable of to handle large and complex data sets [Andrews and Fox, 2007; Philbin, 2007; Agrawal et al., 2005], enabling them to process data models of large numbers of elements or variables with regard to many clustering objectives.

As the definition of the term *cluster* varies in a wide range, the large is the number of different *CA* methods. Following Estivill-Castro [2002] the concepts of the algorithms are based on the utilized cluster model primarily:

- *Connectivity*: This cluster model assumes, that close elements are more related than more distant ones, i.e. a cluster is defined by its inherent maximal connection distance: all element pairs, which are closer to each other than this limit, belong to the same cluster.

- *Centroids*: This approach utilizes a center point for each cluster, implying that each element belongs to that cluster, whose center is closest. The determination of these cluster centers are diverse, e.g. the contained elements' weighted average or the geometric center.

- *Distribution*: The methods of this concept assume, that the elements are distributed in a statistically describable manner, for example in equal or normal distributions. Consequently each target cluster is equivalent to the parameter set of such a statistical distribution.

- *Density*: This often utilized model defines a cluster as a set of elements, which are positioned in an area of uniform density. For example, the density of a single element can be regarded as the average distance of the $k$ nearest neighbor elements. The borders of such a cluster are represented by elements, for which the density drops or raises in comparison to the inner cluster area.

- *Subspacing*: Methods of this strategy analyze subsets of the entire target space independently, recombining the partial results later to the final clusters. The clustering objectives may alter for each subset, i.e. for each element attribute, dependent on the attribute type domain.

- *Graphs*: Element relations like edges within a graph (if available) can also be utilized to define clusters. In general *cliques*[1] or *quasi cliques*[2] form the base concept of clusters in graph based models.

Furthermore, all cluster models can be separated regarding the stringency with which elements are assigned to clusters: $i$) *strict*: each element belongs to exactly one cluster; $ii$) *strict with outliers*: each element belongs to exactly one or to no cluster; $iii$) *fuzzy / overlapping*: each element may belong to more than one cluster, for example expressed by probabilities; and $iv$) *hierarchical*: each element belongs to none or more clusters, where the clusters are organized in a tree-based structure.

The combination of the utilized cluster model and assignment stringency, and mixed forms of both, leads to a large number of different cluster algorithms. As the introduction of all relevant *CA* methods would exceed the frame of this work, only an overview about the available technqiues is provided. According to Soni and Ganatra [2012]; Berkhin [2006] known clustering methods can be categorized coarsely as follows:

1. *Partitioning*: decomposing the data set into partitions (clusters)

   a) *Iterative Relocation Based*: fitting sequentially the unknown parameters (and count) of the clusters to comply with the clustering objectives

   b) *Density Based*: accumulating elements to clusters if they meet density based criteria

   c) *Grid Based*: quantizing the data space to create cells where subsequent clustering operations (e.g. merge, split, ..) are executed on

---

[1] a set of elements, which are completely connected
[2] a set of elements, in which each element is connected to a minimal fraction of the other elements

d) *Subspace Based*: analyzing subsets of the data sets' item feature independently, finally combining the final clusters

2. *Hierarchical*: decomposing the data set into a hierarchy of groups (clusters)

a) *Agglomerative*: starting at the elements as leafs, combining them to increasingly larger clusters (bottom-up)

b) *Dividing*: starting with the entire data set, dividing it to decreasingly smaller clusters (top-down)

The decision which method should by applied to the present clustering task depends on the available data, its size and complexity, the sought insights and/or structure objectives as well as the available computing resources and time [Jain et al., 1999; Jain, 2008; Soni and Ganatra, 2012], i.e. the clustering method has to be chosen very individually.

## 3.1.2 The Standard k-Means Algorithm

The *Standard k-Means Algorithm* (STD K-MEANS Algorithm) is one of the most common base mechanisms to cluster *non supervised*[3] large data sets ([Jain et al., 1999; Estivill-Castro, 2002]) and can be classified as "*centroid based* and *partitioning-iterative relocation based*" [Soni and Ganatra, 2012].

First, in Section 3.1.2.1 the base scheme of the algorithm is introduced, before in Section 3.1.2.2 typical applications of variants of the STD K-MEANS Algorithm are summarized. Subsequently Section 3.1.2.3 illustrates the major advantages and disadvantages of the algorithm and finally Section 3.1.2.4 presents the major enhancements of this *CA* method, developed by several authors so far.

### 3.1.2.1 Base Algorithm Scheme of the k-Means Algorithm

The STD K-MEANS Algorithm awaits a set of elements to cluster and a given cluster count as optimization target. The elements are placed in a $n$-dimensional space and assigned iteratively to the clusters. Conventionally the clusters are spread randomly across the given space and updated with regard to the element positions after element assignment changes, usually by obtaining the arithmetic center of their elements. Hence the clusters "move" through the $n$-dimensional space until stable positions are reached and the algorithm terminates [Lloyd, 1982]. The algorithm's standard scheme is shown in Figure 3.1.

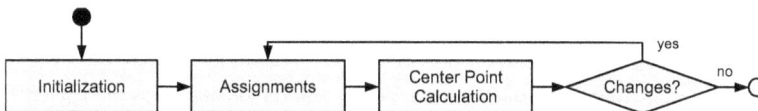

**Figure 3.1:** STD K-MEANS *Algorithm Scheme*; Subsequent to the initialization, which spreads the cluster points randomly, the main algorithm loop is executed: first, the elements are assigned to the nearest cluster point. Second, the positions of all cluster points are recalculated, utilizing the center of their assigned elements, for example the arithmetical or geometrical center. The algorithm terminates when no further cluster point changes occur.

It could by shown that the algorithm's runtime is polynomial in $n$ by Arthur et al. [2009]; Vattani [2011] and its time complexity is $O(n^{dk+1}log(n))$ [Inaba et al., 1994], where $n$ is the number of elements

---

[3]without preliminary structure information or external guidance to determine the underlying data patterns

to cluster, $d$ the dimensionality and $k$ the count of initial clusters. Cluster problems processed by the STD K-MEANS Algorithm can be classified as $np$-hard [Mahajan et al., 2009].

### 3.1.2.2 Fields of Application for k-Means Variants

Adapted versions of the STD K-MEANS Algorithm are applied in various fields [Jain, 2008]:

- *Image Processing*: segmenting images by separating fore- and background as well as objects, utilizing functions which regard pixel intensities and coordinates as distance methods [Mather, 1999; Acharya and Ray, 2005]

- *Data Compression*: more general, the clustering results of the algorithm after processing a data set, e.g. an image, can be used to reduce the total data amount [Oehler, 1995]

- *Pattern Recognition*: detecting underlying data structures, for example, market researches utilize clustering methods to summarize potential customers into different groups for efficient marketing [Punj and Stewart, 1983]

- *Classification*: analyzing unknown data regard known patterns, e.g. astronomical data such as multi-spectral images while combining spatial and spectral information to improve the classification quality [Theiler and Gisler, 1997]

- *Genome Analysis*: special implementations of the STD K-MEANS Algorithm are applied to predict the location and function of specific genes [Heintzman et al., 2007]

- *Document Analysis*: classifying and evaluating documents, particularly textual content of websites [Steinbach et al., 2000]

- *Scheduling:* grouping processes with similar constraints to improve the overall quality and reduce costs during facility scheduling tasks [Topcuoglu et al., 2002]

- *Data Transformation*: improving *PCA* methods [Pearson, 1901; Jolliffe, 2005], which simplify and reduce large data sets to linear combinations [Ding and He, 2004]

### 3.1.2.3 Advantages and Disadvantages

The STD K-MEANS Algorithm was analyzed regarding its strengths and weaknesses by several authors [Jain et al., 1999; Jain, 2008; Soni and Ganatra, 2012]. On the one hand, as the approach is easy to understand, it is inexpensive to implement and maintain the algorithm in own applications without using external libraries. Furthermore, the technique is comparable fast in relation to other *CA*s, which enables it to process large data sets multiple times with different start conditions. It was also shown, that the algorithm is robust, i.e. it terminates with valid results in every case.

On the other hand, there are significant limitations regarding the properties of the processed data: the technique recognizes only convex shapes, due to its centroid based approach, which results always in hypersphere shaped clusters. Furthermore, noisy data can make the cluster identification difficult, or in other word, noise can not be detected. More general the STD K-MEANS Algorithm can not handle hierarchical structures, which would be necessary to identify noise or embedded clusters. If the data contain diffuse cluster boundaries, the algorithm is not able to recognizes them, thus producing unsatisfactory cluster results. The overall quality of result depends strongly on the initial cluster positions, hence several runs with different cluster initializations might be necessary, thus increasing the overall runtime and requiring the selection from multiple solutions.

### 3.1.2.4 Available Enhancements

As shown previously the STD K-MEANS Algorithm and its modifications are frequently used techniques, hence several improvements, known as enhancements, were developed. They can be separated into three classes: $i$) *exact* enhancements, which return exactly the same results when given identical initial conditions, but in less time, $ii$) *heuristic* enhancements, which deliver comparable or better results in comparable or less time, and $iii$) *meta-heuristic* enhancements, utilizing general meta-heuristic approaches, which can be applied to different algorithms.

The following sections provide a brief overview of available and representative STD K-MEANS Algorithm enhancements of the three classes.

**Exact Enhancements:** One of the most important exact improvement is a technique introduced by Elkan [2003]. It is based on the fact, that a cluster problem involves much less clusters than data points, hence it is sensible to calculate and track the distance between all of them. This information enables the algorithm to estimate the required data point – cluster distances, utilizing an upper bound estimation based on the triangular inequality. The estimation can prevent time-expensive distance calculations by excluding the data point early as to distant from the cluster. This method accelerates the STD K-MEANS Algorithm by a large factor for many use cases. In the following this method is referenced as TRIN K-MEANS Algorithm.

Another major enhancement was presented by Pelleg and Moore [1999]. By applying a precalculated *kd*-tree, covering all data points and storing essential metrics, the number of required nearest-neighbor queries, which are performed by the STD K-MEANS Algorithm frequently, is reduced. The *kd*-tree separates the data points iteratively along their dimensional axes (features). In this way a hierarchical structure is created, whose nodes represent the data points beneath. During the cluster assignment phase the metrics of the nodes can be utilized to identify and hence exclude large groups of data points as to distant from the current cluster efficiently. This technique reduces the number of required distance calculations significantly and yields large speedups for well structured and low-dimensional data point distributions.

**Heuristic Enhancements:** The enhancement developed by Kanungo et al. [2002] utilizes the previously introduced *kd*-tree to process the data points preliminarily. In this way it delivers comparable, but not the same results as the standard implementation. In contrast to Pelleg and Moore [1999] the algorithms operate with a set of cluster candidates for each branch of the tree. Starting at the root, every cluster is handled as a candidate. Regarding the hyper rectangles' properties per branch only a subset of the candidates are delegated deeper into the tree until only one cluster, the closest one, remains. This efficient implementation of the STD K-MEANS Algorithm yields a significant speedup and comparable low distortions of the clustering results.

Another approach also attempts to prepare the given data to accelerate the subsequent clustering. The approach of *brFCM* was introduced by Eschrich et al. [2003] and is based on the *fuzzy c-means* algorithm by Bezdek et al. [1984]. In a preliminary step the position vectors are quantized, and similar or identical data points are combined into single ones, thus reducing the total data amount. The resulting compressed data is processed by the *fuzzy c-means* algorithm, which leads to significant speedups at acceptable distortions dependent on the applied problem class.

Similar to that work the approach of Frahling and Sohler [2006] introduces a modification referred to as *CoreMeans*, which is based on so called *core sets*, built while utilizing a precalculated quad tree and integrating several data points each. During the cluster assignment phase the core sets' sizes are increased until they cover all data points. This enables the algorithm to handle different $k$ values efficiently, thus making it useful to determine the optimal value of $k$. The author showed that the implementation has a good scaling behavior and its results diverge only slightly compared to other *k-Means* variants.

As STD K-MEANS Algorithm is limited to numeric data, Huang [1998] developed two enhancements, *k-modes* and *k-prototypes*, to process data with categorical type domains. The first one implements a method to calculate the distance between data points with several categorical attributes and modifies the clusters to fit into the categorical domain. The second method combines the conventional STD K-MEANS Algorithm with the *k-modes* approach to enable the clustering of mixed domain data points. It was shown, that the implementations deliver satisfactory results and scale well with growing data model sizes.

**Meta-Heuristic Enhancements:**   Since the clustering quality of STD K-MEANS Algorithm depends strongly on the start positions of the clusters, Arthur and Vassilvitskii [2007] introduced a technique called *k-means++* to improve these. Here the conventional *k-Means* algorithm is executed multiple times, but instead of seeding the starting positions randomly for each run, they are chosen in a way, so that the distances to already utilized start positions are as large as possible. This heuristic results in more compact and distinct clusters and also a slightly increased overall execution speed.

The well known algorithm *ISODATA* by Ball and Hall [1965] is based on the iterative execution of the STD K-MEANS Algorithm and is able to vary the cluster count. This is done by two operations, which are performed between each *k-Means* run: $i)$ a split operation, which divides a cluster into two new ones, whose standard deviation exceeds a user-defined threshold, and $ii)$ a merge operation, which combines two clusters, whose distance falls below a user-defined threshold. This mechanism enables the algorithm to detect a cluster count, which is close to the optimum, but introduces new parameters which influence the algorithm's behavior and thus the cluster quality greatly.

The *Fuzzy c-Means (FCM)* algorithm by Bezdek et al. [1984] represents another important meta-heuristic enhancement. Instead of the unambiguous assignment data points to clusters, this approach introduces a soft association vector for each data point, which expresses a continuous assignment to each cluster. The technique provides two advantages: First, it caused the cluster results to be less dependent on the cluster's start position, as the continuous association provides more flexibility of the clusters during the assignment phase, in contrast to the strict conventional way. Second, it enables the data points to belong to different clusters with varying intensities, if data points can not be assigned unambiguously.

A similar approach was introduced by Dunn [1974], extending the *ISODATA* algorithm by fuzzy data point – cluster associations. This enhancement enables the algorithm to detect also well separated compact clusters and thus improves the cluster results globally.

The *BIRCH* algorithm by Zhang et al. [1996] requires only one full scan of the database to build a balanced tree. The tree construction is based on STD K-MEANS Algorithm, covering all data points within the given memory limit if possible. Nodes of the tree hold metrics about the underlying data points, whereat not all leafs have to be inserted into the tree. A subsequently applied hierarchical cluster mechanism extracts a user defined number of clusters. This implementation has the advantage to be able to handle noisy data and process particularly large data sets.

Adapted as document clustering technique, Steinbach et al. [2000] presented a bisecting *k-Means*-based approach. In every iteration a cluster is selected, e.g. the largest possible one, and is split into two subclusters by applying the STD K-MEANS Algorithm. This is repeated until the desired count of clusters is reached. The algorithm is proved to deliver better results than conventional methods and scales well regarding its runtime.

To cluster very large databases, Bradley et al. [1998] developed a framework, which exemplary utilizes the STD K-MEANS Algorithm. Since scanning large data bases is expensive, the approach aims to perform only a single scan. Therefore the analyzed and clustered data so far is sorted into three categories: discardable sets, compressible sets and retained sets. These three sets are maintained during the data scan and serve as base for the final result construction. It was shown, that the method outperforms sampling based approaches.

The heuristic *X-Means* by Pelleg and Moore [2000] treats the issue that the optimal count of clusters is not known and the poor runtime scaling behavior of the conventional *k-Means* implementations. The approach executes the STD K-MEANS Algorithm multiple times and optimizes the value of $k$ while utilizing the following strategy: First, it begins at a low value of $k$ and splits suitable clusters until an upper bound is reached. Second, it optimizes the *Bayesian Information Criterion* globally to determine the optimal value of $k$. The single execution of *k-Means* is accelerated by using a preliminary created *kd*-tree, similar to approach of Pelleg and Moore [1999]. Experiments showed that *X-Means* performs faster and better than the STD K-MEANS Algorithm.

## 3.1.3 The Tree Fusion k-Means Approach

The following two Sections 3.1.3.1 and 3.1.3.2 describe firstly the motivation to develop a new variant of the STD K-MEANS Algorithm and secondly introduce the base concept for this novel approach.

### 3.1.3.1 Motivation for a new k-Means Enhancement

It is shown in Section 3.1.2.2 that the *CA* in general and the STD K-MEANS Algorithm in particular have a wide field of application. As presented by several authors, the runtime of the STD K-MEANS Algorithm increases non-linearly with the count of elements to cluster, the count of initial clusters and the count of utilized dimensions [Inaba et al., 1994; Mahajan et al., 2009; Arthur et al., 2009]. Applied to large data sets, this results in impractical long runtimes, despite of the existing enhancements (Section 3.1.2.3). The requirement to process such large data sets arises the question how to reduce the algorithm's computation time at constant qualities of results. For that reason an execution time profiling is applied to identify the bottlenecks of the algorithm.

It quickly becomes clear, that the iterations over all given elements and clusters, to decide whether or not an element has to be assigned to a new cluster, are the main bottleneck. Although enhancements like Elkan [2003] reduce the count of necessary distance calculations, or Pelleg and Moore [1999] utilizes the element distribution, based on a *kd*-tree, while processing large data models, the algorithm requires more than $90\%$ of its runtime to $i)$ evaluate pairs of elements and clusters and $ii)$ maintain its data structures (compare Section 3.3.1). Hence a novel *CA* approach based on the STD K-MEANS Algorithm is required to reduce the overall runtime at unchanged clustering results.

### 3.1.3.2 Base Concept for the Tree Fusion k-Means Algorithm

The base concept of the *Tree Fusion k-Means Algorithm* (TF K-MEANS Algorithm) is to extend the TRIN K-MEANS Algorithm's approach by Elkan [2003] by the utilization of static trees to integrate large numbers of elements and clusters in hierarchical ordered groups to reduce the amount of necessary comparisons. If it can be assured that no cluster of one group can be the closest one of any element within the other group, further comparisons of these groups can be skipped entirely, thus reducing the overall computation effort- It also implies the assumption that the tree building and maintaining process will not consume the time benefit gained through the improved search mechanism.

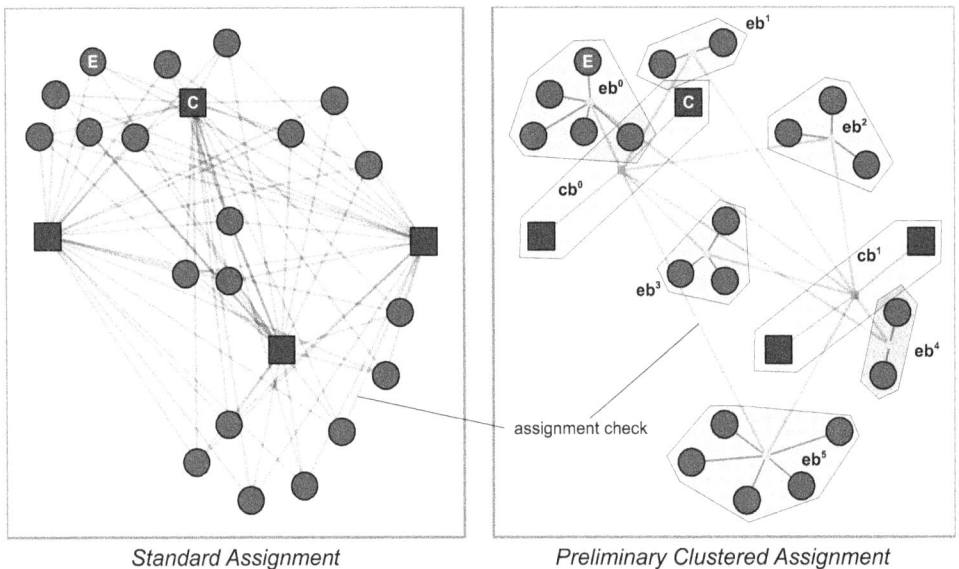

Standard Assignment                              Preliminary Clustered Assignment

**Figure 3.2:** *Base Concept of the* TF K-MEANS *Algorithm*: Example with $20$ elements $E$ (blue) and four clusters $C$ (red), showing the base concept to reduce necessary assignment checks (whether an element belongs to a cluster or not): the standard scheme on the left side requires $80$ checks ($4 \cdot 20$, gray lines); the preclustering approach on the right side introduced six "coarse" element groups $eb^0..eb^5$ and two cluster groups $cb^0$ and $cb^1$, thus reducing the number of element – cluster checks down to $12$ ($2 \cdot 6$, exclusively the checks to refine the group assignments later on).

Figure 3.2 illustrates the base concept of treating coarse preliminary clusters. In contrast to the conventional *k-Means* approach on the left side, which calculates the distance from each element to each cluster to determine the closest element – cluster pairs, the TF K-MEANS Algorithm firstly identifies static coarse groups of elements and clusters. These groups are utilized during the iterative assignment phase to exclude far distant groups from further investigation, saving computation time, i.e. the preliminary grouping is less expensive than the comparison of every element – cluster pair. Furthermore, but not shown in the Figure, the groups can be combined in hierarchical structures (trees) to increase the performance gain even more.

As the enhancement doesn't change the result, but accelerates the process, it can be classified as an *exact* enhancement (compare Section 3.1.2.4). Because the precalculated trees are mostly static, the TF K-MEANS Algorithm is suited for iterative meta heuristics, for example changing the initial cluster positions. Figure 3.3 shows a temporary result of a the TF K-MEANS Algorithm, merging an element and a cluster tree. Although both trees have the same characteristics, they are shown in different styles to visualize the concept to group both elements as well as clusters preliminarily, and process the trees simultaneously.

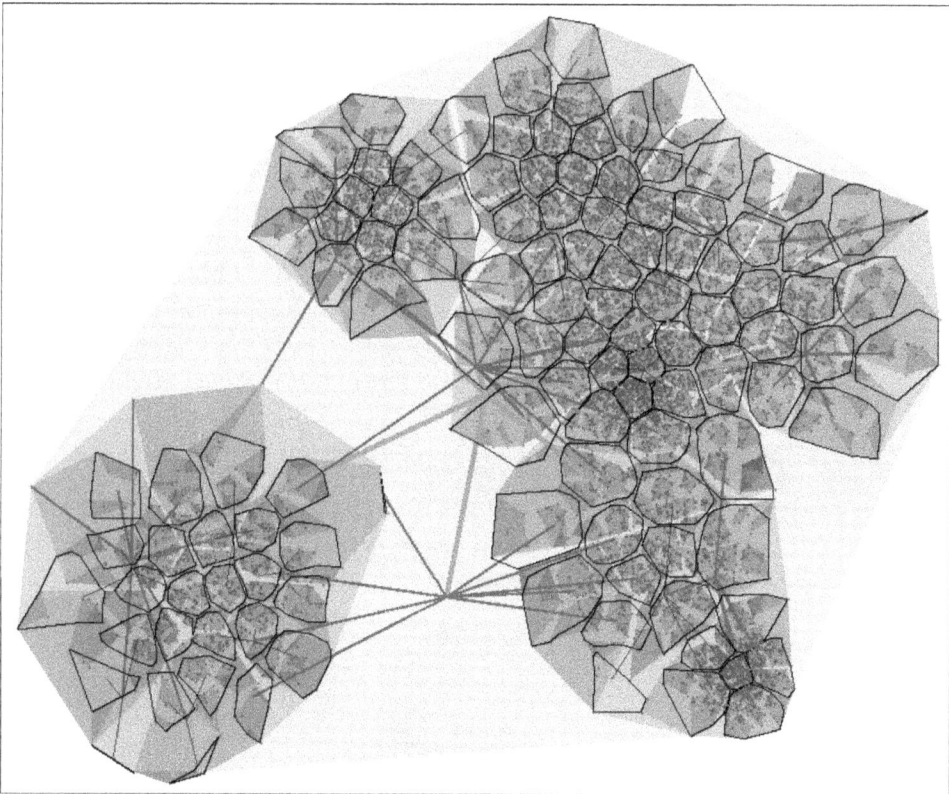

**Figure 3.3:** *Final Result of a CA performed by the* TF K-MEANS *Algorithm,* involving $10k$ elements (blue) and 200 clusters (black polygons). This result was calculated by fusing two different trees: an element tree, forming the elements to element branches (gray transparent polygons) and a cluster tree, forming the clusters to cluster branches (green lines to red cluster centers).

## 3.2 Methods and Materials

In the following the testing and analysis strategy for the proposed *CA* technique (Section 3.1.3.2) is presented. First, Section 3.2.1 describes which algorithms are compared, followed by the introduction of the measurements to evaluate in Section 3.2.2. Subsequently Section 3.2.3 defines the parameter ranges during the testing, i.e. the considered data models and algorithm configurations. The last two Sections 3.2.4 and 3.2.5 describe the designed test scenarios for the final performance analysis and for the development process respectively.

All *CA* algorithms, including existing approaches, were (re-)implemented in C# .NET 4.5 as a transparent and well developed programming language [C#.NET, 2001].

### 3.2.1 Algorithm Comparison

To determine the performance of TF K-MEANS Algorithm, representative runtime tests were executed. Since the algorithm is an exact technique (compare Section 3.1.2.4), its runtimes can be compared to those of the STD K-MEANS Algorithm, to be comparable to measurements of other exact methods, for example those of Elkan [2003]; Pelleg and Moore [1999]. While extensions of the STD K-MEANS Algorithm exist for several platforms, they are extensive to integrate and analyze, which is another reason to limit the comparisons to well researched examples. Quality measurements can be neglected as the method provides exactly the same results as the standard implementation.

Although there exist a wide range of element distribution types or sources, the analysis of the TF K-MEANS Algorithm is restricted to two corner cases of element distribution, covering two extrema (not structured and very well structured), allowing data models of arbitrary sizes and complexities. In contrast to measure and compare runtimes of distinct data models, the performance, especially the scaling behavior and its dependencies are identified by systematic array tests, varying the data model and algorithm parameters step-wisely in specific ranges.

### 3.2.2 Measuring the Algorithm's Results and Performance

To detect the execution acceleration of the TF K-MEANS Algorithm in comparison to the STD K-MEANS Algorithm, the following values are measured or evaluated. Table 3.1 shows an overview of the single measurements, detailed explanation of the measurements can be found below.

| Measurement | Symbol | Description | Relevance |
|---|---|---|---|
| Total Time | $t_{total}$ | total runtime | STD, TF |
| Element Tree Build Time | $t_{build}^{et}$ | time to preconstruct the element tree | TF |
| Cluster Tree Build Time | $t_{build}^{ct}$ | time to construct the cluster tree | TF |
| Runtime | $t_{run}$ | actual runtime | STD, TF |
| Speedup | $q_{speedup}$ | relation of actual runtimes | STD vs. TF |
| Runtime Slope | $slope_{runtime}$ | relative runtime increment | STD, TF |
| Efficiency | $q_{efficiency}$ | processed data model amount per time unit | STD, TF |

**Table 3.1:** *Measurements / Evaluations during the* TF K-MEANS *Algorithm Testing*; Shown are the names and symbols of the single measurements in combination with short descriptions and the relevances for the single algorithms.

The total time span $t_{total}$ represents the total computing duration, starting after the data model creation with the trees' construction and ending with the last *k-Means* iteration (Section 3.3.2.1). In this way it describes the total time costs of the chosen algorithm. The symbols $t^{et}_{build}$ and $t^{ct}_{build}$ describe the time spans to create the element tree and the cluster tree (Section 3.3.2.2) respectively. They occur only once, before the actual algorithm starts, and only when performing the TF K-MEANS Algorithm. Combining the total time span and the tree construction times results in the actual algorithm's runtime $t_{run}$, excluding the time span to construct the element tree, if the TF K-MEANS Algorithm is applied:

$$t_{run} = t_{total} - t^{et}_{build} \tag{3.1}$$

The time span to construct the element tree $t^{et}_{build}$ is excluded from the runtime evaluation, because even if different cluster counts, random seeds or additional algorithm runs (e.g. to evaluate better start position) are applied, the element tree keeps unchanged. Hence it is independent from theses conditions and can be precalculated and stored. In contrast to that the cluster tree build time $t^{ct}_{build}$ is included, as the tree has to be reconstructed after each condition change. As major criteria $q_{speedup}$ is chosen, describing the runtime relation of the STD ($t^{STD}_{total}$) and the TF K-MEANS Algorithm ($t^{TF}_{total}$):

$$q_{speedup} = \frac{t^{STD}_{run}}{t^{TF}_{run}} \tag{3.2}$$

To illustrate the time complexity when processing larger data models, $slope_{runtime}$ is utilized, describing the runtime increment per *MSI*, calculated as average runtime increment (see Formula (2.5)).

Finally, expressing the efficiency of an algorithm, $q_{efficiency}$ is introduced. It represents the processed data model amount per time unit, in this case:

$$q_{efficiency} = \frac{n_{element} \cdot n_{cluster}}{t^{STD,TF}_{total} - t^{TF}_{build}} \tag{3.3}$$

As the TF K-MEANS Algorithm is an exact technique, i.e. delivers exactly the same clustering result as the STD K-MEANS Algorithm, quality measurements can be neglected.

### 3.2.3 Parameters of the Data Models and Algorithm Configuration

To detect the sensitivity of the different evaluated runtimes and the speedup as major criterion, Table 3.2 shows the swept parameters within given ranges:

| Parameter | Symbol | Description | Range |
|---|---|---|---|
| Element Count | $n_{element}$ | number of elements | $100..300k$ |
| Dimensionality | $n_{dimension}$ | position vector size | $2..10$ |
| Data Model Class | $type_{distribution}$ | element distribution method | *ED, GFD* |
| Cluster Count | $n_{cluster}$ | desired number of clusters | $10..30k$ |
| Optimal Branch Size | $size_{opt}$ | branch width | $4..20$ |
| Clustering Strategy | $type_{algorithm}$ | CA algorithm | STD, TF |

**Table 3.2:** *Varied Parameters between Algorithm Runs*; Shown are the names and symbols of the single parameters in combination with short descriptions and typical value ranges during the tests.

The value $n_{element}$ describes the data model size, i.e. the data items which have to be clustered by the algorithms. Besides the cluster count $n_{cluster}$, it is the major parameter which defines the model size. The count of dimensions $n_{dimension}$ describes the position vectors' size, locating elements and clusters within the $n$-dimensional space.

Because of the element distribution scheme within the given space, $type_{distribution}$, influences the behavior and thus the execution time of the *CA* algorithms, two corner cases were chosen: *i*) the *Equal Distribution (ED)* type, spreading the elements randomly and evenly across the given space and *ii*) the *Gaussian Fields Distribution (GFD)* type, grouping the elements to a given number of Gaussian fields, which each spreads the elements normally distributed within an individual range and centered around a random position. In this way two different data model types are available: one provides no spatial structure and the other contains circular, *k-Means*-friendly distributions, i.e. provides significant spatial structures. Figure 3.4 shows example element models for each distribution type, both in two dimensions.

For both types the data models can be scaled, i.e. varying the element and also dimension count, thus providing data models of arbitrary sizes and complexities.

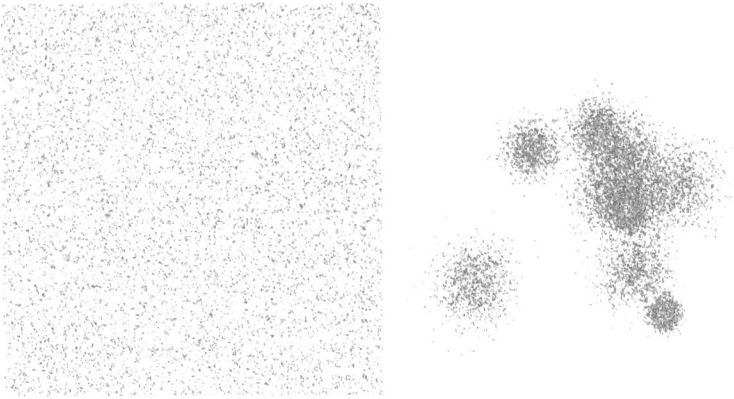

**Figure 3.4:** *Distribution Types;* **Left:** *ED* of $10l$ elements; **Right:** *GFD* of $10l$ elements in ten Gaussian fields of different sizes.

As the *k-Means* algorithm requires a fixed number of clusters (which influences the final clustering behavior and results greatly), the target cluster count parameter $n_{cluster}$ is utilized. Reasonably it is a fraction of the element count $n_{element}$.

Finally the optimal branch size parameter $size_{opt}$ is utilized by the tree building process (Section 3.3.2.2), describing the number of subbranches per branch. In this way it defines the shape of the element and the cluster trees, whereat low values cause lean and deep trees in contrast to wide and flat trees, caused by high values. It is assumed that the tree structure has influence onto the algorithms' runtimes. The parameter $type_{algorithm}$ defines which *CA* algorithm is performed.

### 3.2.4 Exploring the Configuration Space

To test and compare the TF K-MEANS Algorithm (Section 3.2.1), multiple test runs are performed: Each test consists of a parameter configuration (Section 3.2.3), which describes the data model and the algorithm behavior. Due to the large evaluation scatterings, each configuration is applied several times with different random seeds by both the STD and the TF K-MEANS Algorithm. Subsequent to the execution the measurements (Section 3.2.2) are evaluated and stored as well as a result check mechanism verifies the equality of clustering results, and thus the correct behavior of the TF K-MEANS Algorithm.

Because the analysis of the entire parameter space, which is depicted in Table 3.3, would result in a run count of the order of millions and thus impractical run times, the testing is reduced to a fraction of the parameter space, covering regions of interest. Table 3.4 lists the chosen parameter regions as test scenarios, where $n_{seed}$ is always 10, representing a suitable compromise of test run and scattering minimization. The measurement results of these tests are presented in Section 3.3.3.

| Parameter | Description | Range | Step Size | Run Count | Total Run Count |
|---|---|---|---|---|---|
| $n_{element}$ | number of elements | $5k..50k$ | $5k$ | 20 | |
| $n_{dimension}$ | number of dimensions | $2..10$ | 1 | 9 | |
| $type_{distribution}$ | element distribution type | $ED / GFD$ | | 2 | |
| $n_{cluster}$ | number of target clusters | $500..5k$ | 500 | 20 | |
| $size_{opt}$ | tree shape | $4..20$ | 1 | 17 | |
| $type_{algorithm}$ | utilized $CA$ algorithm | STD / TF | | 2 | |
| $n_{seed}$ | number of random seeds | 10 | | 10 | |
| | | | | | 2.448.000 |

**Table 3.3:** *Entire Parameter Space*; Shown are the varied ranges of the described parameter. The total run count of all configurations indicates the size of the entire parameter space.

| ID: Scenario | Varied Parameter | Step Size | Fixed Parameter | Evaluations |
|---|---|---|---|---|
| 1: Performance Evaluation | $n_{element} = 5k..50k$ | $5k$ | $n_{dimension} = 2$ | $t_{total}$ |
| | $n_{cluster} = 500..5k$ | 500 | $size_{opt} = 8$ | $t_{build}^{et,ct}$ |
| | $type_{distribution} = ED / GFD$ | | | $q_{speedup}$ |
| | $type_{algorithm} = $ STD / TF | | | $q_{efficiency}$ |
| 2: Performance Evaluation of Large Data Models | $n_{element} = 25k..300k$ | $25k$ | $n_{cluster} = n_{element}/10$ | $t_{total}$ |
| | $type_{distribution} = ED$ | | $n_{dimension} = 2$ | $t_{build}^{et,ct}$ |
| | $type_{algorithm} = $ STD / TF | | $size_{opt} = 8$ | $q_{speedup}$ |
| 3: Dimension Count Influence | $n_{element} = 2k..20k$ | $2k$ | $n_{cluster} = n_{element}/20$ | $t_{total}$ |
| | $n_{dimension} = 2..10$ | 1 | $size_{opt} = 8$ | $t_{build}^{et,ct}$ |
| | $type_{distribution} = ED / GFD$ | | | $q_{speedup}$ |
| | $type_{algorithm} = $ STD / TF | | | |
| 4: Branch Size Influence | $n_{element} = 5k..50k$ | $10k$ | $n_{cluster} = n_{element}/10$ | $t_{total}$ |
| | $size_{opt} = 4..20$ | 1 | $n_{dimension} = 2$ | $t_{build}^{et,ct}$ |
| | $type_{distribution} = ED / GFD$ | | | $q_{speedup}$ |
| | $type_{algorithm} = $ STD / TF | | | |

**Table 3.4:** *Test Scenarios for the* TF K-MEANS *Algorithm*: Shown are the configurations of the four test scenarios, analyzing the influence of the described parameters in Section 3.2.3 onto the performance indicators out of Section 3.2.2 for the TF K-MEANS Algorithm.

Scenario 1 covers the influence of the data model size (element and cluster count) onto algorithm's performance. By scaling the models up by constant steps, the runtime, speedup and efficiency behavior can be observed. As this scenario treats element and cluster counts independently and thus the practicable maximal model size is limited, Scenario 2 investigates runtimes and speedups while processing very large data models with a fixed element – cluster count ratio. Scenario 3 aims to detect the performance dependency of the dimensionality and Scenario 4 tries to find the link between the speedup and the tree shape. All scenarios, except Scenario 2 due to its redundancy, apply both element distribution types to investigate also their influence onto the measurements.

To suppress the impact of randomness onto the results, the number of runs per configuration is always set to 10, i.e. each data model and algorithm configuration is instantiated and processed 10 times with different random seeds. The resulting measurements serve as statistical base to calculate average values and standard deviations for each configuration set.

The tests were executed in parallel on a *Windows 7 Professional* workstation, hosting an *Intel Core $i7 - 2760QM$* CPU with access to 12 GB of RAM. The results were collected automatically into detailed logs, which serve as the charts' database in Section 3.3.3 for the final analysis.

### 3.2.5 Algorithm Development Procedure

Equal to the iterative DT Framework development approach in Section 2.2.5, the described profiling mechanism was utilized as analysis tool for the TF K-MEANS Algorithm. Additionally a check mechanism was applied to ensure identical results, comparing the final clustering of both the STD. and the TF K-MEANS Algorithm, which were initialized with the same random seed. A subset of tests was chosen to analyze and validate the development progress, shown in Table 3.5.

| Scenario | Varied Parameter | Fixed Parameter | Evaluations |
|---|---|---|---|
| Development | $n_{element} = 1k; 2k; 10k$ | $n_{dimension} = 2$ | $t_{total}$ |
| | $n_{cluster} = 10; 100; 500$ | $size_{opt} = 8$ | $t_{build}^{et.ct}$ |
| | $type_{distribution} = ED \, / \, GFD$ | $n_{seed} = 3$ | $q_{speedup}$ |
| | $type_{algorithm} = $ STD / TF | | |

**Table 3.5:** *Test Configurations utilized during* TF K-MEANS *Algorithm Development*; Shown are the data models, which were clustered and evaluated during the development process to validate and improve the algorithms.

The development tests, containing data models of different sizes and element – cluster count ratios as well as implementing both element distribution types, are processable by both algorithms in a comparable short time, making them suitable for repeated execution.

The development states of the TF K-MEANS Algorithm were applied in the same manner to these data models, as described for the DT Algorithm in Section 2.2.5 to ensure a validated improvement progress.

## 3.3 Results

In the following Section 3.3.1 gives a brief insight in the profiling results as development tool. Subsequently in Section 3.3.2 the actual TF K-MEANS Algorithm is presented, starting with the approach's bases, introducing the main algorithm and enhancements, right through to the final *CA* technique. The measurement results of the final software are illustrated in Section 3.3.3 with respect to the testing strategy previously described in Section 3.2. Finally the results are summarized in Section 3.3.4.

### 3.3.1 Runtime Profiling

To give an example of the evaluated data during the development process, in the following truncated runtime profile logs of both TRIN and the final version of the TF K-MEANS Algorithm are shown in the Tables 3.6 and 3.7. As the algorithms process the same clustering problem, the Tables show the absolute and relative time costs per method and the accordning callers. The complete Tables 6.3 and 6.4 can be found in the appendix Section 6.2.1.

| Method Name | Time Costs | Total Tick Count | Calling Methods | Total Callings | Ratio: Callings |
|---|---|---|---|---|---|
| CheckAtom | 60,76% | 117.274.278 | | | |
| | | | UpdateA2CAssignments | 57.000 | 100,0% |
| UpdateMinA2CDistances | 16,31% | 31.470.388 | | | |
| | | | PerformTrInEq | 57 | 100,0% |
| .. | .. | .. | .. | .. | .. |
| CalculateDistance | 2,92% | 5.626.896 | | | |
| | | | UpdateClusterDistances | 1.524.066 | 78,10% |
| | | | UpdateA2CAssignments | 216.703 | 11,11% |
| | | | CheckAtom | 210.596 | 10,79% |
| .. | .. | .. | .. | .. | .. |

**Table 3.6:** *Example Profiling Log;*
*Data Model: ED, 10k elements onto 500 clusters*
*Algorithm: TRIN K-MEANS Algorithm by Elkan [2003]*
Only the first three entries are shown; the left side presents the time costs of all involved methods, the right side shows the distribution of the according method callings.

This profiling identifies with $61\%$ of the total time costs the method `CheckAtom` as the major bottleneck of TRIN K-MEANS Algorithm. The method encapsulates the mechanism to detect the closest cluster of the current element and is called only by the loop method `UpdateA2CAssignment`, which iterates over all elements. The second important bottleneck is the `UpdateMinA2CDistances` method, calculating the element – cluster distances subsequent to the clusters' reposition. For a detailed explanation of the methods' purpose see Section 3.1.2.1 and 3.1.2.4.

By investigating these parts of the initial implementation the development of the TF K-MEANS Algorithm was started and improved iteratively as described in Section 3.2.5.

| Method Name | Time Costs | Total Tick Count | Calling Methods | Total Callings | Ratio: Callings |
|---|---|---|---|---|---|
| CalculateSquareDistance | 13,52% | 14.067.450 | | | |
| | | | CalculateDistance | 1.574.000 | 100,0% |
| CheckAtom | 11,72% | 12.190.477 | | | |
| | | | UpdateA2CAssignments | 981.608 | 100,0% |
| Filter | 10,94% | 11.380.055 | | | |
| | | | PerformTrInEq | 57 | 0,02% |
| | | | StepDownOrCollect | 307.670 | 99,98% |
| .. | .. | .. | .. | .. | .. |

**Table 3.7:** *Example Profiling Log;*
*Data Model: ED, 10k elements onto 500 clusters*
*Algorithm: TF K-MEANS Algorithm*
Only the first three entries are shown; the left side presents the time costs of all involved methods, the right side shows the distribution of the according method callings.

This Table shows a more equal distribution of the time costs than the previous implementation in Table 3.6, which stands for a less likely occurrence of bottlenecks. Furthermore, the most consuming methods require significantly less CPU ticks than the same methods in the initial profiling log. The development was stopped at this point because no further performance gain was conceivable due to the compactness and the degree of optimization.

In summary the TRIN K-MEANS Algorithm occupies 193.007.629 ticks in total to solve the given problem, the TF K-MEANS Algorithm only requires 104.014.445 ticks, which means a runtime reduction down to $53,9\%$ in this case. As shown in Section 3.3.3, the larger the processed models are, the higher the speedup becomes.

### 3.3.2 Tree Fusion k-Means in Detail

This Section illustrates the relevant components of the TF K-MEANS Algorithm. First, Section 3.3.2.1 gives an overview of the component dependencies and depicts the differences to the conventional STD K-MEANS Algorithm. Subsequently the Sections 3.3.2.2 and 3.3.2.3 adapt the previously introduced techniques for the internal hierarchical data representation (compare Section 2.3.2). After Section 3.3.2.4 illustrates a distance estimation method to accelerate the algorithm, Section 3.3.2.5 introduces one of the key approach to analyze element – cluster relations efficiently, basing the precalculated data trees. Finally Section 3.3.2.6 summarizes methods to maintain the internal data structures, before Section 3.3.2.7 concludes with the final TF K-MEANS Algorithm.

Table 3.8 shows key terms, utilized to describe the data models and the algorithm.

| Name | Symbol | Description | Color |
|---|---|---|---|
| Leaf | $L$ | element or cluster | Dark Green |
| Branch | $b$ | object containing subbranches and/or leafs | Light Green |
| Element | $E$ | object to cluster | Dark Blue |
| Element Branch | $eb$ | object containing subbranches and/or elements | Light Blue |
| Cluster | $C$ | object the elements are assigned to | Dark Red |
| Cluster Branch | $cb$ | object containing subbranches and/or clusters | Light Red |
| Element / Cluster Tree | | hierarchical structure of element/cluster branches | |
| Assignment | | logical relation of an element (branch) to a cluster (branch) | |

**Table 3.8:** TF K-MEANS *Algorithm Legend / Terminology*; Definition and color coding of key terms.

#### 3.3.2.1 Overview of *Tree Fusion - k-Means* Scheme

The following two paragraphs describe the component structure of the TF K-MEANS Algorithm, as well the data conversion during the algorithm process.

**Base Structure:**   The internal dependency structure of the TF K-MEANS Algorithm can be illustrated as stacked algorithm components, each utilizing the functionality of its subordinated parts, as shown in Figure 3.5 on the left side. The algorithm structure utilizes two bases: $i$) the *CDE* technique to avoid unnecessary distance calculations (Section 3.3.2.4) and $ii$) the TRIN K-MEANS Algorithm, an improved version of the STD K-MEANS Algorithm by Elkan [2003]. This serves for the task to build up the static element and cluster trees, as coarse preliminary groupings by the *TB* component (Section 3.3.2.2). Hereon the metric calculation *MLD* is based to evaluate metric data for every branch to relate elements and clusters groups (Section 3.3.2.3). In combination with a lower bound estimation provided by the *CDE*, the *MLIF* technique is formed, which filters groups of clusters with regard the current element group (Section 3.3.2.5). The final *TF* part integrates this filtering component and a mechanism to update the tree metrics, the *TU* component (Section 3.3.2.6), to renew the process of element – cluster assignment checks during a single iteration of *k-Means* algorithm (see 3.3.2.7).

The right side of the Figure represents an extended version of the conventional STD K-MEANS Algorithm, with two distinctive features: $i$) new algorithm steps are introduced (gray), such as the initial static

*TB* processes, the evaluation and update of the branch metric as well as the storage of cluster movement data, and *ii*) the replacement of the standard assignment check technique by the *TF* component (black).

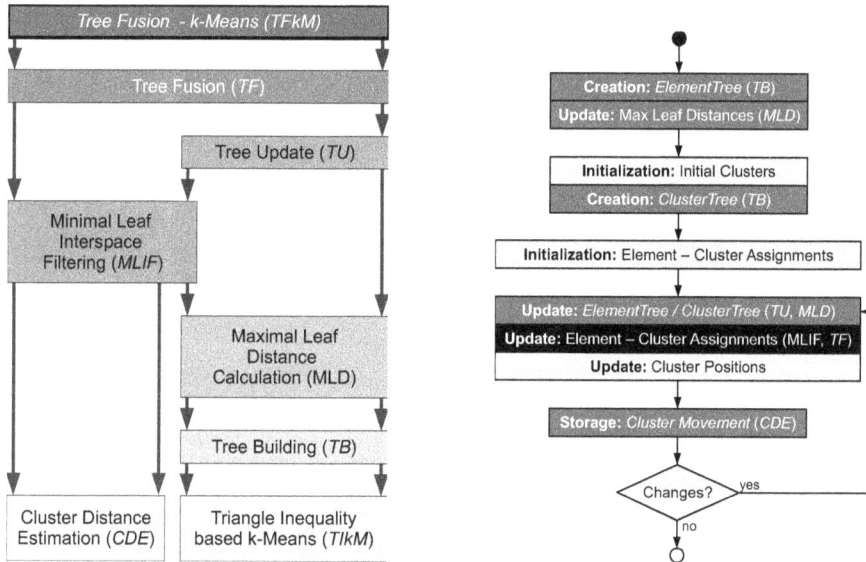

**Figure 3.5:** *Base Structure of the Tree Fusion k-Means Algorithm (*TF K-MEANS *Algorithm)*
    **Left:** Hierarchical component structure of the TF K-MEANS Algorithm.
    **Right:** Overall scheme; First the element tree is created and its *MLD*s are calculated. Subsequently the clusters are initialized and the cluster tree is created. During each iteration both trees are updated and the element – cluster assignments are renewed, fusing both trees. Furthermore, the cluster positions are recalculated and, for performance purpose, the deltas of the positions are stored. The algorithm terminates, like the STD technique, when no further changes occur. Light gray algorithm elements represents unchanged elements from the original STD K-MEANS Algorithm, dark gray elements are newly introduced. The black element – cluster assignment step stands for the major change and heart of the TF K-MEANS Algorithm, integrating all presented elements into a process called *TF*.

**Data Conversion Flow:** Considering a data model, containing elements which should be assigned to a given number of clusters, the data conversion flow of the algorithm is presented in Figure 3.6. The first step is to create and evaluate a static element and a cluster tree based on the given data, which refers to the *TB* and MLD components in Figure 3.5. In each iteration the trees are updated or reconstructed and subsequently fused utilizing the *MLIF* and the *TF* subalgorithms to update the element – cluster assignments efficiently. Finally the positions of the clusters have to be recalculated and hence also parts of the cluster tree (supported by *CDE*), before the main loop can continue.

Note that this illustration utilizes two different but coexisting tree representation styles to visualize the element and the cluster tree simultaneously, depicting the element branches as gray semi-transparent polygons and the cluster branches as green lines with depth-dependent thickness and red dots as cluster centers. Thus no statement about the trees' features is done; they differ only regarding their leaf types (elements or clusters).

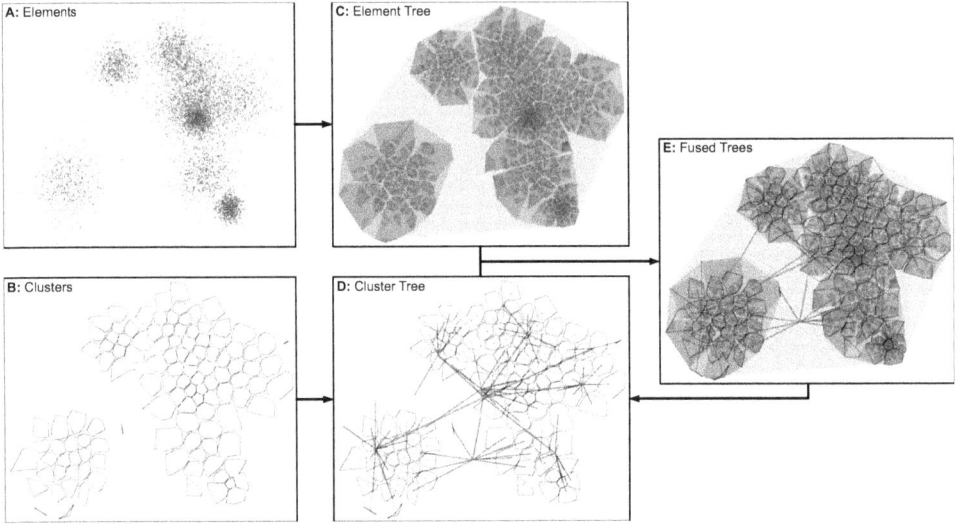

**Figure 3.6:** *Data Conversions Flow within the* TF K-MEANS *Algorithm*;

    **A (data model to cluster):** $10k$ elements (blue dots) in ten Gaussian fields (*DVD* type);

    **B (temporary result):** 200 clusters (black shaped polygons) covering the elements from *A*;

    **C:** element tree shown as semi-transparent gray layers (Section 3.3.2.2), constructed once from *A*;

    **D:** cluster tree whose branches are shown as green lines and cluster centers as red dots (Section 3.3.2.2), constructed from *B*;

    **E:** fusion of both trees, as the essential part of the TF K-MEANS Algorithm to update element – cluster assignments efficiently (Section 3.3.2.7). The loop-back arrow indicates the cluster repositioning and hence the update of the cluster tree branches.

### 3.3.2.2 Tree Building

As described in 3.3.2.1 the static tree creation is one of the fundamental parts of the TF K-MEANS Algorithm. To accomplish this task, the *TB* subalgorithm of the DT concept (Section 2.3.2.3) is utilized to generate the element and cluster tree. While the trees during the DT process are split along their dimensional axes, the data models processed by the TF K-MEANS Algorithm lack the requirements for that feature. Hence the initial cluster branch positions during each split operation, are initialized in the conventional way of the STD K-MEANS Algorithm, distributing them randomly across the given space.

    To illustrate the characteristics of the element and cluster trees, Figure 3.7 shows the rendered examples of element trees based on different element distribution types. The left side shows a balanced tree with a constant count of subbranches for every branch due to the processed *ED* model type. The subbranches in each depth have comparable sizes, only the leaf counts differ slightly, caused by the randomness of the data model and the splitting mechanism. The right side shows an unbalanced tree, consisting of branches with different size in equal depth and variant maximal depths in different regions. In this way the tree represents the irregular structure of the *GFD* data model. Detailed information about the used data models can be found in Section 3.2.3.

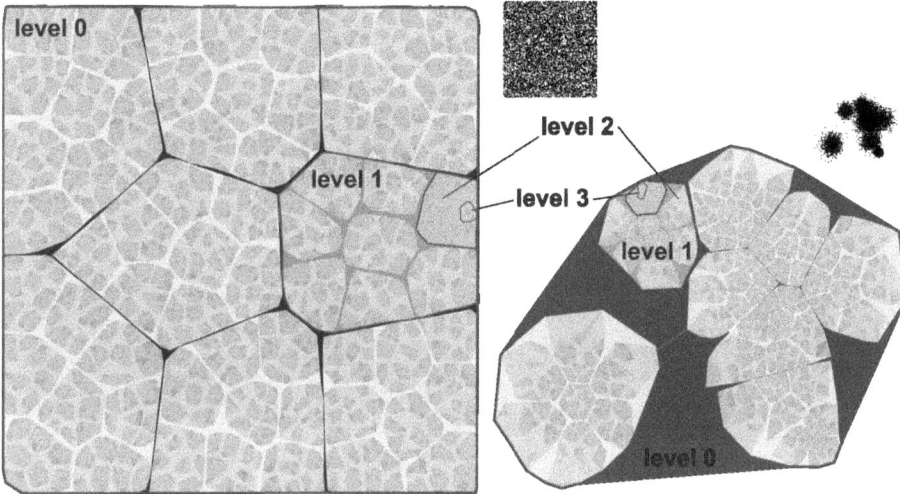

**Figure 3.7:** *Rendered Element Tree Examples*;
  **Both:** $size_{min} = 3$, $size_{opt} = 8$ and $size_{max} = 8$; small pictures with blue dots $\hat{=}$ original element distribution; large pictures with gray semi-transparent polygons $\hat{=}$ trees consisting of branches in various depths; green colored areas indicates branches in the denoted depths
  **Left:** *ED* with $10k$ elements in two dimensions
  **Right:** *GFD* with $10k$ elements in ten Gaussian fields in two dimensions

### 3.3.2.3 Maximal Leaf Distance Calculation

Based on the constructed element or cluster trees (Sections 2.3.2.3 and 3.3.2.2) the *MLD* is utilized as radius of the tree's branches, similar to the usage in the DT Algorithm (Section 2.3.2). The distance to the maximal distant leaf describes a circular area within all other leafs of a branch remain and serves as base for the efficient nearest cluster determination, as explained in Section 3.3.2.5. The calculation scheme is the same as introduced in Section 2.3.2.4.

In addition to the previous examples Figure 3.8 shows more complex and larger structures processed by the *MLD* subalgorithm. The visualization contains two element trees of different distribution type each, built with the *TB* mechanism (Section 2.3.2.3). The *MLD* of every branch is illustrated as a black circle. The elements (leafs) are shown as blue dots and are spread across two dimensions based on the *ED* (left side) and the *GFD* type (right side). Detailed information about the applied data models can be found in Section 3.2.3. The branches of different depths are represented as semi-transparent polygons, enclosing their subordinated items. It can be seen that the *MLD* applied on the geometrical branch center is an appropriate approximation for the extend of the branches, which is relevant for the following subalgorithms, explained in Section 3.3.2.5 and 3.3.2.7.

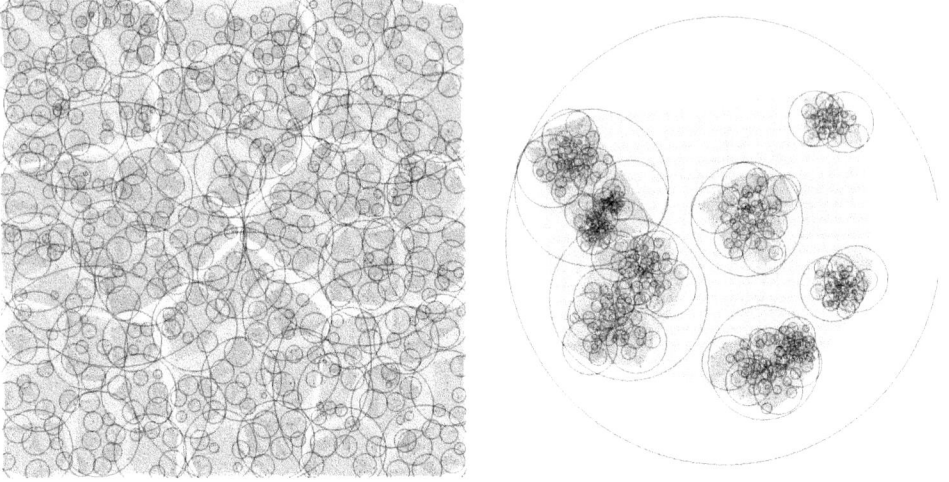

**Figure 3.8:** *Rendered Example of Visualized MLDs for two Element Trees;*
    **Both:** $size_{min} = 3$, $size_{opt} = 8$ and $size_{max} = 8$; blue dots $\hat{=}$ elements; gray semi-transparent polygons $\hat{=}$ branches in various depths, circles $\hat{=}$ MLDs from the branch's geometrical centers;
    **Left:** *ED* with $2k$ elements in two dimensions;
    **Right:** *GFD* with $2k$ elements in ten Gaussian fields in two dimensions;

### 3.3.2.4 Algorithm Enhancement by Cluster Distance Estimation

Similar to the branch distance estimation during the *MLD* calculation, utilizing a path along the hierarchical tree structure, see Formula (2.29), a *Cluster Distance Estimation* (*CDE*) subsequent to occasional cluster position updates can be applied, as suggested by [Elkan, 2003]:

$$dist^{i+m}_{min:e\leftrightarrow c} \geq dist^i_{e\leftrightarrow c} - \sum_{x=0}^{m-1} pos^{i+x,i+x+1}_{delta:c} \tag{3.4}$$

where $dist^{i+m}_{min:e\leftrightarrow c}$ is the minimal distance between an element $e$ and a cluster $c$ after $m$ cluster position updates, starting from update $i$, $dist^i_{e\leftrightarrow c}$ is the calculated distance between $e$ and $c$ during update $i$ and $pos^{i+x,i+x+1}_{delta:c}$ is the cluster position delta of $c$ between update $i + x$ and $i + x + 1$.

This means, storing the "movement" of clusters during the assignment process, enables a prediction of the minimal element – cluster distance, or minimal element branch – cluster branch distance respectively, which saves computation effort. These distances are required to detect whether an element belongs to a cluster, or whether element branches contain potential items for clusters beneath cluster branches (Sections 3.3.2.5 and 3.3.2.7).

The *CDE* is illustrated in Figure 3.9, showing an element, whose distance to a cluster should be estimated based on a former distance calculation and stored cluster position deltas. As only the delta lengths are stored, it is assumed that the cluster approaches through its positions towards the element in a straight way in the worst case. The figure depicts, that the remaining reliable minimal distance after three updates is $dist^{i+3}_{min} = dist^i - pos^{i,i+3}_{delta}$, which can be utilized during subsequent calculations.

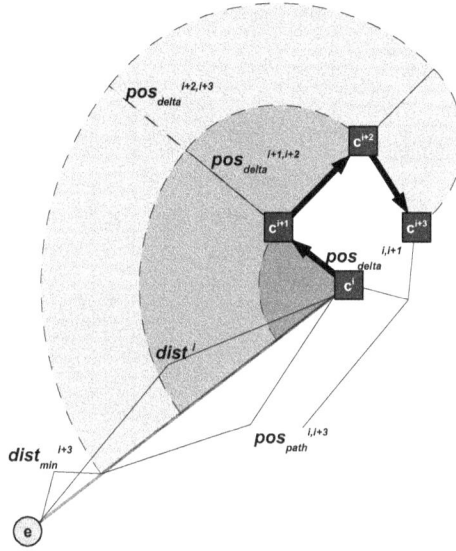

**Figure 3.9:** *Illustration of the Cluster Distance Estimation (CDE);*

The *CDE* assumes a stored element – cluster distance $dist^i$ during a cluster position update $i$ and furthermore, stored cluster position deltas $pos_{delta}^{i+x,i+x+1}$, which occur during the next $x$ cluster position updates.

Visible is an example cluster "movement" for the lower bound estimation: The cluster $c$ (red square) has the position $c^i$ after update $i$, the distance $dist^i$ to the element $e$ (blue circle) and advances through the positions $c^{i+1}$, $c^{i+2}$ and $c^{i+3}$ during the next three position updates. After each update the delta position lengths $pos_{delta}^{i,i+1}$, $pos_{delta}^{i+1,i+2}$ and $pos_{delta}^{i+2,i+3}$ are stored. If the minimal distance between $e$ and $c$ is requested after three updates, the estimation $dist_{min}^{i+3} \geq dist^i - \sum_{x=0}^{2} pos_{delta}^{i+x,i+x+1}$ can be utilized, assuming the cluster moved in a straight way towards the element in the worst case.

This estimation can be applied to element branch – cluster branch relations as well.

#### 3.3.2.5 Minimal Leaf Interspace Filtering

Utilizing the trees as hierarchical data structures, enclosing the elements and clusters (Section 3.3.2.2) and the radius metric *MLD* (Section 3.3.2.3), the following mechanism treats groups (branches) of elements and clusters to detect which element groups contain potential items for the clusters with respect to the iterative *k-Means* cluster reassignment approach.

For each element branch – cluster branch pair, $eb$ and $cb$, the *Minimal Leaf Interspace* ($LI_{min}$) and *Maximal Leaf Interspace* ($LI_{max}$) are defined as the internal minimal and maximal element – cluster interspaces:

$$LI_{min:e\leftrightarrow c}^{eb,cb} = dist^{eb,cb} - dist_{maxleaf}^{eb} - dist_{maxleaf}^{cb}$$
$$LI_{max:e\leftrightarrow c}^{eb,cb} = dist^{eb,cb} + dist_{maxleaf}^{eb} + dist_{maxleaf}^{cb}$$

(3.5)

where $LI_{min:e\leftrightarrow c}^{eb,cb}$ describes the smallest possible distance between an element beneath the element branch and a cluster beneath the cluster branch and further $LI_{max:e\leftrightarrow c}^{eb,cb}$ describes their largest possible distance. Both utilize the precalculated *MLDs* $dist_{maxleaf}^{eb}$ and $dist_{maxleaf}^{cb}$ and the distance $dist^{eb,cb}$ between the branches $eb$ and $cb$. Applying this to an element branch $eb$ and two cluster branches $cb^i$ and $cb^j$, the following bound can be defined:

$$LI^{eb,cb^i}_{max:e\leftrightarrow c} < LI^{eb,cb^j}_{min:e\leftrightarrow c} \tag{3.6}$$

If the equation becomes true, the most distant element – cluster pair within $eb$ and $cb^i$, is closer than the nearest element – cluster pair within $eb$ and $cb^j$. That means $cb^j$ contains no clusters, which are closer to the elements of $eb$, than any cluster beneath $cb^i$. Hence, $cb^j$ can be excluded from further investigation, due to no valid cluster assignment swap would be detectable, as long as $cb^i$ exists. More generally, the term

$$LI^{eb}_{max:e\leftrightarrow c} < LI^{eb,cb}_{min:e\leftrightarrow c} \tag{3.7}$$

applies for every branch $cb$ of the cluster tree to detect whether $cb$ contains potential clusters for $eb$. $LI^{eb}_{max:e\leftrightarrow c}$ is defined as the minimum of the largest element – cluster interspace between $eb$ and every $cb$ out of $k$ cluster branches:

$$LI^{eb}_{max:e\leftrightarrow c} = min\left(LI^{eb,cb^0}_{max:e\leftrightarrow c}, .., LI^{eb,cb^k}_{max:e\leftrightarrow c}\right) \tag{3.8}$$

To illustrate the usage of this upper bound, Figure 3.10 represents different cases, in- and excluding cluster branches based on the Formula (3.7). Shown are an element branch and several cluster branches. One cluster $cb^{ref}$ provides the minimal $LI_{max}$, as limitation for the other cluster branches. The border case whether a cluster branch $cb^i$ can be excluded or not)can be described as:

$$LI^{eb,cb^i}_{min:e\leftrightarrow c} = LI^{eb,cb^{ref}}_{max:e\leftrightarrow c}$$
$$dist^{eb,cb^i} - -dist^{eb}_{maxleaf} - -dist^{cb^i}_{maxleaf} = dist^{eb,cb^{ref}} + dist^{eb}_{maxleaf} + dist^{cb^{ref}}_{maxleaf}. \tag{3.9}$$

The $cb^i$ dependent term can be isolated, resulting in the substitution

$$dist^{eb,cb^i} - LI^{cb^i}_{maxleaf} = dist_{eb,cb_{ref}} + 2dist^{eb}_{maxleaf} + dist^{cb^{ref}}_{maxleaf}$$
$$radius_{limit} = dist^{eb,cb^{ref}} + 2dist^{eb}_{maxleaf} + dist^{cb^{ref}}_{maxleaf} \tag{3.10}$$
$$radius_{limit} = dist^{eb,cb^i} - dist^{cb^i}_{maxleaf}$$

where $radius_{limit}$ describes a circular area, shown as dotted circle, outside of which every $cb^i$ can be neglected. In the example the cluster branch $cb^1$ serves as $cb^{ref}$. It is apparent that $cb^2$ with $radius_{limit} < dist^{eb,cb^2} - dist^{cb^2}_{maxleaf}$ is positioned beyond that area and can be excluded entirely. In contrast $cb^3$, located partially inside $radius_{limit}$, may contain a cluster closer than the most distant cluster beneath $cb^1$ to some element of $eb$, as expressed by $radius_{limit} \geq dist^{eb,cb^3} - dist^{cb^3}_{maxleaf}$.

As Formula (3.7) indicates, given an element branch and a set of cluster branches, firstly the limiting $LI_{max}$ has to be found, and secondly the cluster branches have to be filtered utilizing this upper bound. The illustrated scheme in Figure 3.11 integrates these steps in two stages.

The subalgorithm expects an element branch and a set of cluster branches as input and assumes that

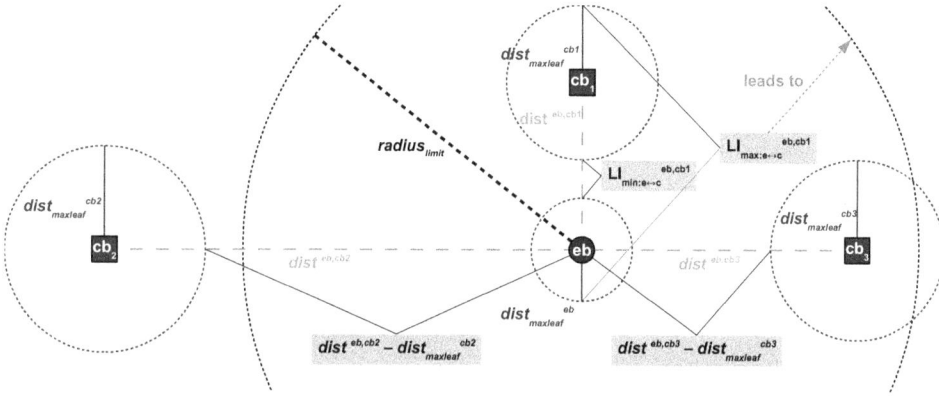

**Figure 3.10:** *Case Analysis of the MLIF;*
To reduce the amount of potential clusters (invisible leafs of cluster branches $cb^1..cb^3$, red squares) for a set of elements (invisible leafs of the element branch $eb$, blue circle), the following estimation is utilized:
A given element branch $eb$ and a cluster branch $cb^1$ have the center distance $dist^{eb,cb^1}$ (dashed line, light gray) and the *MLDs* $dist^{eb}_{maxleaf}$ and $dist^{cb^1}_{maxleaf}$ respectively (dotted circles, blue and red, Section 3.3.2.3). This leads to the $LI_{min}$ and $LI_{max}$ as $LI^{eb,cb^1}_{min:e\leftrightarrow c}$ and $LI^{eb,cb^1}_{max:e\leftrightarrow c}$, i.e. the minimal and maximal element – cluster interspaces between $eb$ and $cb^1$ (see also Formula (3.5)).
Now, if the $LI_{min}$ of another cluster branch $cb^i$ exceeds the current limiting $LI_{max}$ between $eb$ and $cb^1$ (see Formula (3.6)), $cb^i$ can be neglected, because none of its leafs (clusters) can be closer to a leaf (element) of $eb$, than the most distant leaf (cluster) of $cb^1$.
As visual representation of the estimated limitation, the limit $radius_{limit}$, see Formula (3.10), is shown as black dotted circle. Every cluster branch, whose red *MLD* $dist^{cb^1}_{maxleaf}$ circle is outside this limiting area, can be excluded entirely, as too distant from $eb$ in comparison to $cb^1$.
In this case the $LI_{min}$ between $cb^2$ and $eb$ is greater than the limiting $LI_{max}$, i.e. $LI^{eb,cb^1}_{max:e\leftrightarrow c} < LI^{eb,cb^1}_{min:e\leftrightarrow c}$, which is equivalent to $radius_{limit} < dist^{eb,cb^2} - dist^{cb^2}_{maxleaf}$ (denoted in the figure), so $cb^2$ can be neglected.
In contrast, $cb^3$ is inside the limiting radius, which means there can exist a leaf (cluster) within $cb^3$, which is closer than the most distant one within $cb^1$ to $eb$, hence $cb^3$ can not be excluded. Here the estimation is fulfilled: $radius_{limit} \geq dist^{eb,cb^3} - dist^{cb^3}_{maxleaf}$ (denoted in the figure).

the *MLDs* of all branches are up to date. Initially the limiting $LI_{max}$ is set to $\infty$. During the first stage, the mechanism tries simultaneously to find the minimal $LI_{max}$, by checking every cluster branch, and exclude cluster branches, which contain no potential clusters for the elements beneath the element branch, utilizing the upper bound found so far.

For every cluster branch the distance $LI_{min}$ (see Formula (3.5)) is determined. If the cluster branch passes the filter check (Formula (3.6)), it is added to a temporary result list and the $LI_{max}$ is calculated, which may replace the global limiting $LI_{max}$. This approach of simultaneous filtering and updating reduces the computation effort for the second stage, lessening the amount of cluster branches to check.

The second stage processes the remaining cluster branches and compares their $LI_{min}$ to the updated global $LI_{max}$ found during the first stage. All passing cluster branch are the final result of this filtering subalgorithm.

In this way a given set of cluster branches is reduced by sorting out content, which contain no potential clusters for any element within the given element branch. This is accomplished by utilizing only the distances of their geometric centers and precalculated static metrics. Handling large groups of elements and clusters (branches) and involving their static metrics allows to exclude large numbers of clusters efficiently.

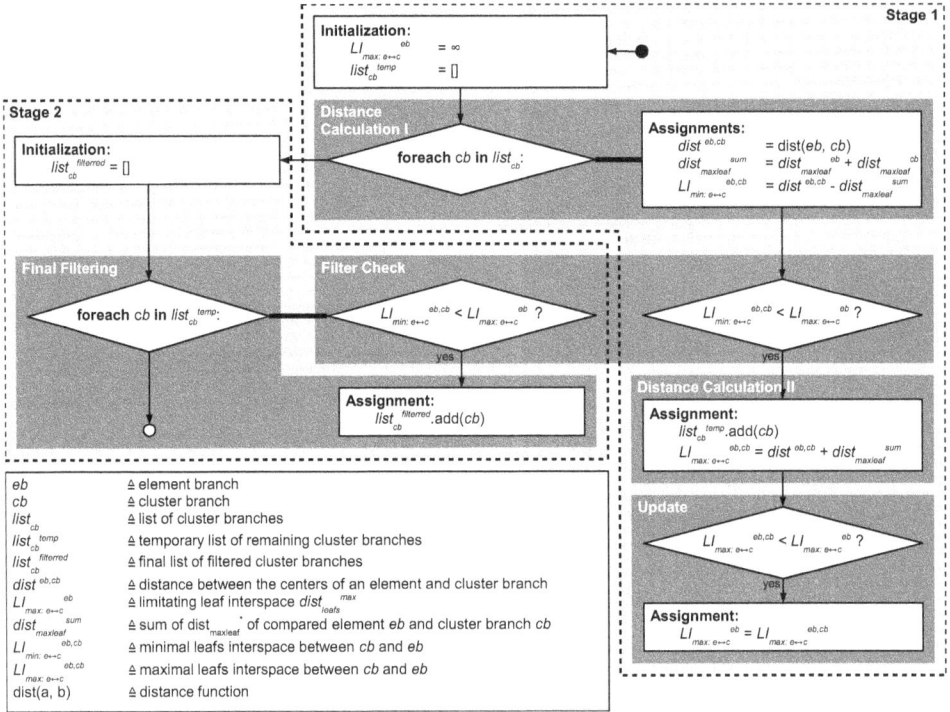

**Stage 1**

**Initialization:**
$$LI_{max:\,e\leftrightarrow c}^{eb} = \infty$$
$$list_{cb}^{temp} = []$$

**Stage 2**

**Initialization:**
$$list_{cb}^{filtered} = []$$

**Distance Calculation I**

foreach $cb$ in $list_{cb}$:

**Assignments:**
$$dist^{eb,cb} = dist(eb,\,cb)$$
$$dist_{maxleaf}^{sum} = dist_{maxleaf}^{eb} + dist_{maxleaf}^{cb}$$
$$LI_{min:\,e\leftrightarrow c}^{eb,cb} = dist^{eb,cb} - dist_{maxleaf}^{sum}$$

**Final Filtering**

foreach $cb$ in $list_{cb}^{temp}$:

**Filter Check**

$$LI_{min:\,e\leftrightarrow c}^{eb,cb} < LI_{max:\,e\leftrightarrow c}^{eb}\ ?$$

$$LI_{min:\,e\leftrightarrow c}^{eb,cb} < LI_{max:\,e\leftrightarrow c}^{eb}\ ?$$

yes

yes

**Assignment:**
$$list_{cb}^{filtered}.add(cb)$$

**Distance Calculation II**

**Assignment:**
$$list_{cb}^{temp}.add(cb)$$
$$LI_{max:\,e\leftrightarrow c}^{eb,cb} = dist^{eb,cb} + dist_{maxleaf}^{sum}$$

**Update**

$$LI_{max:\,e\leftrightarrow c}^{eb,cb} < LI_{max:\,e\leftrightarrow c}^{eb}\ ?$$

yes

**Assignment:**
$$LI_{max:\,e\leftrightarrow c}^{eb} = LI_{max:\,e\leftrightarrow c}^{eb,cb}$$

| | |
|---|---|
| $eb$ | ≙ element branch |
| $cb$ | ≙ cluster branch |
| $list_{cb}$ | ≙ list of cluster branches |
| $list_{cb}^{temp}$ | ≙ temporary list of remaining cluster branches |
| $list_{cb}^{filtered}$ | ≙ final list of filtered cluster branches |
| $dist^{eb,cb}$ | ≙ distance between the centers of an element and cluster branch |
| $LI_{max:\,e\leftrightarrow c}^{eb}$ | ≙ limitating leaf interspace $dist_{leafs}^{max}$ |
| $dist_{maxleaf}^{sum}$ | ≙ sum of $dist_{maxleaf}$ of compared element $eb$ and cluster branch $cb$ |
| $LI_{min:\,e\leftrightarrow c}^{eb,cb}$ | ≙ minimal leafs interspace between $cb$ and $eb$ |
| $LI_{max:\,e\leftrightarrow c}^{eb,cb}$ | ≙ maximal leafs interspace between $cb$ and $eb$ |
| $dist(a,\,b)$ | ≙ distance function |

**Figure 3.11:** *MLIF Scheme;* The goal is to reduce a given set of cluster branches ($list_{cb}$) in relation to an element branch $eb$ so, that only cluster branches remain, whose leafs (clusters) are close enough to the element branch's leafs (elements), to be the potential closest clusters for at least one element of $eb$. The shown mechanism is divided into two stages to combine the search for the global limiting interspace ($LI_{max:e\leftrightarrow c}^{eb}$, Stage 1) and the filtering of the cluster branches ($list_{cb}$, Stage 2).
**Initialization (Stage 1+2):** First, in Stage 1 the limiting $LI_{max}$ ($LI_{max:e\leftrightarrow c}^{eb}$) and a temporary list ($list_{cb}^{temp}$) are initialized. Later on in Stage 2 the final list of filtered cluster branches ($list_{cb}^{filtered}$) is set.
**Distance Calculation I (Stage 1):** For each cluster branch $cb$ of the given $list_{cb}$ the $LI_{min:e\leftrightarrow c}^{eb,cb}$ in relation to the element branch $eb$ is calculated.
**Lower Bound Check (Stage 1+2):** The cluster branch $cb$ passes the check, if its $LI_{min}$ is smaller than the limiting $LI_{max}$ (see Figure 3.10 and Formula (3.6)).
**Distance Calculation II (Stage 1):** If $cb$ passes the filter check, also its $LI_{max}$ is determined and $cb$ is added to the temporary list.
**Update (Stage 1):** If the $LI_{max}$ including $cb$ ($LI_{max:e\leftrightarrow c}^{eb,cb}$) exceeds the global limiting $LI_{max}$ ($LI_{max:e\leftrightarrow c}^{eb}$), it is updated to that value.
**Final Filtering (Stage 2):** Finally the remaining cluster branches are filtered, utilizing the same lower bound check as during the first stage, now with the final $LI_{max:e\leftrightarrow c}^{eb}$.

### 3.3.2.6 Algorithm Enhancement by Tree Updating

During each iteration of the TF k-MEANS Algorithm both the element tree as well as the cluster tree require maintenance to assure the filtering mechanism, as explained in Section 3.3.2.5. In the following the executed update processes are represented. Also further techniques to reduce the computing effort are illustrated.

**Cluster Branch Updates:** After each element reassignment from one cluster to another, both cluster positions have to be updated (Section 3.3.2.1, Figure 3.5). Changing a leaf's position may imply branch position updates up to the tree's root. For that reason every cluster branch $cb$ keeps its geometric boundaries in every dimensions $i \in (0, ..., d)$, $min(p^i_{cb^0_{sub}}, .., p^i_{cb^k_{sub}})$ and $max(p^i_{cb^0_{sub}}, .., p^i_{cb^k_{sub}})$, see Formula (2.25), to improve the position updating speed by checking the position vector $\vec{p}_{cb_{sub}}$ of each subbranch $cb$, against these values. If the new value $p^i_{cb_{sub}}$ of the subbranch is within the stored range of dimension $i$, the position update of $cb$ in $i$ can be skipped. Consequently, if no subbranches of a branch change their position, the branch's position remain and the update process stops at this level. Otherwise the process is executed recursively in a bottom-up manner until the positions of all branches of the cluster tree are up to date.

Simultaneously the distance to the branch's super branch $dist^{cb}_{super}$ and the *MLD* $dist^{cb}_{maxleaf}$ of every cluster branch $cb$, whose position or subbranchs' positions were changed, are recalculated (Section 3.3.2.3). Also with every position update the position delta $pos^i_{delta}$ during iteration $i$ is stored as base for distance estimations later on in the process (Section 3.3.2.4).

**Branch Stable Flags:** To enable the algorithm to recognize altered branches and leafs, which require repeated investigations to detect further reassignment swaps, an *instable flag* was introduced. After an element – cluster assignment swap, the flag is distributed to the following:

- the former cluster of the element (and all its super branches)

- the new cluster of the element (and all its super branches)

- the element itself (and all its super branches)

Marking tree regions which require investigations for assignment swaps in that way, the filtering mechanism (Section 3.3.2.5) can recognize not flagged element and cluster branches as stable and hence negligible with regard to further investigations to accelerate the algorithm.

**Buffered Branch Distances / Element – Cluster Distances:** As shown by Formula (3.5) and in Figure 3.11 the distance $dist^{eb,cb}$ between an element and a cluster branch, $eb$ and $cb$, is an frequently utilized value during the filter process. Based on the instable flag introduced previously, buffered distances are valid as long as involved tree items (branches or leaf) are stable, thus lowering the computation effort.

Similarly the $LI_{min}$ and $LI_{max}$, see Formula (3.5), can be stored, validated and utilized in the same manner, instead of calculating these values repeatedly.

**Estimated Branch Distances / Element – Cluster Distances:** In the case of an invalid buffered distance the following can be applied: utilizing the distance estimation introduced in Section 3.3.2.4, the filter technique in Section 3.3.2.5 can be accelerated further by predicting the distance $dist^{eb,cb}$ similar to Formula (3.4). Consequently $LI_{min}$ and $LI_{max}$ can also be estimated for $m$ updates after iteration $i$ by:

$$\left(LI^{eb,cb}_{min:e \leftrightarrow c}\right)^{i+m} = dist^{eb,cb}_i - dist^{eb}_{maxleaf} - dist^{cb}_{maxleaf}$$
$$\left(LI^{eb,cb}_{max:e \leftrightarrow c}\right)^{i+m} = dist^{eb,cb}_i + dist^{eb}_{maxleaf} + dist^{cb}_{maxleaf} \tag{3.11}$$

based on Formula (3.5), utilizing stored branch distances. Applying these estimations, also the limiting $LI^{eb}_{max:e \leftrightarrow c}$, see Formula (3.6), can be predicted, which enables the algorithm to exclude cluster branches from further investigation without performing the time expensive distance function to calculate the exact values. These preliminary filters work similar to the exact ones and are scheduled prior to them, see Figure 3.11, distance calculation.

**Estimated Limiting Element – Cluster Interspaces:**  As shown in Section 3.3.2.5 the detection of the limiting element - cluster interspace $dist^{eb}_{max:e \leftrightarrow c}$, see Formula (3.7), is an extensive process, requiring the consideration of each potential cluster branch (see Figure 3.11, stage 1). To start with a more appropriate initial value than $\infty$ the following can be applied:

$$\left(dist^{eb}_{max:e \leftrightarrow c}\right)^{i+m} = \left(dist^{eb}_{max:e \leftrightarrow c}\right)^i - \sum_{x=0}^{m-1} pos^{i+x,i+x+1}_{delta:cb^r} \tag{3.12}$$

where $(dist^{eb}_{max:e \leftrightarrow c})^i$ is the limiting element – cluster interspace during iteration $i$, $m$ is the count of position updates since $i$ of the corresponding cluster branch $cb^r$ and $pos^{i+x,i+x+1}_{delta:cb^r}$ are the according position deltas (Section 3.3.2.4). This estimation enables a faster execution of the first stage of the *MLIF* process, see Figure 3.11.

### 3.3.2.7 Tree Fusion / Element - Cluster Assignment Swaps

The name giving heart component of the TF K-MEANS Algorithm is the process which fuses the element and cluster tree to determine the current element – cluster assignment swaps to create the *CA*'s result. While the STD K-MEANS Algorithm detects these swaps by comparing *each* element – cluster pair, the process is accelerated during the TF K-MEANS Algorithm by parsing the precalculated and updated trees (Section 3.3.2.2 and 3.3.2.6) simultaneously and applying the *Minimal Leaf Interspace Filtering (MLIF)* technique (Section 3.3.2.5) onto every element branch and its potential cluster branches.

Figure 3.12 shows the recursive parsing process, checking every element branch and its according cluster branches by applying a filter which reduces the amount of potential cluster branches propagated to the next hierarchy level. The purpose of this mechanism is to minimize the amount of elements and clusters, which require checks regarding their potential assignments, by stepping down in both trees simultaneously and skipping or reducing groups (branches) of clusters for every element group (branch) at every hierarchy level.

The main algorithm starts simultaneously at the roots of both the element tree and the cluster tree. While stepping along the element tree's hierarchy downward, treating only one element branch at one time, all potential cluster branches for the current element branch are considered. If all branches are stable, the process stops the further investigation. Onto tree regions of recent changes a *MLIF* filtering step is applied (Section 3.3.2.5), reducing the count and sizes of potential cluster branches. Based on this filtered set of cluster groups the fusion is continued at every subbranch of the element branch. If the main algorithm reaches a single element leaf, the set of cluster branches is reduced to a single cluster leaf, which represents the new closest cluster. Fusing the trees in this way, a set of element – cluster reassignments is found for the current iteration.

As introduced previously, relevant values can be buffered and estimated, avoiding expensive calculations. Hence Figure 3.12 shows only a simplified scheme, hiding both distance bufferings, buffered distance validations and distance estimations as well as early checks to discard or expand cluster branches (Section 3.3.2.6). These techniques are nested in the filtering and expansion phase.

As expansion check the rule

$$dist^{cb}_{maxleaf} > dist^{eb}_{maxleaf} \qquad (3.13)$$

is implemented, allowing the investigation of branches appropriate of size and number. Too large cluster branches in relation to the element branch's size would prevent the *MLIF* effect, containing always potential nearest clusters for the elements within the element branch. In contrast too small cluster branches would be to numerous, which would undermine the preliminary clustering approach.

After the presentation of the fusion process' scheme, the filtering and expansion sub schemes (Figure 3.12) are demonstrated in an example in Figure 3.13. It is assumed that the fusion process handles currently an element branch, which may consist of more subbranches and leafs. Out of the former fusion step several cluster branches are identified as containing at least one potential closest cluster for at least one element within the element branch.

First (**A**) a *MLIF* step detects the reference cluster branch including the $LI_{max}$ limitation, see Formula (3.8). This results in a limiting radius, see Formula (3.10). Visible in the figure as dotted circle, it shows that one cluster branch is positioned outside the radius, i.e. it contains no cluster closer to any element within the element branch than the most distant cluster out of the reference cluster branch. Thus, this cluster branch can be excluded from further calculations. In contrast another cluster branch is positioned partially inside the limiting radius and therefore it can not be excluded. But all remaining cluster branches are too large in relation to the current element branch (see previous scheme description for further details) and requires expansion, which replaces them by their subbranches.

During the second iteration (**B**), see Figure 3.12, the new cluster branches are taken into account and a further *MLIF* step identifies another cluster branch as closest one, requiring the recalculation of the limiting radius. This reduction disqualifies several cluster branches from further investigation as too distant with respect to the new cluster branch reference.

At this stage no further exclusions are possible, thus the main algorithm is recalled recursively upon every subbranch of the element branch with the currently filtered cluster branch list as initial search set. Note that the limiting radius is not a utilized value during the fusion process, serving only as illustration value. Furthermore, the element branch may be not the only element branch in the shown region of the trees. For the purpose of clarity these branches are invisible.

To illustrate the fusion process further, a second example in Figure 3.14 depicts the simultaneous descent in both trees, disassembling the trees in disjoint assignment sets to reduce the computation effort. While Figure 3.13 presents the filtering and expansion techniques, this example illustrates the recursive components of the scheme in Figure 3.12.

The example consists of several elements and clusters, embedded in precalculated and analyzed trees

**Initialization**

Assignments:
$eb$ = root of element tree
$list_{cb}$ = root of cluster tree
$map_{e \to c}^{swaps}$ = {}

$eb$ is instable or ∃ $cb$ in $list_{cb}$ is instable ?

Assignment:
$list_{cb}^{filtered}$ = $list_{cb}$

Legend:
| | |
|---|---|
| $eb$ | ≙ current element branch |
| $cb$ | ≙ cluster branch |
| $list_{cb}$ | ≙ list of cluster branches |
| $list_{subbranches}^{b}$ | ≙ subbranches of a branch $b$ |
| $list_{cb}^{remaining}$ | ≙ temporary list of filtered cluster branches |
| $list_{cb}^{filtered}$ | ≙ list of cluster branches, filtered regarding their minimal leaf distances |
| $map_{e \to c}^{swaps}$ | ≙ map of swaps (element → new cluster assignment) |
| $dist_{maxleaf}^{b}$ | ≙ maximal leaf distance from the center of a branch $b$ |
| $MinimalLeafInterspaceFiltering(eb, list_{cb})$ | ≙ function to perform the *Minimal Leaf Interspace Filtering* for a given element branch and list of cluster branches |

**Recursion**

Recursive Call:
$eb$ = $eb_{sub}$
$list_{cb}$ = $list_{cb}^{filtered}$

**Filtering**

while ∃ $cb$ in $list_{cb}^{filtered}$ is instable and ∃ exp. occured

Assignment:
$list_{cb}^{remaining}$ = $MinimalLeafInterspaceFiltering(eb, list_{cb}^{filtered})$
$list_{cb}^{filtered}$ = []

foreach $eb_{sub}$ in $list_{subbranches}^{eb}$     no

$eb$ is a leaf ?     yes

**Expansion**

foreach $cb$ in $list_{cb}^{remaining}$

Cluster Switch:
$map_{e \to c}^{swaps}[eb]$ = $list_{cb}^{filtered}[0]$

$dist_{maxleaf}^{cb}$ > $dist_{maxleaf}^{eb}$ ?

no →  Update:
$list_{cb}^{filtered}$.add($cb$)

yes →  Expansion / Update:
$list_{cb}^{filtered}$.add($list_{subbranches}^{cb}$)

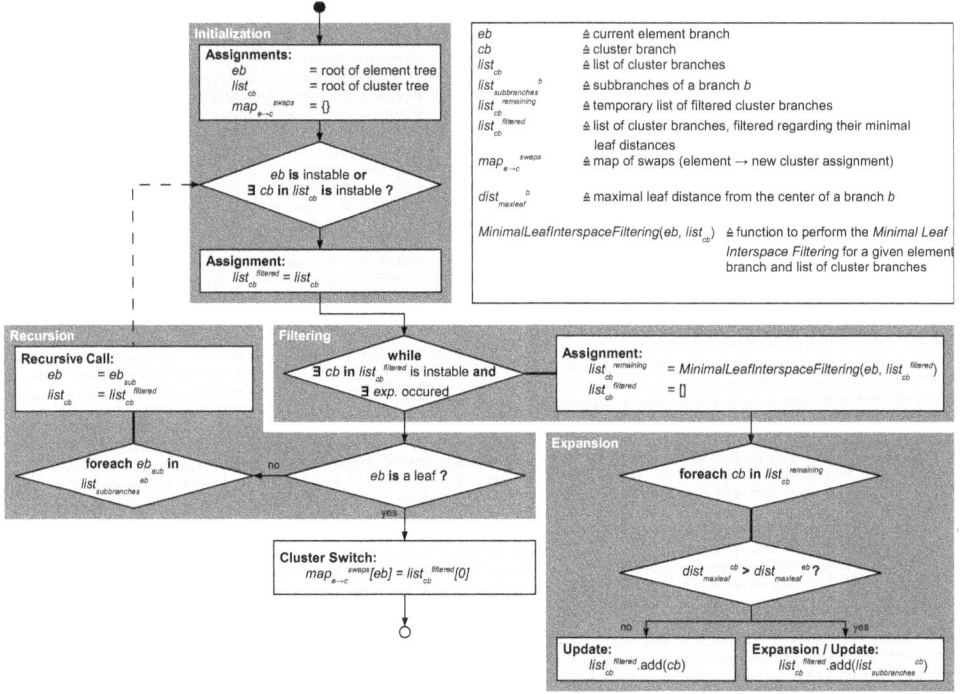

**Figure 3.12:** *Tree Fusion (TF) Scheme as Heart Component of the* TF K-MEANS *Algorithm*;
The purpose is to determine element – cluster assignment swaps, depending on changed clusters positions, just like the STD K-MEANS Algorithm. In this case the precalculated element and cluster trees are utilized to reduce the computation effort.
**Initialization:** The element tree's root became the current element branch $eb$, the cluster tree's root fills the list of the current cluster branches $list_{cb}$. The result map of assignment swap $map_{e \to c}^{swaps}$ is initialized empty. Subsequently during every recursive step along the element tree branches, $eb$ and all current cluster branches in $list_{cb}$ are checked for being instable. If none instability occurs, the entire element branch can be skipped, hence nothing has changed and the algorithm terminates. Otherwise a temporary cluster branch list $list_{cb}^{filtered}$ is initialized with the current cluster branch list.
**Filtering:** As long as expansions occur and instable cluster branches appear, the following loop is executed: $list_{cb}^{filtered}$ is reduced, using the *MLIF* mechanism (Section 3.3.2.5).
**Expansion:** Each item of $list_{cb}^{remaining}$ is checked for being too large in comparison to $eb$, i.e. if its *MLD* $dist_{maxleaf}^{cb}$ is greater than the *MLD* $dist_{maxleaf}^{eb}$ of the element branch, the cluster branch $cb$ is replaced by its sub cluster branches in the $list_{sub\ branches}^{cb}$. The Filtering / Expansion process stops, if no cluster branch in $list_{cb}^{filtered}$ is too far away or too large in relation to the element branch $eb$.
**Recursion:** If $eb$ is not a leaf, for each element subbranch $eb_{sub}$ the fusing main algorithm is recalled recursively with the filtered cluster branches $list_{cb}^{filtered}$ as the current cluster branches $list_{cb}$.
**Cluster Swap:** If $eb$ is a leaf, the swap is stored in $map_{e \to c}^{swaps}$, hence at this point $list_{cb}^{filtered}$ contains only one item (the closest cluster).

(Sections 3.3.2.2 and 3.3.2.3). To clarify the process, both the elements as well as the cluster are distributed in only one dimensions (vertically) at which the trees are nested into each other usually, but are shifted (horizontally) in this case to show the assignment relations. The purpose is to determine a map of elements → clusters, representing the elements and their closest cluster to extract the swaps, if the closest cluster differ from the current ones. Note that this example shows a compressed algorithm flow,

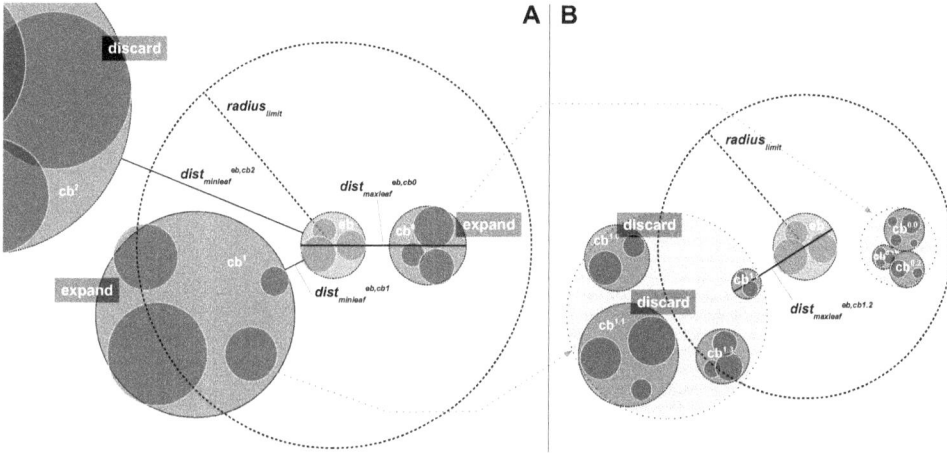

**Figure 3.13:** *TF Example;*

**A** illustrates a fraction of a fusing element and cluster tree. One element branch $eb$ (blue circle), which contains numerous elements in subbranches (small blue circles), and three cluster branches $cb^0..eb^2$ (large red circles), which also contain numerous clusters in subbranches (small red circles) is given. A former *MLIF* step (Section 3.3.2.5) detects the limiting (minimal) $LI_{max}$ between $eb$ and $cb^0$, $dist_{maxleaf}^{eb,cb^0}$. Hence $dist_{maxleaf}^{eb,cb^0} < dist_{minleaf}^{eb,cb^2}$ (see Formula (3.6)) shows, that $cb^2$ is too distant to contain a cluster which is closer than the maximal distant cluster within $cb^0$ ($cb^2$ is positioned completely outside $radius_{limit}$, see Formula (3.10)), and can be discarded. Both $cb^0$ and $cb^1$ are positioned inside or touch the limiting radius ($dist_{maxleaf}^{eb,cb^0} \geq dist_{minleaf}^{eb,cb^{0,1}}$), but are too large in relation to $eb$ ($dist_{maxleaf}^{eb} < dist_{maxleaf}^{cb^{0,1}}$) and will be expanded (split up into their subbranches).

**B** shows the same scenario after the discard of $cb^2$ and the expansion of $cb^0$ and $cb^1$. A repeated *MLIF* step (see Figure 3.12) identifies $cb^{1,2}$ as the closest cluster branch, which reduces $radius_{limit}$. Therefor $cb^{1,0}$ and $cb^{1,1}$ can be discarded, because they contain no clusters close enough for the elements in $eb$. All remaining cluster branches are close and small enough to remain in the cluster list for the recursive down-step to the subbranches of $eb$.

**Note:** This example shows the reduction of possible element – cluster pairs, considering only pre-constructed containers (branches) with their positions and sizes, which reduces the computation effort. Possible further element branches in the area are not shown, because the *TF* main algorithm processes only one element branch at one time.

usually only one pair of an element branch and a cluster branch set is processed at one time.

As explained in Figure 3.12, the fusion process is started at the roots of both trees (**A**). At this level no exclusions are possible, hence the main algorithm advances to the next subelement branches and expanses the cluster branches (**B**). Here the filtering has to perform the comparisons of all element branches and all cluster branches. The *MLIF* step detects the potential cluster branches for the given element branches. In other words, at this early stage some cluster branches are recognized as containing no potential clusters for any element within certain element branches. Thus the problem is divided into disjoint assignment sets, skipping the consideration of numerous branch pairs. At level three (**C**) the main algorithm continues independently of both sets, descending again into the element tree and expanding the cluster branches, to identify more independent assignment sets, which reduce the computation effort even further. At the bottom levels of the trees (**D**) the leafs (elements and clusters) are checked in the same way as the STD K-MEANS Algorithm proceeds, but based on significantly reduced candidate amounts. This means for every element only a small fraction of all existing clusters has to be checked to find the closest cluster.

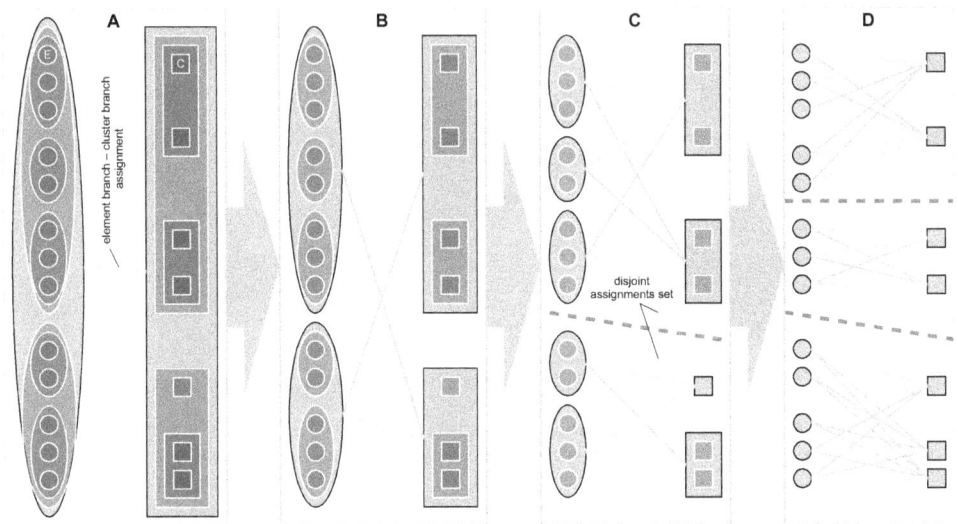

**Figure 3.14:** *TF Dividing Example*;

Illustrates the fusion of an element with a cluster tree, resulting in distinct element – cluster assignments for a single iteration. Instead of comparing all element – cluster pairs one by one, the approach compares bundles (branches) of elements and clusters and skip them entirely if possible to reduce the computation effort.

**A:** Given is an element tree (blue) with $13$ elements $E$ and a cluster tree (red) with $7$ clusters $C$, both in $4$ layers (blue and red ellipses), which both spread in one dimension only. The main algorithm is initialized with the assignment (yellow) of the entire element tree to the cluster tree at the first level.

**B:** At the second level the subbranches of both tree roots are compared regarding distances and sizes (Section 3.3.2.5, gray and yellow lines). This stage reveals, that the lower cluster branch is too distant to contain potential clusters for the elements of the upper element branch (the same applies to the upper cluster and the lower element branch), therefor the further comparison of these combinations can be skipped. The process continues independently with the yellow marked assignments only.

**C:** After finding this two disjoint assignments sets, the third levels of both trees within the sets are compared. In the upper half two element branches are assigned to one cluster branch, the remaining element branch to the second cluster branch. In the lower half no disjoint assignments can be found.

**D:** At the lowest level only elements and clusters remain and are processed similar to the STD scheme, which can now operate in three independent sets and detects the final element – cluster assignments for this iteration. In this example the count of necessary comparisons could be reduced from $13 \cdot 7 = 91$ to $5 \cdot 2 + 3 \cdot 2 + 5 \cdot 3 = 31$.

### 3.3.3 Result and Performance Measurements

The performance of the previously described TF K-MEANS Algorithm (Section 3.3.2) is evaluated by utilizing the testing strategy as described in Section 3.2. The first Section 3.3.3.1 reflects the runtime behavior of both the STD as well as the TF K-MEANS Algorithm. As this proposed new *CA* method yields identical results in comparison to the STD K-MEANS Algorithm, the Sections 3.3.3.2 until 3.3.3.5 consider the speedup dependencies of several model and algorithm parameters. Finally Section 3.3.3.6 investigates the efficiency issue, i.e. the computed problem amount per time span.

### 3.3.3.1 Runtime Dependency of the Data Model Type and Size

In the following representative runtime measurements of *test scenario* 1 are illustrated (Section 3.2.4, Table 3.4).

**Single Runtimes Dependencies of the Element and Cluster Counts:**  To depict the runtime (see Formula (3.1)) improvement of the TF K-MEANS Algorithm in comparison to the standard implementation, the Figures 3.15 and 3.16 show charts of the average runtimes of both algorithms. During the analysis *ED* data models of two dimensions were processed and scaled up by holding the cluster count constant and increasing the element count and vice versa. The resulting models were instantiated multiple times with different random seeds and processed by both algorithms. Note that for the STD K-MEANS Algorithm applies the $y$-axis on the right side, for the TF K-MEANS Algorithm the left $y$-axis.

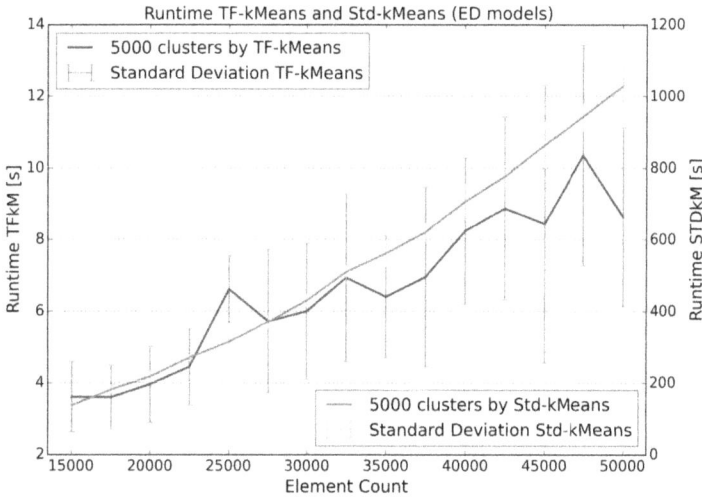

| Algorithms | Data Model Class | Element Count | Cluster Count | Dimensions Count |
|---|---|---|---|---|
| $(type_{algorithm})$ | $(type_{distribution})$ | $(n_{element})$ | $(n_{cluster})$ | $(n_{dimension})$ |
| STD (blue) / TF (green) | ED | $15k..50k$ | $5k$ | 2 |

**Figure 3.15:** *Runtimes ($t_{total}$) of the TF and the STD K-MEANS Algorithms applied to ED Data Models, dependent on Element Count*

The runtimes dependent on the element count behave slightly non-linearly and thus are comparable, but differ in two orders of magnitude. In contrast to that the cluster dependent runtime of the TF K-MEANS Algorithm starts at a comparable high value, but saturates with higher cluster counts. That means the difference of the runtimes' orders of magnitude raises from one to two. The cluster dependent runtime of the STD K-MEANS Algorithm is similar to the element dependent one. Additionally the TF K-MEANS Algorithm shows a higher runtime scatter in both cases, visualized by the standard derivation, which may be caused by the suitability of the current utilized underlying trees, defined by their tree parameters.

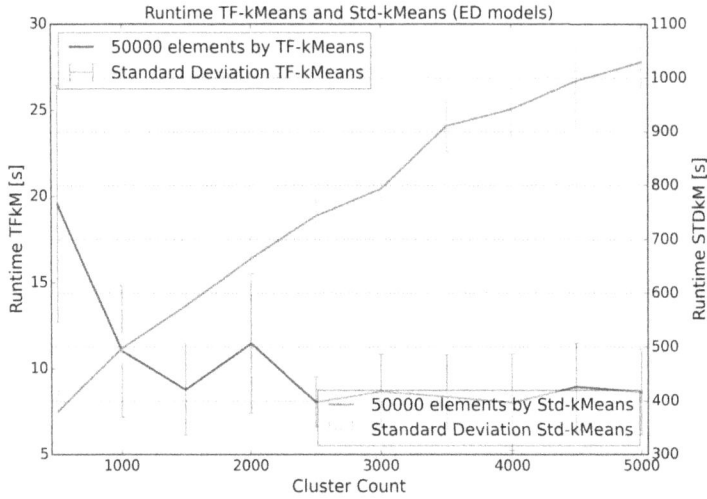

| Algorithms | Data Model Class | Element Count | Cluster Count | Dimensions Count |
|---|---|---|---|---|
| $(type_{algorithm})$ | $(type_{distribution})$ | $(n_{element})$ | $(n_{cluster})$ | $(n_{dimension})$ |
| STD (blue) / TF (green) | ED | $50k$ | $500..5000$ | 2 |

**Figure 3.16:** *Runtimes ($t_{total}$) of the* TF *and the* STD K-MEANS *Algorithms applied to ED Data Models, dependent on Cluster Count*

A similar behavior is observable for processed *GFD* data models in the Appendix in Section 6.2.2.1.

**Multiple Runtimes Dependencies of the Element and Cluster Counts:** Figure 3.17 illustrates runtimes for multiple cluster counts dependent on growing element counts for both algorithms to compare. The selected data models are similar in type and parameter ranges to the models previously described . Note that this chart utilizes a logarithmic $y$-axis.

It is obvious that the runtimes' order of magnitude varies from one (small data models) to two (large data models). The runtime's element count dependency is equivalent to the behavior shown previously. While increased cluster counts cause higher runtimes of the STD K-MEANS Algorithm, they have only observable influence onto the runtimes of the TF K-MEANS Algorithm for small model sizes.

More runtime measurement charts, including cluster count dependencies with multiple element counts in *ED* and *GFD* data models can be found in the Appendix in Section 3.3.3.1.

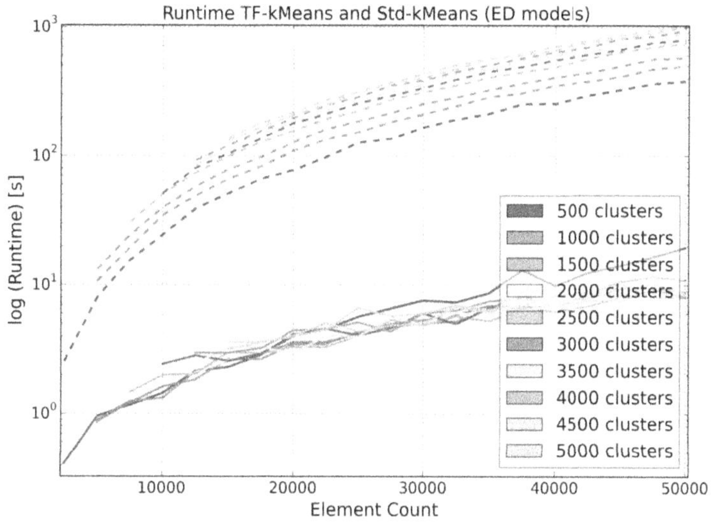


| ***Algorithms*** *($type_{algorithm}$)* | *Data Model Class ($type_{distribution}$)* | *Element Count ($n_{element}$)* | *Cluster Count ($n_{cluster}$)* | *Dimensions Count ($n_{dimension}$)* |
|---|---|---|---|---|
| STD (dashed) / TF (solid) | *ED* | $5k..50k$ | 500..5000 | 2 |

**Figure 3.17:** *Runtimes ($t_{total}$) of the* TF *and the* STD K-MEANS *Algorithms applied to ED Data Models, dependent on Element Count*

### 3.3.3.2 Speedup Dependency of the Data Model Type and Size

In the following representative speedup measurements of *test scenario* 1 are illustrated (Section 3.2.4, Table 3.4).

**Single Speedup Dependency of the Element Counts:** To present the performance gain more clearly, Figure 3.18 shows the speedup (see Formula (3.2)) of the TF in relation to the STD K-MEANS Algorithm. The processed data models are identical to the previously described ones in Section 3.3.3.1, utilizing a fixed cluster count and a growing element count. In addition to the standard deviation, which illustrates the runtime scatter, a moving average ($n = 4$, $\tau = 2$) as smoothed speedup course is shown.

Despite of the large scatter, it is observable that the speedup increases with growing data model size from around 40 up to 130. The smoothed trend shows a virtually linear speedup slope in the selected range. The same behavior is observable during measurements with a fixed element and a growing cluster count as well as while processing *GFD* data models, shown in the Appendix in Section 6.2.2.2.

**Multiple Speedup Dependencies of the Element and Cluster Counts:** Figure 3.19 illustrates the speedup for several cluster counts dependent on a growing element count, comparing both algorithms. The processed data models are similar in type and parameter ranges to the previously described ones.

It is obvious that the speedup increases nearly linearly with growing element counts, but saturates in dependence of the utilized cluster count. That means the higher the element and cluster counts are, the higher is the speedup until the element – cluster count ratio exceeds a certain limit, $\frac{n_{element}}{n_{cluster}} > 30$ in this

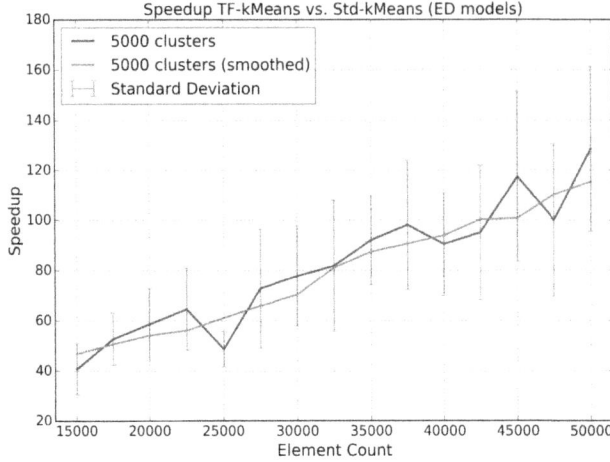

| **Algorithms** | Data Model Class | Element Count | Cluster Count | Dimensions Count |
|---|---|---|---|---|
| $(type_{algorithm})$ | $(type_{distribution})$ | $(n_{element})$ | $(n_{cluster})$ | $(n_{dimension})$ |
| STD / TF | ED | $5k..50k$ | $5k$ | 2 |

**Figure 3.18:** *Speedup ($q_{speedup}$, normal and smoothed) of the TF vs. the STD K-MEANS Algorithm applied to ED Data Models, dependent on Element Count*

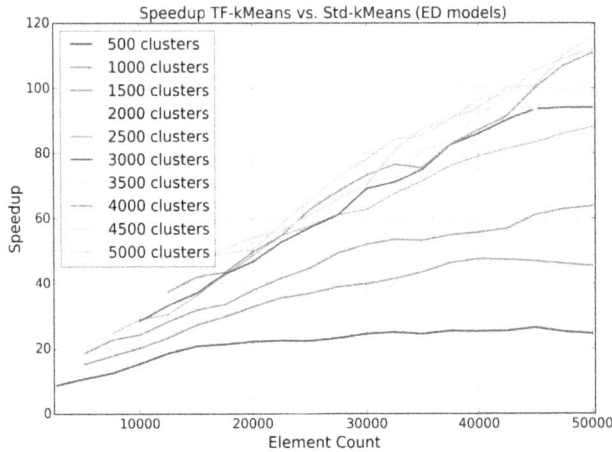

| **Algorithms** | Data Model Class | Element Count | Cluster Count | Dimensions Count |
|---|---|---|---|---|
| $(type_{algorithm})$ | $(type_{distribution})$ | $(n_{element})$ | $(n_{cluster})$ | $(n_{dimension})$ |
| STD / TF | ED | $5k..50k$ | $500..5000$ | 2 |

**Figure 3.19:** *Speedup ($q_{speedup}$, smoothed) of the TF vs. the STD K-MEANS Algorithm applied to ED Data Models, dependent on Element and Cluster Count*

case. Similar charts, showing the same speedup behavior dependent on element and cluster count as well as while processing *GFD* data models, can be found in the Appendix in Section 6.2.2.2.

To integrate the speedup dependencies of the element and cluster counts into a single summarizing up chart, Figure 3.20 shows two surface plots of the speedup for both distribution types. Both the element and cluster counts were incremented by constant steps and for each resulting grid point both algorithms were executed multiple times, utilizing the same seed. The $z$-value stands for the according average speedup. Furthermore, the surface profiles are projected to the element count / speedup layer left, and the cluster count / speedup layer right to further illustrate the speedup's parameter dependencies. Also, these projections can be found in the previous section. The element / cluster count layer contains a heat map of the resulting speedup, which is equivalent to the $z$-value.

The surface charts show a clear trend of speedup increases for growing data model sizes (element and cluster count), whereupon the model distribution types have no influence. Observable is a slight diagonal ridge of the speedup surface plots, indicating an optimal element / cluster count ratio for the TF K-MEANS Algorithm ($\frac{n_{element}}{n_{cluster}} \approx 10$ in this case).

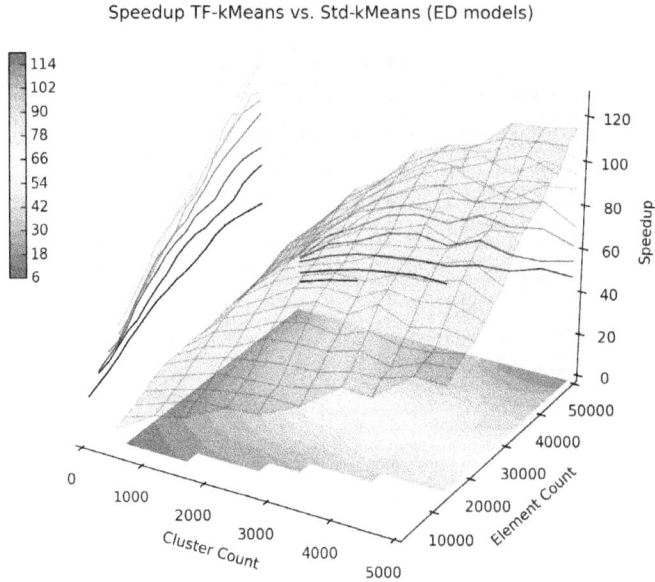

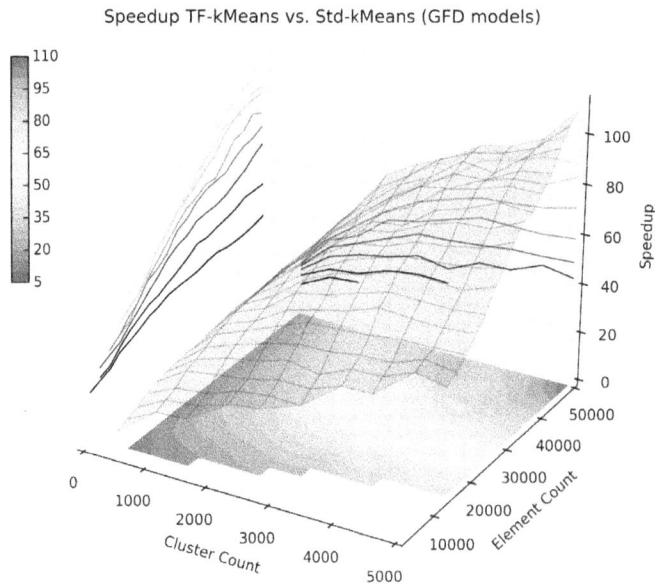

| Algorithms | Data Model Class | Element Count | Cluster Count | Dimensions Count |
|---|---|---|---|---|
| $(type_{algorithm})$ | $(type_{distribution})$ | $(n_{element})$ | $(n_{cluster})$ | $(n_{dimension})$ |
| STD / TF | ED (top), GFD (bottom) | $5k..50k$ | $500..5000$ | 2 |

**Figure 3.20:** *Speedup ($q_{speedup}$, smoothed) of the TF vs. the STD K-MEANS Algorithm applied to ED and GFD Data Models, dependent on Element and Cluster Count*

### 3.3.3.3 Speedup and Runtime Slope Dependency of Large Scale Data Models

In the following representative runtime slopes and speedup measurements of *test scenario* 2 are illustrated (Section 3.2.4, Table 3.4).

**Single Speedup Dependency of the Data Model Size:**  While processing large data models by the STD and the TF K-MEANS Algorithm, the runtime and speedup scaling behavior were determined. During the analysis *ED* data models were utilized, which sizes ranged up to $300k$ elements, and which cluster counts were set to $\frac{1}{10}$ of the current element count, i.e. up to a maximum $30k$ clusters. The models were spread in two dimensions and instantiated multiple times with different seeds. Subsequent to the processing of the STD and the TF K-MEANS Algorithm, the average speedup was evaluated, presented in Figure 3.21.

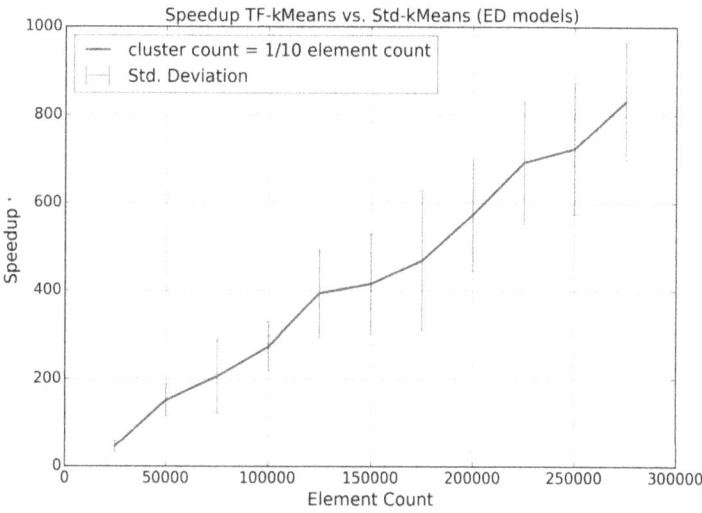

| **Algorithms** | Data Model Class | Element Count | Cluster Count | Dimensions Count |
|---|---|---|---|---|
| $(type_{algorithm})$ | $(type_{distribution})$ | $(n_{element})$ | $(n_{cluster})$ | $(n_{dimension})$ |
| STD / TF | ED | $5k..300k$ | $\frac{n_{element}}{10}$ | 2 |

**Figure 3.21:** *Speedup ($q_{speedup}$) of the TF vs. the STD K-MEANS Algorithm applied to ED Data Models, dependent on Element and Cluster Count*

The chart shows a nearly linear speedup slope from around 25 to 850 for growing data model sizes. The larger the model size is, the higher the speedup and its scatter. Charts of the underlying runtimes can be found the Appendix in Section 6.2.2.3. Large scale tests with *GFD* data models are not performed due to the comparable algorithm behavior for both distribution types as shown in the previous Sections 3.3.3.1 and 3.3.3.2.

**Single Runtime Slope Dependency of the Data Model Size:**  Figure 3.22 illustrates the runtime slopes as runtime complexities of both the STD and the TF K-MEANS Algorithms. According to Formula (2.5) the charts show the runtime increments for linear growing data models, utilizing the same runtime data as the previous chart. The slope is evaluated for each model size step (dashed lines) and smoothed, utilizing a moving average (solid lines, $n = 4$, $\tau = 2$). Note that the charts refer to two different $y$-axis (left side: TF, right side: STD).

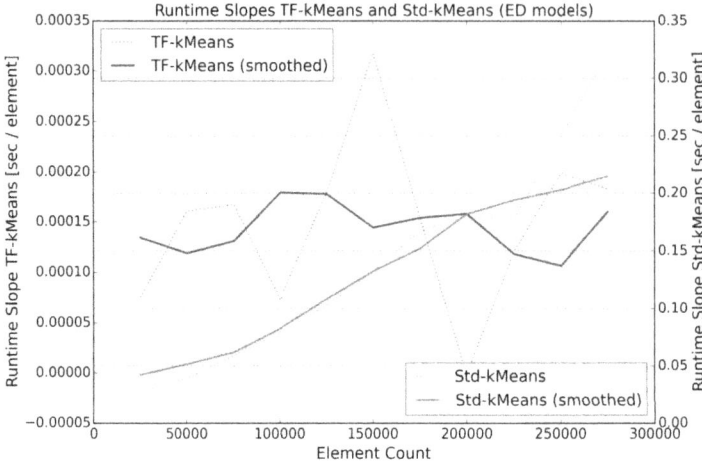

| Algorithms | Data Model Class | Element Count | Cluster Count | Dimensions Count |
|---|---|---|---|---|
| $(type_{algorithm})$ | $(type_{distribution})$ | $(n_{element})$ | $(n_{cluster})$ | $(n_{dimension})$ |
| STD / TF | ED | $5k..300k$ | $n_{element}/10$ | 2 |

**Figure 3.22:** *Runtime Slopes (slope$_{runtime}$, averaged (dashed) and smoothed (solid)) of the* TF *vs. the* STD K-MEANS *Algorithm applied to ED Data Models, dependent on Element and Cluster Count; see Formula (2.5)*

First, it is observable that the runtime slopes of both algorithms differ in more than two orders of magnitude, i.e. the STD K-MEANS Algorithm's runtime is increased by a factor of $> 100$ in comparison to the runtime of the TF K-MEANS Algorithm. Second, and more important, for increasing the data model sizes, the STD K-MEANS Algorithm's runtime increment per model size grows approximately linearly from about $0.03\frac{sec}{element}$ to $.32\frac{sec}{element}$. In contrast, the TF K-MEANS Algorithm's runtime slope remains relatively constant at about $0.0002\frac{sec}{element}$.

This allows the prediction of an approximately linear runtime increment for the TF K-MEANS Algorithm, i.e. a doubled data model size will result only in a doubled runtime, despite of the quadratic problem character. On the other side, due to its inconstant slope, the runtime of the standard implementation will exceed acceptable runtimes at a certain point, regardless of the available computational power.

### 3.3.3.4 Speedup Dependency of the Dimension Count

In the following representative speedup measurements of *test scenario* 3 are illustrated (Section 3.2.4, Table 3.4).

The Figures 3.23 and 3.24 present the speedup dependency of the utilized dimension count for both *ED* as well as *GFD* data models. The models were varied by scaling up their sizes with respect to a constant element / cluster count ratio of 20 and by altering the dimensionality, i.e. the count of available dimensions.

Based on different seeds the data models were instantiated multiple times and processed by both algorithms to yield average runtimes. The resulting speedups (see Formula (3.2)) are shown for both model distribution types, utilizing logarithmic $y$-axes.

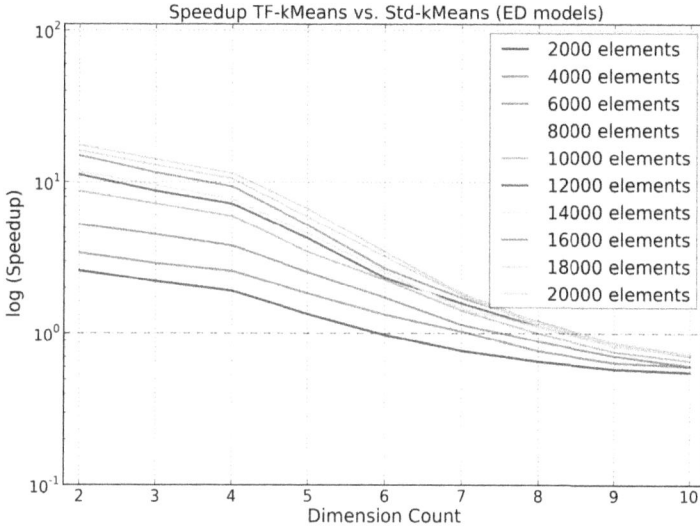

| Algorithms | Data Model Class | Element Count | Cluster Count | Dimensions Count |
|---|---|---|---|---|
| $(type_{algorithm})$ | $(type_{distribution})$ | $(n_{element})$ | $(n_{cluster})$ | $(n_{dimension})$ |
| STD / TF | ED | $2k..20k$ | $n_{element}/20$ | $2..10$ |

**Figure 3.23:** *Speedup ($q_{speedup}$, smoothed) of the* TF *vs. the* STD K-MEANS *Algorithm applied to ED Data Models, dependent on Element, Cluster and Dimension Count*

The chart shows different speedup behaviors for the different data model types. For *ED* data models the speedup falls below 1 at a certain point, which is dependent on the data model size and dimension count. Small data models ($n_{element} = 2k$, $n_{cluster} = 100$) hit the limit already at six dimensional axes, large data models ($n_{element} = 20k$, $n_{cluster} = 1k$) are affected of this for eight and nine dimensions. The larger a model is, the higher dimension counts are manageable for the TF K-MEANS Algorithm before executing slower than the standard implementation.

In contrast to that, *GFD* data models results in a speedup independent from the dimension count. Growing data model sizes cause higher speedups even for high-dimensional models. Charts of single runtimes and speedups dependent on dimension counts can be found in the Appendix in Section 6.2.2.4.

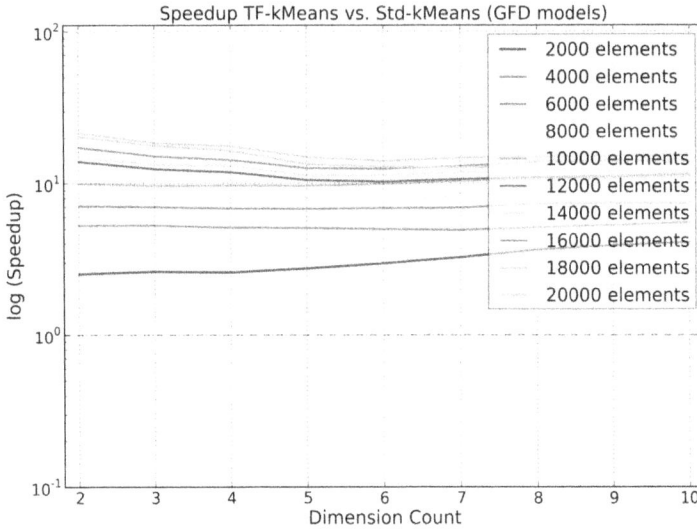

| Algorithms | Data Model Class | Element Count | Cluster Count | Dimensions Count |
|---|---|---|---|---|
| $(type_{algorithm})$ | $(type_{distribution})$ | $(n_{element})$ | $(n_{cluster})$ | $(n_{dimension})$ |
| STD / TF | GFD | $2k..20k$ | $n_{element}/20$ | $2..10$ |

**Figure 3.24:** *Speedup ($q_{speedup}$, smoothed) of the* TF *vs. the* STD K-MEANS *Algorithm applied to GFD Data Models, dependent on Element, Cluster and Dimension Count*

### 3.3.3.5 Speedup Dependency of the Branch Size

In the following representative speedup measurements of *test scenario* 4 are illustrated (Section 3.2.4, Table 3.4).

To measure the influence of the tree's width and depth onto the speedup, the $n_{branch}$ parameter was varied while processing data models of different sizes. A detailed description of the parameter can be found in Section 3.3.2.2. Figure 3.25 shows the smoothed speedup (moving average, $n = 4, \tau = 2$) dependent on the subbranch count per branch for different *ED* data models. The element / cluster count ratio was set to a constant of 10.

It is obvious that the branch size has almost no influence onto the speedup while processing small data models ($n_{element} < 20k$, $n_{cluster} < 2k$), for large data models the speedup decreases slightly for growing numbers of subbranches. Data models of $50k$ elements and $5k$ clusters exhibit a constant speedup reduction of about 20% for branch sizes larger than 13. The same behavior occurs while processing *GFD* data models, as shown in the Appendix in Section 6.2.2.5.

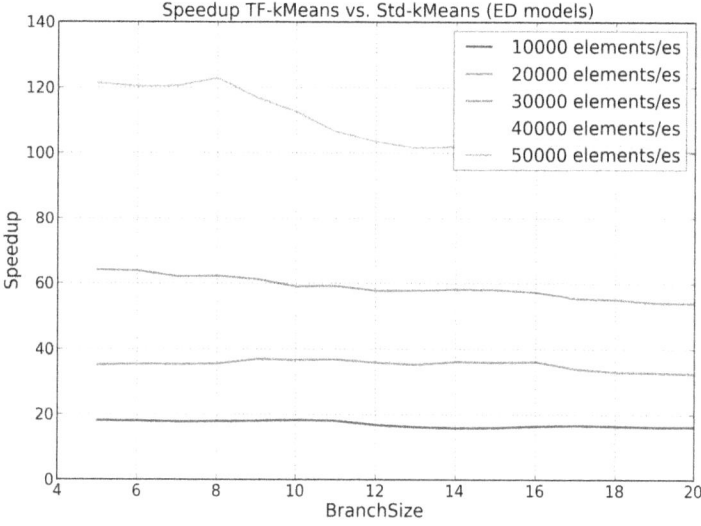

| Algorithms | Data Model Class | Element Count | Cluster Count | Dimensions Count | Branch Size |
|---|---|---|---|---|---|
| $(type_{algorithm})$ | $(type_{distribution})$ | $(n_{element})$ | $(n_{cluster})$ | $(n_{dimension})$ | $(size_{opt})$ |
| STD / TF | ED | $10k..50k$ | $n_{element}/10$ | 2 | $4..20$ |

**Figure 3.25:** *Speedup ($q_{speedup}$, smoothed) of the* TF *vs. the* STD K-MEANS *Algorithm applied to ED Data Models, dependent on Element and Cluster Count, as well as Branch Size*

### 3.3.3.6 Efficiency Dependency of the Data Model Size

In the following representative efficiency measurements of *test scenario* 1 are illustrated (Section 3.2.4, Table 3.4).

As an expression of the efficiency of an algorithm, the $q_{efficiency}$ ratio was introduced (compare Formula (3.3)). It describes the processed data model amount per time unit. While varying element and cluster counts, *ED* and *GFD* data models were utilized to determine the efficiency of both *CA* methods. Each data model was instantiated multiple times with different seeds and processed by both the STD as well as the TF K-MEANS Algorithm. Figure 3.26 shows smoothed surface plots (moving average, $n = 2, \tau = 1$) for both algorithms, where the $z$-value stands for the according efficiency. Note that the $z$-axis is scaled to *Mega Problem Size Units* (mPSU), whereat in this case at problem unit is defined as $n_{element} \cdot n_{cluster}$.

Besides the fact that the TF K-MEANS Algorithm has a larger efficiency compared to the standard implementation of up to two orders of magnitude, for both algorithms applies that the efficiency is increased for growing element counts, whereat the slope of the TF is higher than the one of the STD K-MEANS Algorithm. In contrast, growing cluster counts cause a falling efficiency for the standard implementation. However, for the TF K-MEANS Algorithm, this causes a constant efficiency for small element counts ($n_{element} < 3k$) and a slightly increasing efficiency for larger element counts.

Additional efficiency measurements for *GFD* data models with a similar behavior can be found in the Appendix in Section 6.2.2.6.

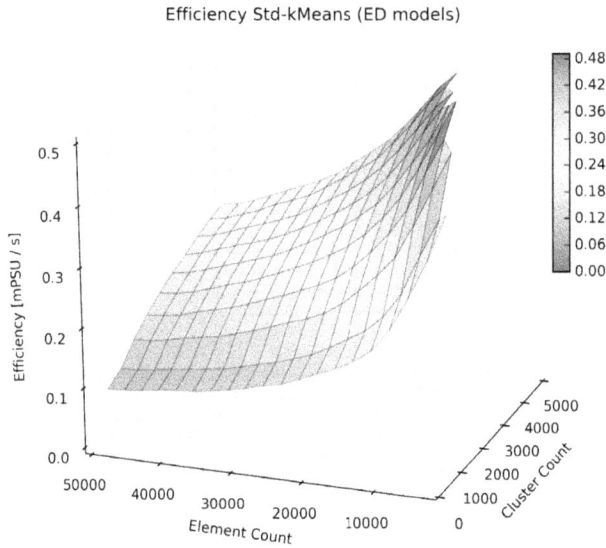

Efficiency Std-kMeans (ED models)

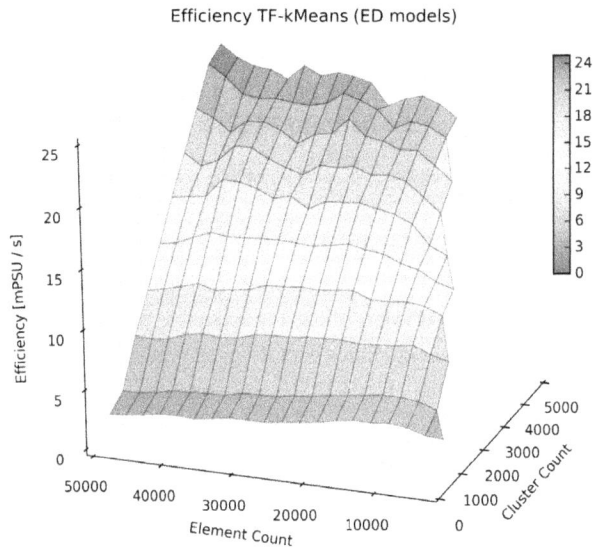

Efficiency TF-kMeans (ED models)

| Algorithms | Data Model Class | Element Count | Cluster Count | Dimensions Count |
|---|---|---|---|---|
| $(type_{algorithm})$ | $(type_{distribution})$ | $(n_{element})$ | $(n_{cluster})$ | $(n_{dimension})$ |
| STD (top) / TF (bottom) | ED | $5k..50k$ | $500..5000$ | 2 |

**Figure 3.26:** *Efficiency ($q_{efficiency}$, smoothed) of the* STD *and the* TF K-MEANS *Algorithm applied to ED Data Models, dependent on Element and Cluster Count;* Efficiency is defined as *PSU* per *TU*, see Formula (3.3).

### 3.3.4 Summary of the Insights regarding the new Clustering Algorithm

This Section gives a brief overview of the newly developed TF K-MEANS Algorithm in Section 3.3.4.1 and its performance evaluations in Section 3.3.4.2. The resulting insights are compared to existing *k-Means* enhancements in Section 3.3.4.3.

#### 3.3.4.1 Algorithm Scheme

The TF K-MEANS Algorithm as an exact enhancement of the STD K-MEANS Algorithm was presented. It's purpose is to accelerate the standard implementation at constant, unchanged clustering results. The method is based on precalculated data structures (namely trees), whose nodes represent groups of elements as well as clusters and store important metrics. Subsequently these information are utilized to compare element and cluster groups in an efficient way to exclude pairs of large groups from further comparison early, thus reducing the computation effort. The development was based on iterative profiling runs (Section 3.3.1), processing predefined data models for evaluation and analysis purposes (Section 3.2.5).

The algorithm is decomposed into several sub methods (Section 3.3.2.1), creating the tree structures by using the TRIN K-MEANS Algorithm (Section 3.3.2.2), calculating node metrics such as the *MLD* (Section 3.3.2.3), estimating frequently utilized values (Section 3.3.2.4), filtering cluster groups regarding a given element group by using the previously calculated metrics and estimated values (Section 3.3.2.5), updating the data structures (Section 3.3.2.6), and finally executing the filter mechanism recursively at every node of the element tree, to obtain the final element – cluster assignments (Section 3.3.2.7) as same result as the conventional approaches, but faster in execution.

#### 3.3.4.2 Measurement Analysis

To establish a general testing strategy, test scenarios were composed, which allow the creation of artificial data models of arbitrary size and dimensionality, covering the corner cases of element distributions (*ED* and *GFD*), and which are executable in practical times (Section 3.2.4, Table 3.4). In this way the new TF K-MEANS Algorithm was compared to the STD K-MEANS Algorithm, evaluating the runtime performances, following other authors, and providing comparable analyses [Elkan, 2003; Pelleg and Moore, 1999]. Hence the measurements represent meaningful comparative indicators for the algorithm's performance. Furthermore, the utilization of artificial data is a reasonable approach to evaluate the algorithm's qualities of results and runtime complexities systematically, as it provides the flexibility to create data models of arbitrary complexity and to trace the measurements, while increasing the model sizes gradually.

The collected data show, that the TF K-MEANS Algorithm requires significantly less time in comparison to the STD approach, when processing the same data models (Section 3.3.3.1). While the runtime's scaling behavior is comparable regarding the count of elements, it saturates for growing cluster counts, if the TF K-MEANS Algorithm is applied in contrast to the standard implementation.

The speedup, dependent on element and cluster counts, increases for growing model sizes and becomes optimal for distinct the element / cluster count ratio (Section 3.3.3.1). The element distribution type has no influence onto the speedup.

Large scale tests proved virtually linear speedup slopes, i.e. a maximal factor of $850$ in comparison to the STD K-MEANS Algorithm at a data model size of $275k$ elements and $27.5k$ clusters (Section 3.3.3.3). The time complexity analysis of the large scale tests revealed a constant runtime slope for the TF K-MEANS Algorithm in contrast to the standard implementation, implying linearly increasing runtimes for linearly growing data model sizes.

Applying high-dimensional data models revealed a speedup sensitivity to the dimensional count in dependence of the utilized distribution types (Section 3.3.3.4). *ED* data models are processed slower by the TF K-MEANS Algorithm, if exceeding a certain dimensional count. In contrast, the speedup is independent from the spatial magnitude while processing *GFD* data models.

The branch size has only a slight influence onto the performance (Section 3.3.3.5). The speedup decreases minimally for larger branches, indicating that small numbers of subbranches are sufficient.

The analysis of the algorithms' efficiency (Section 3.3.3.6) shows an increases for growing element counts for both algorithms, where the TF K-MEANS Algorithm exhibits a larger slope. In contrast, growing cluster counts decrease the STD's efficiency, for the TF K-MEANS Algorithm it remains at least constant.

### 3.3.4.3 Performance Comparisons

The runtime performance of the TF K-MEANS Algorithm can be compared to the introduced enhancements of other authors (Section 3.1.2.4). Since no runtime scaling behavior tests are available for the publications, comparisons regarding the runtime slopes are not possible.

The exact approach of Elkan [2003] reaches a maximal speedup of $351$ for two-dimensional *GFD* data models with $100k$ elements and $100$ clusters. For less significant structures the speedup drops. The TF K-MEANS Algorithm proves a comparable speedup of around $300$ for *ED* and *GFD* data models with $100k$ elements, but $10k$ clusters (Section 3.3.3.3) and thus outperforms the pure triangular inequality based algorithm for large data models. At very high-dimensional data models, i.e. $n_{dimension} = 1k$, the method by Elkan [2003] is about three times faster for *ED* data models than the standard implementation in contrast to the TF K-MEANS Algorithm, whose speedup falls below $1.0$ for high dimension counts. For *GFD* data models the speedup behavior is comparable.

The exact $kd$-tree based enhancement by Pelleg and Moore [1999] shows a maximal speedup of $177$ for $300k$ elements and $5k$ clusters in two dimensions with an element distribution based on astronomical data. According to the large scale tests (Section 3.3.3.3) and the observed saturation (Section 3.3.3.2) the TF K-MEANS Algorithm has a speedup of about $400$ for comparable *ED* data models, hence executes approximately three times faster the $kd$-tree based approach.

In general, the heuristic approaches show, that the loss of cluster quality of is comparable low in relation to the performance gain. This aspect is neglected since the TF K-MEANS Algorithm is an exact technique, i.e. yields identical results as the STD K-MEANS Algorithm. The performance comparison is focused on the speedup, the scaling behavior and the distribution dependencies, if data are available. The method of Kanungo et al. [2002] utilizes sampled data models to examine the cluster result quality. Nevertheless it was shown that a speedup of $25$ is possible even for small data models, which is slightly faster than the exact TF K-MEANS Algorithm. The quantization approach of Eschrich et al. [2003] exhibits a comparable low speedup of up to $306$ for data models based on infra-red images of $400k$ elements, but a large speedup of up to $2346$ for data models based on magnetic resonance images of $65k$

elements. Both models include three dimensions and aim at a fixed and comparable low cluster count. These measurements are based on comparisons to the *FCM* algorithm by Bezdek et al. [1984]. The algorithm introduced by Frahling and Sohler [2006] is not directly comparable to the STD K-MEANS Algorithm. Due to the intention to enable the clustering of different data type domains, no comparable runtime tests exist.

The majority of the published *k-Means* meta heuristics aim at objectives such as to determine a suitable value of $k$, to handle very large databases, to detect noise or to identify ambiguities. Hence these publication provide no useful data to compare with the TF K-MEANS Algorithm.

## 3.4 Discussion

Besides its role as *CA* component within the TRACS-MOO (compare Section 1.4) the TF K-MEANS Algorithm can be regarded as a stand-alone *k-Means* variant. The advantages and disadvantages in general are discussed in Section 3.4.1, followed by a brief summary of potential fields of applications in Section 3.4.2. Finally conceivable further improvements of the algorithm are considered in Section 3.4.3.

### 3.4.1 Advantages and Disadvantages

As the TF K-MEANS Algorithm is an exact enhancement of the STD K-MEANS Algorithm, it shows the same *k-Means* characteristics as presented in Section 3.1.2.3. Additionally a significant runtime speedup in comparison to the standard algorithm could be proven, which outperforms other exact enhancements presented so far [Elkan, 2003; Pelleg and Moore, 1999], in particular for large data models. The speedup grows linearly with the model size and is independent from the distribution type, except when processing high-dimensional models with less or no internal structure (Sections 3.1.2.4, 3.3.3 and 3.3.4.2).

On the other hand the TF K-MEANS Algorithm is a comparatively complex technique, making reimplementations expensive (Section 3.3.2). Due to the precalculation of the tree structures the memory consumption seems to be linear but clearly larger than for the STD K-MEANS Algorithm. This issue could limit the processable data model size at systems of low memory and requires further investigation to allow precise statements. As no detailed scaling tests of other exact techniques are available, comparative scaling behavior evaluations would require the reimplementation of these methods or the provision of corresponding measurements.

### 3.4.2 Applications

The TF K-MEANS Algorithm scales well for data models with large numbers of elements and clusters, and also with high dimensionalities, if the data provides distinct cluster structures. Hence it is suitable or a base for every application of that characterization (Section 3.1.2.2), e.g.:

- *Image Processing:* combining large numbers of pixels to comparatively many clusters, in particular high resolution images with depth information [Mather, 1999; Acharya and Ray, 2005]

- *Overall Compression Techniques:* reducing the size large data sets to detect and/or cover redundancies [Oehler, 1995]

- *High Granular Pattern Recognition:* merging large numbers of data items (e.g. consumers) to a fixed number of groups with similar properties [Punj and Stewart, 1983; Theiler and Gisler, 1997]

- *Data Preparation:* precluster large databases to accelerate search and store queries [Heintzman et al., 2007]

- *Base for Agglomerative Cluster Analysis:* providing a mechanism to combine sets of data items or to identify group information for the selection processes [Soni and Ganatra, 2012]

### 3.4.3 Further Enhancements and Improvements

The tree building process (Section 3.3.2.2) utilizes the TRIN K-MEANS Algorithm to create the branches. As the process performs only one cluster seeding to minimize the overhead, it is possible that unfavorable branch assignments slow down the filter process. As compact branches assist the exclusion mechanism, it could be investigated whether it is worth to repeat subbranch creation steps, i.e. vary the initial branch positions, to obtain higher tree qualities at the expense of longer setup times.

Furthermore, the execution time could be reduced by the parallelization of the tree operations. After leaving the root nodes, the processed trees remain unchanged and the methods are executed independently (compare Section 3.3.2.2 - 3.3.2.7). This approach suggests a big speed gain, but would complicate the algorithm further.

Another strategy is the application of meta heuristics to the TF K-MEANS Algorithm. This covers the methods introduced in Section 3.1.2.4, i.e. optimizing the initial cluster positions [Arthur and Vassilvitskii, 2007] or the cluster count [Ball and Hall, 1965; Steinbach et al., 2000; Pelleg and Moore, 2000] during several iterations, as well as enhancing the TF approach to treat very large data models e.g. with sampling methods [Zhang et al., 1996; Bradley et al., 1998]. Due to the early exclusion approach of the TF K-MEANS Algorithm, only fuzzy-based methods are difficult to apply [Bezdek et al., 1984; Dunn, 1974].

# 4 Applications

In the following chapter the application of the proposed TRACS-MOO approach (Section 1.4) is demonstrated with two examples: First, the *Segmentation*[1] issue is considered [Shapiro and Stockman, 2001], in which biological inspired networks of neurons and synapses have to be transferred to specialized neural hardware, that process and simulate the dynamic behavior of the networks. Second, a generalized *Pattern Recognition*[2] approach is presented [Jain et al., 2000], identifying density peaks in multi-dimensional data sets.

Both examples comprise the data model challenges (large, complex, diverse) as described in Section 2.1.1 and conflicting optimization objectives, that characterize *MOO* problems. The presentation of their solutions should illustrate the proof of the proposed *MOO* concept, than the detailed investigation of the specialized issues. Hence the measurements are limited to the verification of the quality of result and the time complexity.

The next Section 4.1 introduces the application of the TRACS-MOO approach as well as the chosen examples within that scope. Subsequently the Sections 4.2 and 4.3 present the resulting algorithms, the analysis strategies and the performance measurements for both application examples. Finally Section 4.4 reflects the results in the context of the *MOO* in general.

## 4.1 Introduction to the new Approach's Application

The novel concept of TRACS-MOO integrates the initial data model transposition via a *MDS* to a low-dimensional uniform space and the subsequent clustering and interpretation, utilizing adapted common *CA* techniques. The generalized scheme of this approach is illustrated in Figure 4.1. The *MOO* problem is transformed to a *CA* problem [Pulido and Coello Coello, 2004; Taboada and Coit, 2007] and processed as such in two steps. The introduced DT Framework (Section 2) serves as *MDS* technique to perform the model transposition, enabling the subsequent application of adapted common *CA* methods (e.g. the TF K-MEANS Algorithm in Section 3) to group the elements in a way, that the cluster assignments represent for the sought pareto front.

The following sections describe the two examples of the segmentation (Section 4.1.1) and pattern recognition (Section 4.1.2), represented by the *Neuron Placement* and the *Peak Detection* task respectively.

---

[1]decomposing a set of elements into (a fixed number of) segments, whose components are affine according to given criteria
[2]equivalent to a center detection of an unknown number of clusters according to a specified distinctness

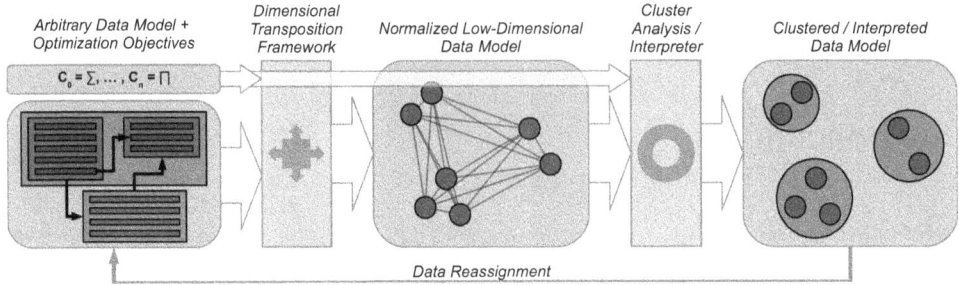

**Figure 4.1:** *Applied* TRACS-MOO *Approach*; A given arbitrary data model is transposed to a normalized low-dimensional representation with regard to the optimization objectives, utilizing the DT Framework (Section 2). Subsequently this model and the objective integrating result is analyzed by an adapted *CA* technique, e.g. the TF K-MEANS Algorithm (Section 3). The outcome can be assigned back to the original data elements, representing a (set of) pareto-optimal solution(s).

## 4.1.1 Segmentation as Neuron Placement

In recent years the research on brain functionality and the search for novel computing techniques initiated the development of so-called *neural hardware*. These hardware systems model the elements of biological neural networks, namely neurons and synapses, as electronic units, encapsulating their electrical behavior. By manufacturing them as integrated circuits, it is possible to create large arrays of artificial neural elements to simulate and investigate biological inspired neural network models for the purpose of deeper insights in brain activities.

Reputed and well known representatives of this fields are the *Spikey* Project [Schemmel et al., 2006], the *FACETS*[3] and the *BrainScaleS*[4] Systems [Schemmel et al., 2010], and, most recently, the *Human Brain Project* (HBP) [HBP, 2012], which all aim to combine large numbers of neural grids (electronic dendrites with adjacent synapse blocks, connected via communication networks) to simulate artificial neural networks of more than $10^6$ neurons.

The intention to simulate biological inspired models on these systems raises the challenge to map a given model description, including the network's topology and elements' parameters, to the neural hardware, taking into account all soft and hard hardware constraints, e.g. capacities, parameter ranges and connection limitations. According to Brüderle et al. [2011] the complete mapping process can be divided coarsely into $i$) the *Neuron Placement*, $ii$) the *Routing* and $iii$) the *Parameter Transformation*. Investigations revealed, that the *Neuron Placement* is one of the most crucial step regarding overall mapping runtime and mapping quality, as it has to consider the entire system and its result influences the reachable quality of the subsequent steps and thus the final mapping significantly [Brüderle et al., 2011; Wendt, 2007].

The term *Neuron Placement* stands for the neuron assignment of the biological model to the neuron blocks of the neural hardware, so that a set of given target functions is optimized. For example, such objectives are:

- *Neuron Loss*: minimizing the number of not realized neurons, e.g. caused by inappropriate adja-

---

[3]Fast Analog Computing with Emergent Transient States
[4]Brain-Inspired Multiscale Computation in Neuromorphic Hybrid Systems

cent synapse types, over occupancies or parameter mismatches

- *Synapse Loss*: minimizing the number of not realized synapses, e.g. caused by insufficient feed-in capacities, routing limitations or parameter mismatches

- *Parameter Matching*: maximizing the similarity of shared neuron / synapse parameters, e.g. influenced by inconvenient neural element grouping

- *Hardware Utilization*: maximizing/minimizing the occupied hardware amount; maximal utilization may increase losses (due to feed-in and routing capacities), minimal utilization may increase parameter mismatches

User-defined preferences determine the priority of these optimization objectives and thus the pareto-optimal result. As the final target function evaluation is scheduled to the completion of the total mapping process (i.e. subsequent to the parameter transformation), the neuron placement has to operate with heuristics, which estimate the function values.

Common biological models utilize the concept of *populations* and *projections* as illustrated in Figure 4.2. A population encapsulates a certain number of neurons along with stochastic information of the neuron parameters. Instantiating a population means to create the specified number of neurons and randomly determine their parameters within the given ranges. Similarly projections connect a source and a target population with a constant probability $p$ and contain also stochastic synaptic parameter data. Establishing a projection is equivalent to connect all possible pairs of source and target neurons of the previously instantiated populations with a chance of $p$. Such a single connection represents a synapse, which also receives a set of randomly determined parameters, based on the projection's specification [Ehrlich and Schüffny, 2013]. In this way, the model of a biological neural network is created from a population network.

On the opposite side the abstract modeling of the neural hardware can be regarded as a hierarchy of hardware elements, whose cores of constant sizes are formed by a synapse and a neuron block. The configurable synapse block contains the pulse amplifying or depressing synapses, which feed the pulse accumulating neurons in the also configurable neuron block, thus representing the synapse-neuron relations. The element behavior of both blocks is determined by shared and individual parameters. With each super-ordinated hierarchy level, the hardware forms a larger grid, that interconnects its components, utilizing a configurable pulse communication network, thus realizing the neuron-synapse relations. Each element has limited capacities and parameter ranges, reflecting the technological boundaries of the neural hardware [Brüderle et al., 2011; Wendt, 2007]. Figure 4.3 shows an example of a neural hardware model with three levels of hierarchy.

For example with respect to the *FACETS* project, the cores represent the so-called *HICANN* ICs[5], the core grid stands for a single wafer and the grid system is equivalent to the entire wafer system [Schemmel et al., 2010].

An immediate *CA* of the original biological model to find appropriate neuron groups for later placement would have only a small chance of success, as the according model system is $i$) high-dimensional, $ii$) unscaled and domain-different, $iii$) unable to reflect the given objectives or the hardware model characteristics, and thus is unsuitable for clustering (Section 2.1.1).

---

[5]High Input Count Analog Neural Network

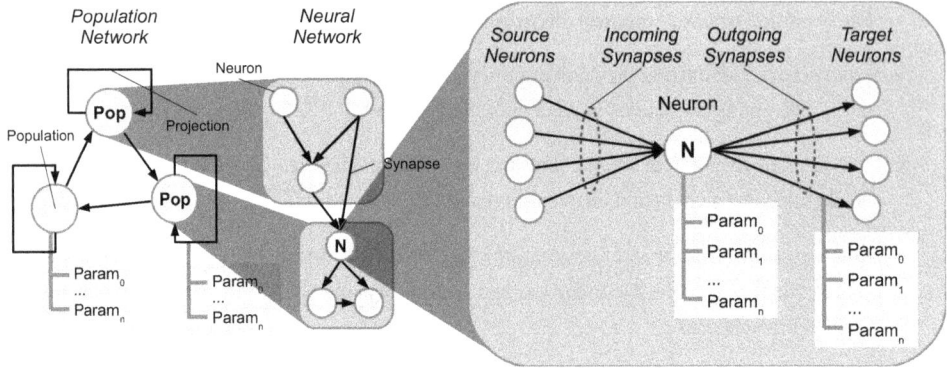

**Figure 4.2:** *Generalized Biological Model*; On top level, within the *Population Network* (left side), the model consists of neuron groups (*Populations*) of different sizes, which are connected via directed assignments (*Projections*). Each population may contain several individual parameters and their stochastic ranges ($Param_{0..n}$), describing the encapsulated subpopulations or neurons. Besides optional synaptic parameters, each projection represents a constant probability $p$ to connect a single neuron of the source population with a single neuron of the target population. Establishing the neurons with their parameters and the projections as synapses between neuron pairs, leads to the *Neural Network* (middle), which has to be mapped to the neural hardware. A closer look at a single neuron (right side) shows its mapping relevant individual information, such as the set of *Incoming* and *Outgoing Synapses*, as synaptic fan-in and fan-out respectively, together with the instantiated parameters.

Following the introduced *MOO* strategy in Section 1.4 and Figure 4.1, the approach to optimize the prioritized objectives described above, while assigning the neurons of the biological neural network to the neuron blocks of the neural hardware, is illustrated in Figure 4.4. Both the model and the objectives are transposed to an uniform low-dimensional equivalent, utilizing the DT Framework (Section 2). A subsequent *CA* technique, based on the alternating execution of the TF K-MEANS Algorithm (Section 3) and a balancing method, divides the transposed model into affine neuron groups with sizes equal to the targeted neural cores. The optimization of the mapping objectives is implicit, due to the weighted transformation process, positioning neurons closely, which are affine with regard to the optimization objectives. In the following this method is referred as SEGMENTATION ALGORITHM (SEGM Algorithm).

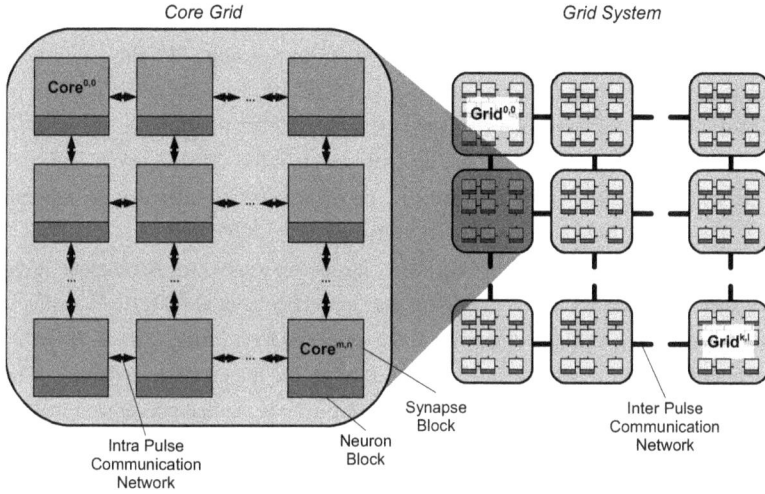

**Figure 4.3:** *Generalized Neural Hardware*; The base element of the neural hardware is the *Core*, consisting of a *Synapse Block*, feeding the dendrites of the *Neuron Block*. Synapses and dendrites can be combined synapse-neurons circuits of various sizes, the counts of which are equal per core. Multiple cores are arranged in a *Core Grid*, as $n \times m$ 2D array, which interconnects adjacent cores via communication channels. This *Intra Pulse Communication Network* establishes neuron-synapse connections over several cores, thus emulating the topology of the simulated biological neural network. On top level, within the *Grid System*, $k \times l$ core grids are connected via an *Inter Pulse Communication Network*, transporting neural spike pulses beyond the borders of a single core grid.

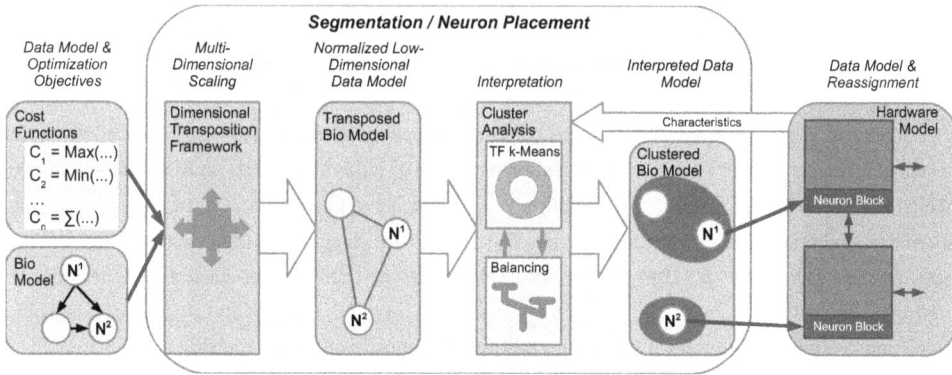

**Figure 4.4:** *Neuron Placement Base Scheme*; First, the SEGM Algorithm imports the *Bio Model* in combination with the *Cost Functions* to be optimized. The DT Framework (Section 2) converts these data to a *Transposed Bio Model*, where the low-dimensional spatial distances correspond with the neurons' affinities, i.e. the similarities of the synaptic interconnectivity and the behavior defining parameters. Then, the following *CA* loop is executed, until the quality gain saturates: the TF K-MEANS Algorithm (Section 3) clusters the transposed model with respect to the characteristics of the targeted hardware model. A subsequent *Balancing* mechanism ensures, that the numbers of elements per cluster don't exceed the cores' capacities. Iteratively the clustering refines the segmentation and the balancing fixes over occupancies, until the quality gain falls below a given threshold. Finally the elements (neurons) of the *Clustered Bio Model* are assigned to the neuron blocks cluster-wisely, which completes the neuron placement.

### 4.1.2 Pattern Recognition as Peak Detection

In the context of the automated data analysis, the *Pattern Recognition* (PR) plays an important role to retrieve integrating information from unstructured data sets. Its task is to detect patterns (regularities, repetitions or similarities), i.e. common characteristics, which determine distinct categories. These analyses allow general statements about the entire data set as well as the classification of new, unknown data items, and are usually based on clustering methods. The *PR* can be considered as an important sub field of the *Knowledge Discovery in Databases* (KDD) [Bishop et al., 2006].

Most recently the Nobel price 2014 of chemistry has been awarded for a technique to visualize objects smaller than the wavelength of the utilized laser light [Betzig et al., 2014]: Exposing said objects to light of very low energy multiple times provides a set of different images, which show the reflections of only few molecules and is subsequently combined to enable the creation of images of much higher resolutions than the original data. For this purpose *PR* methods applied specialized image processing algorithms identify the position of single molecules as centers of the blurred dots in every photo shoot. The combination of all these positions yields the final image with a resolution of $nm$ scale. This technique allows for deeper insights in structures, where the localization of single molecules is relevant. For example the higher image resolution reveals the positions of single proteins within cells during exchange processes, i.e. processes of the metabolism [Bar-On et al., 2012].

A further typical application is the processing of measurement data from physical, chemical, biological or other scientific experiments. As each real system returns noisy data, the *PR*'s purpose is to identify single distinguishable results, thus making the measurements meaningful and comparable. For example, the research on the biological DNA replication mechanisms by Cheng et al. [2014] requires information about the state of the enzymes, which copy segments of the DNA. Due to the natural scattering of the emitted light, the state measurements yield several overlapping Gauss distributions, whose centers are equivalent to the sought states. The utilization of *PR* methods reveals the states in their number and values as cluster centers, thus identifying the states.

More practical applications of the *PR* can be found in the field of image analysis [Liu et al., 2007], retrieving the existence, validation or classification of known or unknown objects within provided images, which is relevant for tasks like face recognition, medical cancer detection, astronomical data processing, as well as in robotics and for self-driving cars. The detection of categories during categorization or classification tasks [Kittler et al., 1998] is also accomplished by the utilization of clustering algorithms, determining distinct object groups, e.g. different consumer classes on the base of single consumer data. The same applies for speech analyses [Rabiner, 1989] in the acoustic domain, requiring a reliable database to compare the preprocessed audio data with. Both the database as well as the audio preprocessing are based on the automated extraction of general characteristics, provided by methods of the *PR*.

With regard to one of the *PR*'s major field of application, the *KDD*, many of the provided data for a *PR* task can be characterized as follows:

- *Unknown Pattern Count:* In dependence of the chosen granularity and pattern definition the data contain different numbers of patterns. The patterns may overlap or contain each other.

- *Unknown Pattern Shape and Sizes:* Similar to the introduction in Section 3.1.1, the pattern's definition determines the results of the *CA*s, and thus those of the *PR* significantly. Dependent on the specified pattern type, the sought groups alter in size, shape and density course.

- *Noise Existence:* Taking into account or neglecting the existence of noise, e.g. data items, which are not assignable to any found pattern, also influences the final *PR* result.

In consequence the parametrization of the *PR* for the unsupervised data analysis (*unsupervised learning*[6]) is defined by $i$) the pattern or cluster definition (e.g. centroid, density or distribution based, see also Section 3.1.1) and $ii$) the granularity, i.e. a specification of the patterns' distinction. With regard to the previous introduced data characteristics, the *PR*'s objectives can be described as:

1) the identification of clusters, i.e. the patterns in count and features including the competitive sub-objectives:

   $a$) the minimization of the cluster count, to ensure the result's meaningfulness

   $b$) the maximization of the pattern specification matching (e.g. minimal distances to the according cluster centers or maximal density of each cluster) to retrieve patterns of the sought type

2) the data assignment to these clusters

Figure 4.5 illustrates the tasks of the *PR* as an example: Provided raw data should be grouped to reveal the internal structure and group characteristics, i.e. the patterns, in this case the centers of and the data assignment to the found clusters.

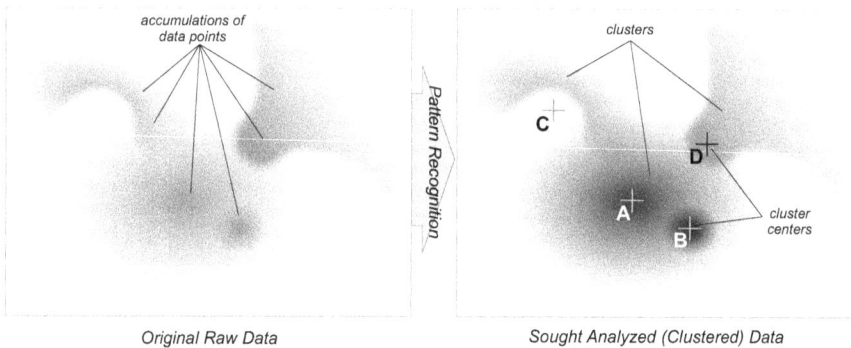

Original Raw Data          Sought Analyzed (Clustered) Data

**Figure 4.5:** *Example of a PR task in the field of unsupervised data analysis*; Shown is an unknown data set, spread in a two-dimensional system (left side). The gray shapes represent large groups of data items, which are located according to their features. In this example the subsequent *PR* utilizes a centroid-based cluster definition and should identify the pattern count and centers to group the data items. Here (right side) the *PR* may reveal for clusters, their locations $A..D$ within the feature space and assign the data items accordingly, shown as colorization.

Formally countless methods can be associated to the *PR* [Bishop et al., 2006], starting with techniques of the *Regression Analysis* to estimate relationships between variables, or *Neural Networks* for classification tasks. Furthermore, graphical approaches as *Bayesian Networks* or *Markov Random Fields* allow for the computer-aided detection of covered data dependencies and decision making. Finally, but

---

[6]In contrast to *supervised learning*, the term *unsupervised learning* means the absence of known labels, i.e. previously defined categories, and thus implies the detection of these.

not completing the enumeration, the large field of the *CA* also belongs to the *PR*, detecting groups of similar data for abstraction [Xu et al., 1998]. In the following the focus is on the latter subcategory.

As already stated in Section 2.1.1 current data sets are large, high-dimensional as well as diverse, and thus not suitable for an immediate *CA*. With regard to the *PR*'s objectives as described previously, the following concept is proposed to automatically detect clusters (patterns) and their representative center points as peaks within unstructured data sets: First, as shown in Figure 4.6, the original raw data is transposed to an uniform space, thus minimizing its dimensionality and diversity as well as smoothing the shapes of the sought clusters. Second, a clustering algorithm determines the count, the locations and the data assignments of the clusters before the model is mapped back to the original system. In this way the for- and backward transposition allows for the development and utilization of a generalized clustering method, which's purpose is to identify clusters of arbitrary count, shapes and sizes within large and complex data sets.

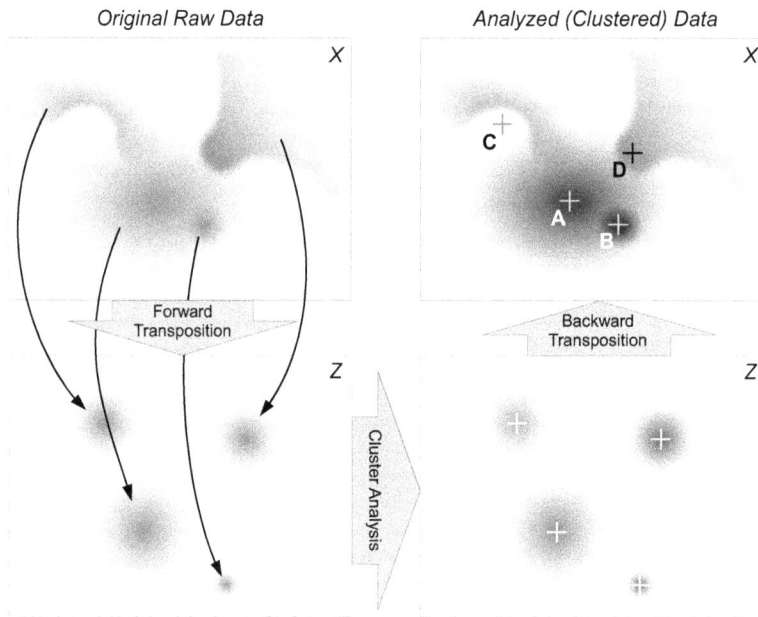

**Figure 4.6:** *Concept of the* PD *Algorithm processing Large-Scale Data Sets*; The provided raw data in a hetero-geneous high-dimensional space $X$ are transposed to an uniform low-dimensional space $Z$, enabling a general *CA* to subsequently detect the clusters, i.e. their representing center points (peaks), shown as crosses. The final backward transposition maps the clusters back to the original data space, thus yielding the *PR*'s result as cluster centers and assigned data items (colorization).

The concept's scheme is illustrated in Figure 4.7, which shows the data models transformation by the DT Framework and the subsequent *CA* as an agglomerative clustering algorithm, whose purpose is to combine the data items to increasingly growing clusters, starting with the leafs, until the cluster qualities fall below a given limit or the tree's root is reached. This should enable the algorithm to detect the clusters without information about their count, sizes, distributions and shapes. In the following this method is referred as *Peak Detection* Algorithm (PD Algorithm).

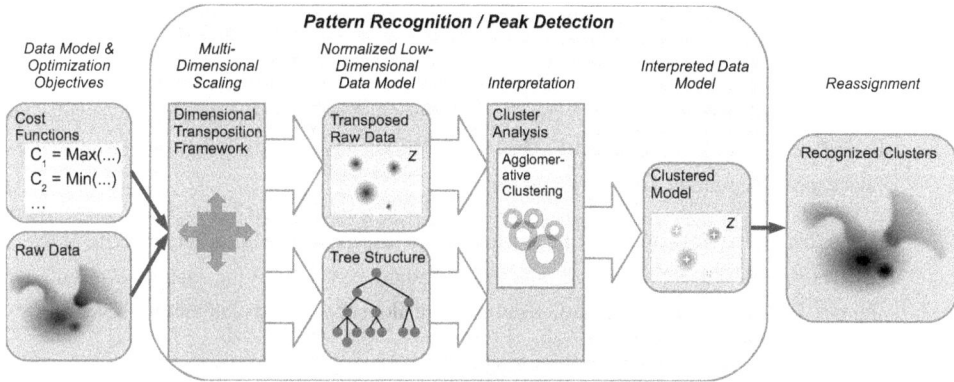

**Figure 4.7:** PD *Algorithm Base Scheme*: The given data model, i.e. the data items and the *PR*'s objectives (e.g. the cluster definition), is delegated to the DT Framework (Section 2), which transforms it to a normalized low-dimensional equivalent and provides a tree structure, embedding the transformed model. Both the transposition result as well as the hierarchical representation are utilized by the subsequent *CA*, a generalized agglomerative clustering algorithm, to group the data items to clusters, which are represented by a center point each (centroid based approach). The reassignment to the original data system yields the sought cluster centers as density peaks.

## 4.2 Application to the Segmentation Issue as Neuron Placement

The next two Sections 4.2.1 and 4.2.2 firstly describe the analysis strategy regarding the utilized data models, the measurements as well as the testing methods, and secondly the obtained results including new algorithms and performance analyses for the segmentation issue.

### 4.2.1 Methods and Materials

In the following Section 4.2.1.1 firstly considers the general objectives of the analysis strategy. Subsequently the Sections 4.2.1.2 and 4.2.1.3 describe the utilized data models, algorithm configurations as well as parameter spaces, and the measurements for the performance analysis respectively. Finally Section 4.2.1.4 concludes with the applied test and development scenarios.

### 4.2.1.1 Proof of Concept

To validate the quality of result of the application example, to ensure the approach's capability to also treat prospective problem sizes, and to keep the measurements meaningful, the following strategy is chosen to evaluate the presented concept: For testing purposes artificial, random based and sufficiently complex example problems of arbitrary size are created, from which the ideal results are known. This offers the following advantages:

- *Relative Quality Evaluation*: By defining the quality of result of the known ideal result as 1, the algorithm's outcome can be compared against this reference. Accumulating the weighted relative differences of all optimization objectives leads to a relative overall difference, and thus to a relative overall quality. The more the algorithm's relative quality of result is below 1 or the relative deviation is above 0 respectively, the more it differs from the ideal result, i.e. the more inadequate it is. In this way an uniform and unitless quality measurement method is specified. For more details see also Section 4.2.1.3.

- *Runtime Complexity Evaluation*: Increasing the processed problem size step-wisely enables the measurement of the time complexity. The slope of the evaluated runtimes allows for predictions of the expectable runtime for even larger tasks.

- *Scatter Reduction*: Instantiating and processing the same model with different seeds provides sufficient statistical data for average values and standard deviations of the qualities of results and runtimes. Involving these information reduces the random dependence and thus the scattering.

- *Validation and Comparability*: The detailed data model descriptions can be utilized to recreate the artificial data models, to (re-)evaluate these or other *MOO*, *MDS* and *CA* approaches, whose results are easily comparable in consequence. In this way an independent database approach for algorithm investigations is provided.

In the following this investigation strategy is applied to the neuron placement application.

### 4.2.1.2 Data Models and Algorithm Configuration

In the following the utilized biological and hardware models are described, as well as the resulting mapping objectives and the conclusions for the required model transposition.

**Biological Model:** As introduced in Section 4.1.1 neural networks can be modeled by utilizing populations and projections (Figure 4.2). To create a data model, which satisfies the requirements defined in Section 4.2.1.1, the following concept is applied: Given a 2D array of populations, projections can be established between each population pair within the array. The connection probability between two populations $P^i$ and $P^j$ is set to:

$$prob_{proj}^{P^i,P^j} = \frac{prob_{base}}{dist^{P^i,P^j}} \tag{4.1}$$

where $prob_{base}$ is the global base probability of connection and $dist^{P^i,P^j}$ the array distance between both populations, i.e. each neuron has a chance to be connected to any other neuron, and this chance decreases with growing distances. In this way a random based synaptic network is created. Additionally each population consists of several subpopulations, dividing the neurons in different parameter domains: At the first level, the population is split evenly by the boolean synaptic *type* (*exhibitory* or *inhibitory*), followed by two shared neural integer parameters $v_0$ and $v_1$, which evenly divides the population further by applying their minimal and maximal values. At the bottom level the neurons differ only in the individual parameters $(x, y)$, which are not relevant for the placement. In other words, each population of the array is divided into 8 ($2^3$) core populations, whose neurons share an unique parameter set of *type*, $v_0$ and $v_1$.

Hence the grid population and model size, i.e. their neuron counts are

$$size_{grid\,population}^{Bio} = 8 \cdot size_{core\,population}^{Bio}$$
$$size_{system}^{Bio} = size_{grid\,population}^{Bio} \cdot \left(n_{grid}^{Bio}\right)^2 \tag{4.2}$$

where $n_{grid}^{Bio}$ is the length of the quadratic population array and $size_{core\ population}^{Bio}$ represents the count of neurons within a core population. The scheme of the artificial biological model is illustrated in Figure 4.2.

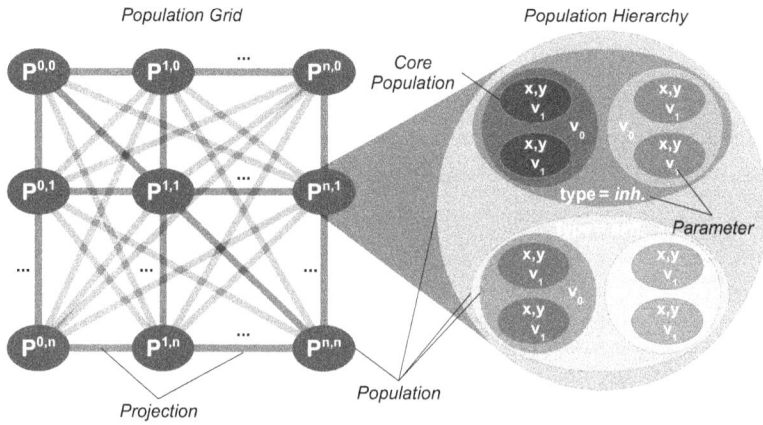

Population Grid          Population Hierarchy

Projection

Population

**Figure 4.8:** *Artificial Biological Model of Arbitrary Size*; On top level the model consists of $n \times n$ populations ($P^{0,0}..P^{n,n}$), arranged in a quadratic array (*Population Grid*). All populations are completely inter-connected by projections, whose connection probabilities are determined by the according distances within the grid (Formula (4.1), the distance between two adjacent grid points is set to 1). A closer look at population $P^{n,1}$ reveals its inner structure: in a tree like manner the shared parameters *type*, $v_0$ and $v_1$ split up the population into several subpopulations:
*Level 0*: two *type* groups of inhibitory (top) and exhibitory (bottom) synapses
*Level 1*: two *parameter* groups with $v_0 = v_0^{min}$ (left) or $v_0 = v_0^{max}$ (right)
*Level 2*: two *parameter* groups with $v_1 = v_1^{min}$ (top) or $v_1 = v_1^{max}$ (bottom)
At the bottom, each population of the grid consists of 8 core populations with a unique configuration from $[type = inh., v_0 = v_0^{min}, v_1 = v_1^{min}]$ (top left) to $[type = exh., v_0 = v_0^{max}, v_1 = v_1^{max}]$ (bottom right).

The expansion, i.e. the creation of neurons, including their random parameters, for each population and the application of the projections to these neurons leads to the concrete neural network, the biological model. The resulting features per neuron, which are shown in Table 4.1 have to be considered by the SEGM Algorithm during the neuron placement.

**Hardware Model:** To ensure a possible, but difficult to find ideal neuron placement and to stress the SEGM Algorithm, the targeted neural hardware model is designed so, that it corresponds to the given biological model: According to the introduced generalized neural hardware model (Section 4.1.1, Figure 4.3), the width and height of the grid system is set to:

$$n_{grid}^{HW} = n_{grid}^{Bio} \tag{4.3}$$

i.e. equal to the size of the population grid. Furthermore, the neuron capacity of a single hardware core grid is set to the size of a grid population (Formula (4.2)):

$$capacity_{grid}^{HW} = size_{grid\ population}^{Bio} \tag{4.4}$$

| Name | Symbol | Description | Unit: [Range] |
|------|--------|-------------|----------------|
| fan-in 0 | $fan_{in}^{P^0}$ | fraction of $P^0$, which serves as neural sources | float: $[0.0..1.0]$ |
| .. | .. | .. | .. |
| fan-in $n$ | $fan_{in}^{P^n}$ | fraction of $P^n$, which serves as neural sources | float: $[0.0..1.0]$ |
| fan-out 0 | $fan_{out}^{P^0}$ | fraction of $P^0$, which serves as neural targets | float: $[0.0..1.0]$ |
| .. | .. | .. | .. |
| fan-out $n$ | $fan_{out}^{P^n}$ | fraction of $P^n$, which serves as neural targets | float: $[0.0..1.0]$ |
| neuron type | $type$ | type of the incoming synapses, hence the neuron type | bool: $[inh./exh.]$ |
| parameter 0 | $v_0$ | shared neuron behavior parameter | int: $[0..10]$ |
| parameter 1 | $v_1$ | shared neuron behavior parameter | int: $[0..100]$ |

**Table 4.1:** *Neuron Features*; Shown are the resulting neuron features of the biological model, which have to be considered by the SEGM Algorithm.

i.e. with a constant core capacity ($capacity_{core}^{HW}$), the number of cores per grid is determined by:

$$n_{cores\ per\ grid}^{HW} = \frac{capacity_{grid}^{HW}}{capacity_{core}^{HW}} \tag{4.5}$$

In this way the hardware model has exactly the same neural capacity as the biological model has neurons, i.e. the ideal placement has no tolerance, there exists only one single solution.

**Neuron Placement Objectives:**   With regard to the described optimization objectives during the neuron placement introduction (Section 4.1.1) the cost functions can be formulated as:

$$deviation_{routing} = \sum_{grid}^{system} \sum_{core}^{grid} \sum_{neuron\ i,j}^{core} \sum_{proj}^{(i,j).projections} |prob_{proj}^i - prob_{proj}^j| \cdot capacity_{core}$$

$$incompatibility_{synapses} = \sum_{grid}^{system} \sum_{core}^{grid} \sum_{neuron\ i,j}^{core} |type_{synapse}^i - type_{synapse}^j| \tag{4.6}$$

$$distortion_{parameters} = \sum_{grid}^{system} \sum_{core}^{grid} \sum_{neuron\ i,j}^{core} \sum_{param}^{(i,j).parameters} |param^i - param^j|$$

The first function *deviation*$_{routing}$ describes the estimated differences of the synaptic fan-in and fan-out for all cores within the hardware model. Deviations regarding the neural sources and targets of a hardware core means, that the subsequent routing algorithm has to establish more routing connections, than if all neurons of the hardware core have the same sources and targets. More routing implies the utilization of more routing capacities, and thus, if the capacities are depleted, synapse losses. Hence minimizing the routing deviations is a suitable strategy to minimize the later occurring synapse loss.

The second term *incompatibility*$_{synapse}$ stands for the type differences of the synapses within a hardware core. As a hardware core can realize only one synapse type, differing types per hardware core cause the loss of all incompatible synapses, and in consequence the loss of their according neurons. Thus, assuring that only neurons with equally typed synapses are mapped to a single hardware core, ensures the realization of them.

The third term $distortion_{parameters}$ represents the difference of shared parameters within a hardware core. For technological reasons neurons of the same hardware core share some behavior determining parameters, i.e. minimizing their deviation implies a later satisfactory neuron parametrization, and thus a matching behavior.

To test the SEGM Algorithm as neuron placement method, only the following data is provided:

- $N$, the set of neurons, including parameters and synaptic connections, i.e. the biological neural network, without the population structure

- $capacity_{grid}^{HW}$, the neuron capacity of a single hardware core grid, i.e. an array of several neural cores

- $capacity_{core}^{HW}$, the neuron capacity of a single hardware core

The algorithm's objective is to minimize the neuron placement costs (Formula (4.6)) by an adequate clustering of $N$ into groups of $capacity_{grid}^{HW}$ and each of these into subgroups of $capacity_{core}^{HW}$, for the later assignment to the hardware cores.

**Model Transposition:** To enable the SEGM Algorithm to cluster the provided, potential large and high-dimensional biological model, it has to be transposed by the embedded DT Framework to a low-dimensional equivalent. For this purpose the following property compare functions (Section 2.3.2.1) are applied:

$$Func_{aff}(\overrightarrow{frac_{rnd}}^i, \overrightarrow{frac_{rnd}}^j) = \frac{\sum_k^{\overrightarrow{frac_{rnd}}}(1 - |^{scl}prob_k^i - {}^{scl}prob_k^j|) \cdot size_k}{\sum_k^{frac_{rnd}} size_k}$$

$$Func_{aff}(single^i, single^j) = 1 - |^{scl}value^i - {}^{scl}value^j|$$

(4.7)

The first function $Func_{aff}(\overrightarrow{frac_{rnd}}^i, \overrightarrow{frac_{rnd}}^j)$ compares two random fraction maps $i$ and $j$. A single random fraction $k$ describes a set of items of the $size_k$ and a probability $prob_k^i$ as a random fraction of this set. The second function $Func_{aff}(single^i, single^j)$ calculates the affinity of two single numerical values ($value^i$ and $value^j$). The prefix $scl$ indicates a scaled property value, which is performed by a self adaptation technique to utilize the complete affinity range $[0..1]$. More details about the self adaptation can be found in the Appendix in Section 6.3.1.1.

Finally the neuron features of the biological model (Table 4.1) can be covered by these compare functions, as shown in Table 4.2. The incoming and outgoing synaptic connections are modeled as random fraction maps, neuron parameters as single numerical values. In addition the element weight (Section 2.3.2.2) is extended to an affinity specific weight to enable a hierarchical segmentation. A detailed explanation can be found in the Appendix in Section 6.3.1.1.

The minimization of the cost functions (Formula (4.6)) and thus the optimization of the objectives (Formula (4.9)) is performed implicitly, as the affinity based transposition enables the clustering of group elements, which would correspond to the mapping objectives.

| Original Features | DT Framework Property Type | DT Framework Priority | Description |
|---|---|---|---|
| $fan_{in}^{P^{0..n}}$ | random fraction map ($\overrightarrow{frac_{rnd}}$) | 1.0 | incoming synapse fan-in |
| $fan_{out}^{P^{0..n}}$ | random fraction map ($\overrightarrow{frac_{rnd}}$) | 1.0 | outgoing synapse fan-out |
| $type$ | numerical value (*single*) | 0.2 | neuron type |
| $v_0, v_1$ | numerical value (*single*) | 0.1 | shared neuron parameters |
| $x, y$ | numerical value (*single*) | 0.01 | individual neuron parameters |

**Table 4.2:** *Neuron Features as* DT *Framework Element Properties*; Shown are the neuron features modeled by utilizing the introduced affinity functions (Formula (4.7)).

### 4.2.1.3 Measuring Results and Performance

As explained in Section 4.2.1.1 the measurements focus is on the *relative deviation* as difference from the ideal result and the *runtime*. Table 4.3 shows an overview of the single measurements, detailed explanations can be found below.

| Measurement | Symbol | Description | Unit |
|---|---|---|---|
| Deviation of Property 0 | $deviation_{prop^0}$ | mapping deviation of property 0 for all hardware cores | - |
| .. | .. | .. | .. |
| Deviation of Property $l$ | $deviation_{prop^l}$ | mapping deviation of property $l$ for all hardware cores | - |
| Total Deviation | $deviation_{total}$ | overall mapping deviation | - |
| DT Runtime | $t_{run}^{DT}$ | DT execution time | *seconds* |
| DT Runtime Slope | $slope_{runtime}^{DT}$ | relative DT runtime increment | *second/MSI* |
| TF Runtime | $t_{run}^{TF}$ | TF execution time | *seconds* |
| TF Runtime Slope | $slope_{runtime}^{TF}$ | relative TF runtime increment | *second/MSI* |
| *Balancing* Runtime | $t_{run}^{Balancing}$ | *Balancing* execution time | *seconds* |
| *Balancing* Runtime Slope | $slope_{runtime}^{Balancing}$ | relative *Balancing* runtime increment | *second/MSI* |
| Total Runtime | $t_{run}^{total}$ | total execution time | *seconds* |

**Table 4.3:** *Measurements / Evaluations the during* SEGM *Algorithm Testing;* Shown are the names and the symbols of the single measurements in combination with short descriptions and the according units.

The sizes and characteristics of the biological and the hardware model (Section 4.2.1.2) are chosen such, that the neural network can be mapped 1 : 1 to the neural hardware, if segmented ideally. In this context ideal segmentation means, that every final mapping cluster, i.e. the assignment to a single hardware core, contains only neurons, which are identical regarding their properties (topological connections and parameters). To further restrict the algorithm's degree of freedom, the biological model size requires the utilization of the entire hardware capacities.

Hence a single neuron pair's difference within a cluster stands for a deviation from the ideal. The overall deviation of a property *prop* can be described as:

$$deviation_{prop} = \frac{\sum_k^{clusters} \sum_{i,j}^{k.elements} |value_{prop}^i - value_{prop}^j|}{count_{cluster} \cdot (size_{cluster})^2} \tag{4.8}$$

where $k$ represents all resulting mapping clusters (hardware core assignments), $i$ and $j$ two elements, neurons respectively, within $k$. The expression $value_{prop}^{i,j}$ stands for the according property values of both elements, whereas $count_{cluster}$ and $size_{cluster}$ describe the size of the result, and thus the count of possible comparisons.

As stated before, the mapping cost functions with regard to $l$ properties are:

$$\overrightarrow{cost}_{mapping} = \begin{pmatrix} deviation_{prop^0} \\ \cdots \\ deviation_{prop^l} \end{pmatrix} \tag{4.9}$$

i.e. a set of single relative differences $deviation_{prop}$. Summarizing the deviations to a single expression such as:

$$deviation_{total} = |\overrightarrow{cost}_{mapping}| = \frac{\sum_{prop}^{properties}(deviation_{prop} \cdot priority_{prop})}{\sum_{prop}^{properties} priority_{prop}} \tag{4.10}$$

provides the total deviation or the mapping costs $deviation_{total}$, where $prop$ is one of all mapping relevant properties, and $priority_{prop}$ the user-defined global priority of $prop$.

In this way the single and overall qualities are determined. The evaluation of the runtimes $t_{run}$ of the DT Algorithm (Section 2), the TF K-MEANS Algorithm (Section 3) and the *Balancing*, as well as their according time complexities $slope_{runtime}$, are performed similarly to the DT and TF runtime measurements (Sections 2.2.2 and 3.2.2).

### 4.2.1.4 Performance Evaluation

To validate the results and evaluate the performance of the SEGM Algorithm regarding quality and runtime, the following tests are performed:

The grid size of the artificial biological model (Section 4.2.1.2, Figure 4.8) is $n_{grid}^{Bio} = 5$, modeling a $5 \times 5$ population grid with a base connection probability of $prob_{base} = 0.5$. To enlarge the neural network, the size of all core populations ($size_{core\ population}^{Bio}$) is increased gradually, so that also the size of the grid populations ($size_{grid\ population}^{Bio}$) grows evenly.

Consequently the according hardware model (Sections 4.1.1 and 4.2.1.2) has to consist of also $5 \times 5$ core grids, each with a neuron capacity ($capacity_{grid}^{HW}$) equal to the grid population size. With respect to the constant neuron capacity of the hardware cores ($capacity_{core}^{Bio}$), enlarging the biological model implies a growing core count per hardware grid ($n_{cores\ per\ grid}^{Bio}$, Formula (4.5)).

Table 4.4 shows the varied and constant model parameters (Section 4.2.1.2) in combination with the algorithm configurations and the measurements (Section 4.2.1.3).

Applying the minimum of these parameters is synonymous with a neural network of 3.200 neurons and 21.917 synapses, which should be mapped to a hardware of 128 neurons and 877 feed-in synapses on average per core grid (16 cores per grid). The maximum would generate a neural network of 51.200 neurons and 350.665 synapses, which should be mapped to a hardware of 2048 neurons and 14.027 feed-in synapses on average per core grid (256 cores per grid).

The placement is performed by the SEGM Algorithm (Section 4.2.2.1), embedding the DT Framework and the TF K-MEANS Algorithm. The DT Framework's degree of accuracy ($f_{SA} = 0.97$) is increased compared to the tests in Section 2.2.4, due to the required finer granularity of the transposed hierarchical result. The target dimension count is set to 2, minimizing the $CA$ complexity. The target cluster count of

| Scenario | Varied Parameter | Step Size | Fixed Parameter | Measurements |
|---|---|---|---|---|
| Large Scale Neuron Placement | $size^{Bio}_{core\ population} = 16..256$ <br> -> $size^{Bio}_{system} = 3.2k..51.2k$ <br> -> $n^{HW}_{cores\ per\ grid} = 16..256$ | 16 | DT Framework: <br> $f_{SA} = 0.97$ <br> $n_{dimension} = 2$ <br> Neural Network Model: <br> $n^{Bio}_{grid} = 25$ <br> $prob_{base} = 0.5$ <br> Hardware Model: <br> $n^{HW}_{grid} = 25$ <br> $capacity^{HW}_{core} = 8$ | $deviation_{fan-in, fan-out}$ <br> $deviation_{type}$ <br> $deviation_{v_0, v_1}$ <br> $deviation_{total}$ <br> $t_{DT}$ <br> $t_{TF}$ <br> $t_{Balancing}$ <br> $t_{total}$ |

**Table 4.4:** *Test Scenarios for the* SEGM *Algorithm Testing*; Shown is configurations of the SEGM Algorithm test scenario, to analyze the influence of the data models' characteristics (Section 4.2.1.2) onto the quality and performance indicators (Section 4.2.1.3).

the TF K-MEANS Algorithm (Section 3.2.4) are determined by the number of core grids ($n^{HW}_{grid}$) and the number of hardware cores per grid ($n^{HW}_{cores\ per\ grid}$).

To reduce the effects of scattering, each test is instantiated and processed multiple, i.e. 10 times, utilizing different random seeds, providing measurements data, which are statistically useful.

The tests were carried out in parallel on a *Windows 7 Professional* workstation, hosting an *Intel Core* $i7 - 2760QM$ CPU with access to 12 GB of RAM. The results were collected automatically into detailed logs, which serve as the charts' database for the final analysis in Section 4.2.2.2.

## 4.2.2 Results

Presenting the results of TRACS-MOO's application to the segmentation issue, firstly Section 4.2.2.1 describes the developed algorithms and adaptations, and secondly Section 4.2.2.2 shows an overview of the performance analysis. Finally Section 4.2.2.3 summarizes and discusses the obtained insights.

### 4.2.2.1 The Segmentation Algorithm

First, this Section introduces the general scheme of the SEGM Algorithm and second, describes the base of the balancing concept before the applied algorithm enhancements are briefly presented.

**General Scheme:** As the other components of the SEGM Algorithm have already been introduced (the DT Framework in Section 2 and the TF K-MEANS Algorithm in Section 3), this Section focuses on the components' integration and adaptation techniques to enable the algorithm to perform the neuron placement task, as described in Section 4.1.1, following the concept shown there in Figure 4.4.

First, the model transposition, performed by the DT Framework, creates an uniform and low-dimensional equivalent of the provided data model (neural network, Section 4.2.1.2), where affine elements (neurons) are positioned close to each other. Element affinity in this context means the similarity of features with regard to the provided cost functions (Formula (4.6)). This representation enables centroid-based *CA* methods, such as *k-means* based algorithms, to structure the given data model, forming groups of affine elements (for the later hardware assignment). As the number of elements and target clusters can be large, the accelerated TF K-MEANS Algorithm is utilized. As the data model has to be assigned to different stages of the hardware model, the segmentation is performed in several hierarchical steps.

Figure 4.9 shows the main scheme of the SEGM Algorithm. At each level the current segment of the transposed data model is split into clusters of equal size and the segmentation process continues recursively until the cluster core size is reached. Each segmentation step is performed by the *Single Segmentation Algorithm* (SISEGM Algorithm): The TF K-MEANS Algorithm clusters the elements of the current segment, and the subsequent *Balancing Algorithm* (BLC Algorithm) ensures that no cluster remains overcrowded. A detailed description of the BLC Algorithm can be found below. The sequence is repeated until the cluster density gain saturates. The overall density of a set of clusters is determined as:

$$density_{total} = \sum_{k}^{clusters} \sum_{i}^{k.elements} dist^{i \leftrightarrow k} \tag{4.11}$$

where $k$ is one of the clusters, $i$ an element within $k$ and $dist^{i \leftrightarrow k}$ the distance of $i$ to the center of $k$. Hence the relative density difference between two SISEGM Algorithm iterations can be described as:

$$diffdensity_{rel}^{iteration\ i} = \frac{density_{total}^{iteration\ i}}{density_{total}^{iteration\ i-1}} \tag{4.12}$$

where $density_{total}^{iteration\ i}$ is the density after iteration $i$, and $density_{total}^{iteration\ i-1}$ represents the density of the iteration before. In this way a sequence of clustering and size balancing is performed for each segment of the data model until the density improvement falls below a given threshold and the segment is divided further or the algorithm terminates.

The clusters of each hierarchy level are assigned to the according hardware elements. Assuming a hardware model as described in Section 4.1.1, Figure 4.3 the hierarchical clustering would consist of two levels. The resulting top level clusters are assignable to the core grids, the second level clusters would correlate with the neural cores themselves. In the following a rendered example a segmented data model illustrates the process and the result of the SEGM Algorithm. Table 4.5 specifies the utilized artificial neural network and the target hardware system as described in Section 4.2.1.2, as well as the algorithm configuration. The rendered segmentation result of the transposed and clustered biological model is shown in Figure 4.10, visualizing the neuron group creation and assignment to the hardware core grids as well as the later subgroup creation for the cores within the hardware grids.

| Name | Symbol | Value |
|---|---|---|
| *Biological Model* | | |
| core population size | $size_{core\ population}$ | 32 neurons |
| system size | $size_{system}$ | 6.400 neurons |
| grid count | $n_{grid}$ | 25 |
| base connection probability | $prob_{base}$ | 25 |
| *Hardware Model* | | |
| neuron capacity per core | $capacity_{core}$ | 8 neurons |
| core count per grid | $n_{cores\ per\ grid}$ | 32 |
| grid count | $n_{grid}$ | 25 |
| *SEGM Algorithm* | | |
| DT Framework: *SA* Factor | $f_{SA}$ | 0.97 (100 iterations) |
| DT Framework: target dimension count | $n_{dimension}$ | 2 |
| TF K-MEANS Algorithm: cluster count level 0 | $k_0$ | $n_{grid} = 25$ |
| TF K-MEANS Algorithm: cluster count level 1 | $k_1$ | $n_{cores\ per\ grid} = 32$ |

**Table 4.5:** *Data Model Parameters and Algorithm Configuration for the Rendered Example in Figure 4.10*

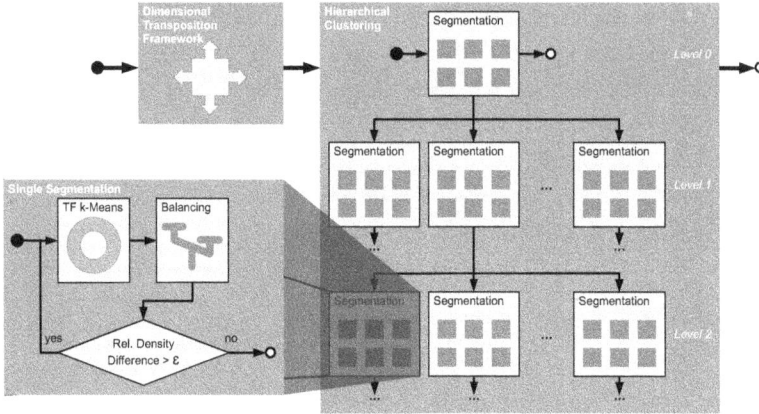

**Figure 4.9:** SEGM *Algorithm Scheme*; First, the DT Framework transforms the data model (neural network) including the optimization objectives (mapping cost functions) to a low-dimensional equivalent. Subsequently the hierarchical segmentation process starts: On top level (*Level* 0) the elements (neurons) of
the data model are divided into a given number (e.g. the number of hardware grids) of equal-sized (e.g.
the hardware grid capacity) groups. Each of this group is divided further by the same segmentation
mechanism, now applying other target counts (e.g. the number of cores within a grid) until no further
splits are required.

A closer look at the segmentation process (bottom left) shows the loop sequence of the introduced
TF K-MEANS Algorithm, which performs the actual clustering, and a BLC Algorithm, which ensures
that no cluster is overcrowded. This sequences is executed until the quality gain falls below a given
threshold $\epsilon$.

The result of the SEGM Algorithm is a cluster tree, whose clusters contain the same number of elements (or subclusters) per hierarchy level and are maximally dense (i.e. optimized with regard to
optimization objectives).

**Concept of Balancing:** Although large parts of the SEGM Algorithm could be abstracted and are
already available, it embeds the essential BLC Algorithm, which extends the TF K-MEANS Algorithm
to consider cluster capacity limitations. For this purpose the approach of *virtual gravity* is introduced.
The single virtual gravity between two clusters $i$ and $j$ is defined as:

$$gravity_{virtual}^{i,j} = \frac{|size^i - size_{opt}| \cdot |size^j - size_{opt}|}{dist^{i,j}} \tag{4.13}$$

where $size^{i,j}$ are the current element counts of both clusters, $size_{opt}$, is the optimal cluster size, or the
cluster capacity respectively, and $dist^{i,j}$ is the spatial (Euclidean) distance between cluster $i$ and $j$. It
follows, that the total virtual gravity is:

$$gravity_{virtual}^{total} = \sum_{i}^{clusters_{oversized}} \sum_{j}^{clusters_{undersized}} gravity_{virtual}^{i,j} \tag{4.14}$$

where $clusters_{oversized}$ is the set of oversized (containing more elements than their capacity allow) and
$clusters_{undersized}$ is the set of undersized (containing less elements than their capacity allow) clusters. In
this way the global degree of cluster imbalance can be described, taking into account the size differences
to the optimum and the distances of unbalanced clusters. Reducing one of these factors implies the
decrease of the total virtual gravity, which is equivalent to the BLC Algorithm's objective.

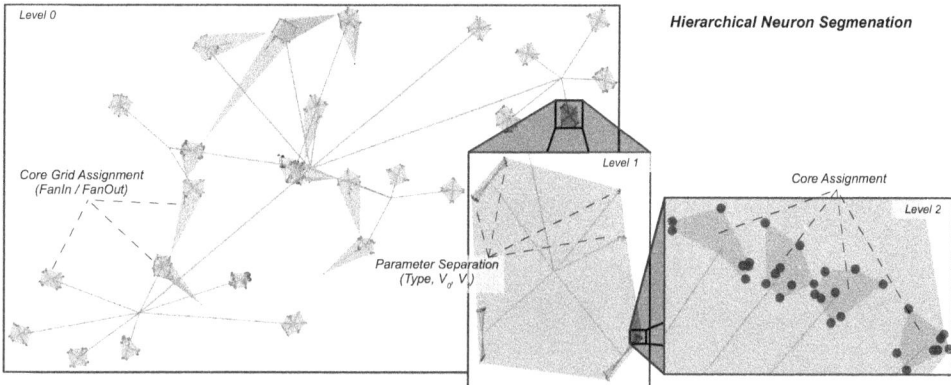

**Figure 4.10:** *Rendered Segmentation Example*; Shown is a transposed and segmented artificial neural network mapped to a neural hardware model as specified in Table 4.5.
**Level 0:** As the target hardware model consists of 25 core grids, the (via the DT Framework) transposed neural network is divided by the SEGM Algorithm into the same number of clusters (gray polygons). The green lines represent the underlying tree structure of the TF K-MEANS Algorithm. With respect to the synaptic connections ($fan_{in/out}^{P^{0...n}}$), the neurons of each cluster of this hierarchy level are assigned to one of the core grid of the neural hardware model.
**Level 1:** On this level the remaining 256 neurons of each core grid are separated regarding the shared neuron parameters ($type$, $v_0$, $v_1$), forming 8 groups (blue dots at the end of the green lines) of equal parameters sets.
**Level 2:** At the bottom level the 32 neurons differ only in the irrelevant parameters for the mapping, that are $x$ and $y$, and the clustering groups them into 4 groups of 8 neurons (blue dots) each. These final core clusters (dark gray polygons) are assigned to the neuron blocks of the neural hardware cores, representing the final neuron mapping.

The scheme of the BLC Algorithm is shown in Figure 4.11 and can be summarized as follows: While oversized clusters exist, the algorithm's task is to find an *element – target cluster* pair, which satisfies the following requirements:

- the *element* is within an *oversized cluster*

- the *target cluster* contains less elements than the *oversized cluster*

- assigning the *element* to the *target cluster* causes:

  - no overlapping with neighbor clusters (intersecting boundaries, i.e. misshaped clusters)

  - the decrease of the total virtual gravity (Formula (4.14))

- the *element – target cluster* distance is minimal with regard to other valid pairs

After the determination of the best element – target cluster pair, a swap is performed, assigning the element to the target cluster. This implies the update of both clusters regarding their weight, position, element – center distances and *MLDs*. The procedure is repeated as long as oversized clusters remain.

This approach ensures that the oversized clusters are reduced step by step. The direct element assignment to undersized clusters may create clusters, which stretch over several other clusters, thus disregarding the objectives of dense and internal affine final clusters. Hence the swapping has to be performed step-wisely, passing clusters between the over- and undersized clusters. To avoid overlappings with inconvenient positioned neighbor clusters, a triangular inequality based overlapping check validates potential element – target cluster pairs. To establish a global guide mechanism and suppress element oscillations, the total virtual gravity approach ensures, that the swappings are performed in the direction of undersized clusters.

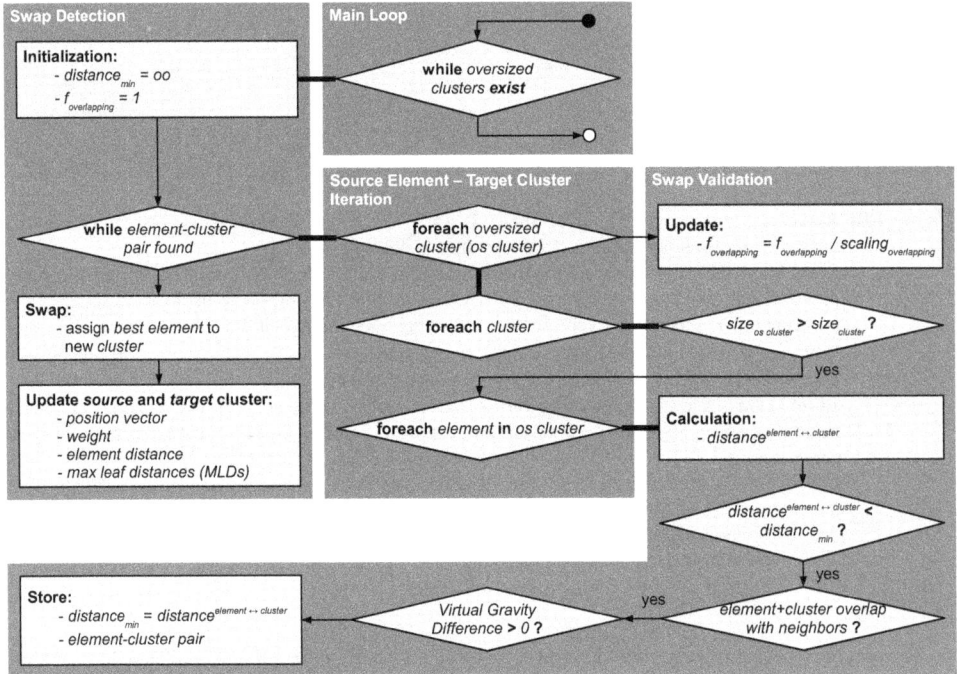

**Swap Detection**

Initialization:
- $distance_{min} = oo$
- $f_{overlapping} = 1$

**Main Loop**

while *oversized* clusters **exist**

while *element-cluster* pair found

Swap:
- assign *best element* to new *cluster*

Update *source* and *target* cluster:
- position vector
- weight
- element distance
- max leaf distances (MLDs)

**Source Element – Target Cluster Iteration**

foreach *oversized* cluster (*os cluster*)

foreach *cluster*

foreach *element in os cluster*

**Swap Validation**

Update:
- $f_{overlapping} = f_{overlapping} / scaling_{overlapping}$

$size_{os\ cluster} > size_{cluster}$ ?

yes

Calculation:
- $distance^{element \leftrightarrow cluster}$

$distance^{element \leftrightarrow cluster} < distance_{min}$ ?

yes

element+cluster overlap with neighbors ?

yes

Virtual Gravity Difference > 0 ?

yes

Store:
- $distance_{min} = distance^{element \leftrightarrow cluster}$
- element-cluster pair

**Figure 4.11:** *Scheme of the* BLC *Algorithm;*
> **Main Loop:** The algorithm is executed, as long as overcrowded, i.e. oversized clusters remain.
> **Swap Detection:** Each iteration is initialized by reinitializing the search data, subsequently the search for the best element – cluster pair starts. After it is found, the element is assigned to its new cluster, causing an update (position, weight, inferior distances, *MLD*) of both involved clusters (former and new cluster of the element).
> **Source Element – Target Cluster Iteration / Swap Validation :** During the search for the best swap, all oversized clusters are compared to all smaller clusters. For each of this cluster pair the element within the oversized cluster is detected, which fulfills the swapping criteria: *i*) being closer than former found swap elements, *ii*) causing no overlappings with neighbor clusters and *iii*) decrease the total virtual gravity (Formula (4.14)). If the element passes all checks, it is stored as the current best swap. If no element is found, the overlapping factor is decreased, allowing a higher degree of overlapping of neighbor clusters, and the swap search is repeated.

To clarify the reassignment scheme of the BLC Algorithm, Figure 4.12 illustrates a balancing example in several steps. Starting with over- and undersized, as well as balanced clusters, the closest elements of oversized clusters are swapped to adjacent clusters, according to the requirements as described previously, until no capacity limits are exceeded any more. In other words, the less crowded clusters move to dense regions, occupying elements and displacing oversized clusters, causing them to transfer elements.

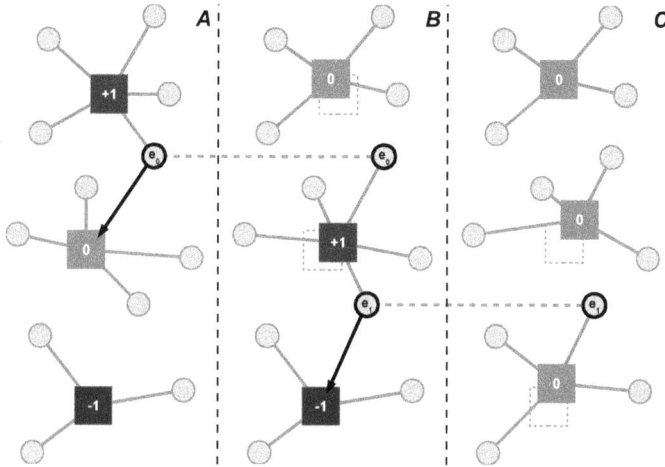

**Figure 4.12:** *Balancing Concept of the* SEGM *Algorithm;* Shown is the step-wise reassignment strategy of the BLC Algorithm to equalize the cluster sizes. Elements are presented as circles, clusters as colored squares (*red* = oversized, *green* = balanced, *blue* = undersized) with an internal number, which indicates the difference to the balanced size.
**A:** Given are three clusters of different sizes, the required cluster size is for. The upper cluster ($+1$) contains five elements, i.e. one more than required. The lower cluster ($-1$) has only three elements and thus one too few. Only the middle cluster (0) is already balanced. Now the BLC Algorithm checks all oversized clusters and determines the best element, which can be assigned to a smaller cluster, so that over- and undersized clusters move closer to each other. In this case the element $e_0$ is assigned to the middle cluster.
**B:** The element $e_0$ is reassigned, the cluster sizes and also their positions are updated. The upper, former oversized cluster becomes balanced and its center shifts upward due to the now missing $e_0$. The middle cluster becomes oversized and moves upward too, due to the added element $e_0$. Performing the next iteration the BLC Algorithm determines $e_1$ as suitable for the next reassignment.
**C:** After the element swap and the cluster updates, which shift both cluster slightly upward, all clusters are balanced and the algorithm terminates.

**Enhancements:** As later measurements show, the BLC Algorithm becomes the most runtime causing component, if processing large models. In the following the most important runtime saving algorithm enhancements are introduced briefly:

- *Data Buffering:* As the detection of the best element swap requires different data and most of it remains constant most of the time, it is conceivable to buffer the following data:

    - *Distances:* spatial distances of element – cluster and cluster – cluster pairs (relevant for swap detections, virtual gravity (difference) calculations, overlapping avoidance, triangular inequality based estimations, nearest neighbors searches)

    - *Virtual Gravity:* single values of cluster – cluster pairs (relevant for virtual gravity (difference) calculations)

    - *Element – Target Cluster Pairs:* valid, but unconsidered swap data (potentially relevant during later iterations)

    - *Nearest Neighbors:* spatial environment of clusters (relevant for swap detections, overlapping avoidance)

     – *Unbalanced Clusters:* lists of over- and undersized clusters (relevant for swap detection)

- *Tree Based Nearest Neighbor Search:* With increasing cluster counts, the number of cluster comparison grows quadratically, hence an improved method to detect cluster neighborhoods is required. For this purpose the exported data trees of the TF K-MEANS Algorithm are utilized to collect the $k$ nearest neighbors of a cluster efficiently. Inverting the tree's *MLD* approach (Section 2.3.2.4), the search can exclude large numbers of clusters without calculating the distances to them.

- *Calculation of the Virtual Gravity Difference:* Following the algorithm scheme in Figure 4.11, each swap validation implies the calculation of the virtual gravity difference. Hence only a fraction of all included single virtual gravity values change after a swap, a difference calculation method recalculates only the values influenced by the current swap.

- *Triangular Inequality Checks:* Similar to the numerous abort checks of the TF K-MEANS Algorithm, many pair considerations can be aborted early, applying triangular inequality estimations to buffered distance data.

- *Data Invalidation:* Subsequent to each swap, an invalidation mechanism marks all invalid buffered data, enabling it for recalculation on demand.

### 4.2.2.2 Measurement Results

In the following representative measurements of the previously described *test scenario* (Section 4.2.1.4, Table 4.4) are illustrated.

    The first charts in Figure 4.13 show the relative mapping deviations in dependency of the model size. The count of neurons per core population was gradually increased, the resulting neural network was mapped to a hardware model of a capacity equal to the biological model's size, so that an ideal transposition is possible (all relative deviation would become 0, see Section 4.2.1.2 for detailed explanations). The charts show the single deviations of the neuron features within a cluster, i.e. a neural hardware core. The single deviation is introduced in Formula (4.8) in Section 4.2.1.3. *Fan In* and *Fan Out* symbolize the totality of the incoming and outgoing synaptic connections, *Type* stands for the inherent neuron type, and *V0* and *V1* represent two shared behavior neuron parameters. The less the deviation, the more equal the neurons within the neural cores are to each other, which corresponds to the optimization objectives.

    First, the Charts shows, that the most important features *Fan In* and *Fan Out* are mapped with a maximal deviation below 0.005, i.e. on average 99.95% of the neural network topologies are transferred to the hardware, hence the connection characteristics are retained. These feature deviations are equivalent to the *deviation$_{routing}$* cost function (Formula (4.6)) and signifies the potential synapse loss during the subsequent routing steps. Second, the deviations of the remaining features *Type*, *V0* and *V1* start at a relative difference of 0.05, which decreases to a minimum of 0.01. The features' mapping deviation correspond to the *incompatibility$_{synapses}$* and the *distortion$_{parameters}$* cost functions, which signify the loss of neurons due to synapse type incompatibilities and neuron behavior distortion implied by mismatching shared parameters respectively. The observable (local) minimum at $n_{neuron} = 40k$ is probably caused by inconvenient model sizes in relation to the branch sizes of the underlying tree structures.

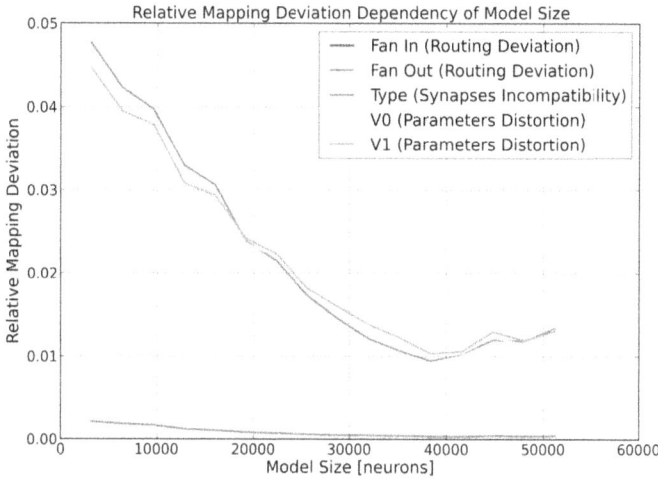

| Algorithm | Dimensions Count | SA Factor | Core Population Size |
|---|---|---|---|
| SEGM Algorithm | $n_{dimension} = 2$ | $f_{SA} = 0.97$ | $size^{Bio}_{core\ population} = 16..256$ |

**Figure 4.13:** *Single Mapping Deviations of the* SEGM *Algorithm, dependent on Data Model Size.*

The largest test case is equivalent to a neural network of $51.200$ neurons, divided into $25$ synaptical connected populations, with $8$ subpopulations each, placed onto a hardware of also $25$ core grids, wherein each grid consists of $256$ cores, each with a capacity of $8$ neurons.

In conclusion it can by said, that the neurons of the given biological models are placed virtually ideally onto the hardware models regarding the provided and transformed cost functions. Independent from the model sizes, the networks' topologies are reestablished, exhibiting only low deviations. The parameter mapping distortions decrease slightly with enlarged models, but also remain also in small magnitudes.

Following the test specification, Figure 4.14 shows the runtimes of the SEGM Algorithm during the same mapping task. As illustrated in Section 4.1.1 the SEGM Algorithm consists of three components, whose runtimes and runtime slopes are presented as charts in dependency of the model sizes. The solid lines stand for averaged runtimes of $i$) the DT Algorithm, $ii$) the TF K-MEANS Algorithm and $iii$) the BLC Algorithm, introduced in the Sections 2.3.2, 3.3.2 and 4.2.2.1 respectively. The dashed lines represent the according slopes and thus the runtime complexities. It should be noted, that the diagram utilizes two different logarithmic $y$-axes: the runtime in *seconds* (left side) and the runtime slope in $\frac{seconds}{neuron}$ (right side).

First of all it becomes apparent, that the runtime of the actual *CA*, the TF K-MEANS Algorithm, is two orders of magnitudes below the runtimes of the other two components, i.e. the clustering's runtime plays only a minor role for the overall runtime. The model transposition and the capacity balancing occupy the majority of the computation effort, so that while processing the largest models, the BLC Algorithm requires about $82.1\%$ and the DT Algorithm about $17.7\%$ of the runtime, the TF K-MEANS Algorithm executes in $0.2\%$ of the total time span.

This leads to the second and more important insight: while the runtime slopes of both the previously introduced components, the DT and the TF K-MEANS Algorithm, remain comparable constant at about $0.5 \cdot 10^{-1}$ and $< 10^{-3} \frac{seconds}{neuron}$ respectively, indicating linear runtime increments for linear growing models,

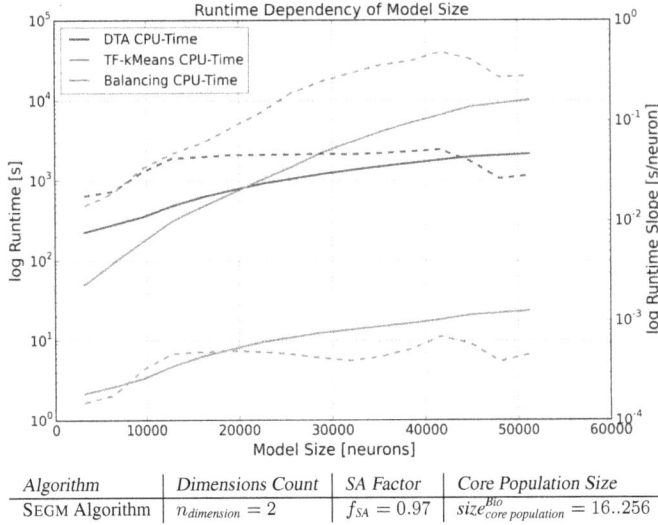

Runtime Dependency of Model Size

| Algorithm | Dimensions Count | SA Factor | Core Population Size |
|-----------|------------------|-----------|---------------------|
| SEGM Algorithm | $n_{dimension} = 2$ | $f_{SA} = 0.97$ | $size^{Bio}_{core\,population} = 16..256$ |

**Figure 4.14:** *Single Runtimes of the* SEGM *Algorithm, dependent on Data Model Size;* solid lines represent the runtimes (left logarithmic $y$-axis), dashed lines stand for their slopes (right logarithmic $y$-axis)

the slope of the BLC Algorithm is clearly not constant, but raises by a factor of ten from $< 0.5 \cdot 10^{-1}$ to $0.5 \cdot 10^{0} \frac{seconds}{neuron}$. The effect is also observable at model sizes of $20k$ neurons, where the former faster BLC Algorithm levels up to the DT Algorithm and becomes the slowest component of the SEGM Algorithm and thus dominant with regard to the overall runtime.

The detailed results can be summarized in Figure 4.15, which shows the overall relative mapping deviation in combination with the total runtime in dependency of the model sizes. The base of these charts is formed by the tests presented above.

The overall mapping deviation is defined in Formula (4.10) and the chart reveals, that the placement quality is close to the theoretical optimum. For smaller models the deviation is about 0.01 and decreases slightly with growing model sizes to a minimum of 0.004 for models of $40k$ neurons, i.e. with respect to the feature priorities the mapping result matches to 99.6% with its original neural network. The chart's course suggests, that the SEGM Algorithm is also capable to process larger models with low mapping deviations.

However, the runtime behavior restricts the processable model size due to its non-linear complexity. Caused by the BLC Algorithm, the SEGM Algorithm's overall runtime clearly exhibits a runtime, whose slope is comparably flat, but not constant, and thus increasing the execution time spans to impractical orders of magnitudes for certain models sizes, depending on the executing system.

Finally it can be said, that models with less than $100k$ neurons are processable by the SEGM Algorithm on the given test system (see Section 4.2.1.4). The approach is capable to reduce the provided cost functions close to the ideal result, but the processing of models beyond the said size will require significantly more than 10 hours (estimated), thus hindering the algorithm's application to larger problems without further enhancements of the BLC Algorithm.

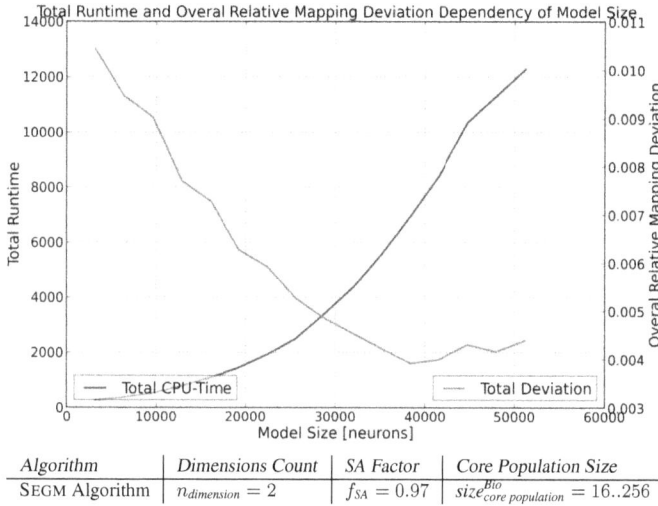

| Algorithm | Dimensions Count | SA Factor | Core Population Size |
|-----------|------------------|-----------|----------------------|
| SEGM Algorithm | $n_{dimension} = 2$ | $f_{SA} = 0.97$ | $size_{core\ population}^{Bio} = 16..256$ |

**Figure 4.15:** *Total Runtime and Total Mapping Deviation of the* SEGM *Algorithm, dependent on Data Model Size; the runtime is assigned to the left y-axis, the mapping deviation to the right y-axis.*

### 4.2.2.3 Results Summary

The former introduced *MDS* component, the DT Framework (Section 2), and the TF K-MEANS Algorithm as *CA* technique (Section 3) were applied to a *MOO* task to validate their usability in that context. The approach follows the TRACS-MOO concept introduced in Section 1.4 where it was postulated, that *MOO* tasks can be converted to a form, so that pareto-optimal solutions or the pareto-optimal front can be found by subsequent adapted *CA*s.

As illustrated in Section 4.1.1 the utilized *MOO* task is the challenge to configure a neural hardware system optimally, given a biological neural network model to simulate. Divided in several steps, this mapping process starts with the most crucial step, the neuron placement. Its underlying *MOO* task is to find a proper neuron – hardware assignment with regard to several conflicting optimization objectives, e.g. the preservation of the network's topology and a suitable parameter matching.

The presented solution is to utilize the DT Framework to transpose the biological model to a low-dimensional form, wherein the neurons are positioned with regard to their affinities, thus expressing the target functions. The subsequent application of the TF K-MEANS Algorithm in combination with a balancing mechanism determines size-limited clusters of affine neurons, which are equivalent to the sought neuron – hardware assignment. The utilization of the existing components requires only the implementation of the compare functions and the provision of the data access, i.e. the defined interfaces (Sections 2.3.2.1 and 3.1.2). As an enhancement the BLC Algorithm was added to enable the TF K-MEANS Algorithm to reflect cluster capacities. The resulting SEGM Algorithm is presented in Section 4.2.2.1.

The algorithm's analysis is based on artificial biological and hardware models of arbitrary size, which were designed in a manner, so that exactly one ideal solution exists. The deviations from this known ideal placement served as quality indicators. In this way and in combination with runtime measurements the performance of the SEGM Algorithm was evaluated. The analysis methods and model descriptions can be found in Section 4.2.1.

The measurement results in Section 4.2.2.2 proved, that the SEGM Algorithm is a proper tool to perform the neuron placement step during the mapping of models of up to $100k$ neurons, executed on the available test system. The mapping distortion was independent from the model size and corresponded with at least 99.0% to the ideal mapping. In spite of the DT Framework' and the TF K-MEANS Algorithm's linear time complexities, the balancing component exhibited inconstant runtime slopes, thus slowing down the overall process for larger models. While models of $50k$ neurons could be placed in about 3 hours on the test system, models of twice that size would require an estimated runtime of at least 10 hours of which the majority would occupy the BLC Algorithm.

## 4.3 Application to the Pattern Recognition Issue as Peak Detection

The next two Sections 4.3.1 and 4.3.2 firstly describe the analysis strategy regarding the utilized data models, the measurements as well as the testing methods, and secondly the obtained results including new algorithms and performance analyses for the *PR* issue.

### 4.3.1 Methods and Materials

In the following Section 4.3.1.1 firstly considers the general objectives of the analysis strategy. Subsequently the Sections 4.3.1.2 and 4.3.1.3 describe the utilized data models, algorithm configurations as well as parameter spaces, and the measurements for the performance analysis respectively. Finally Section 4.3.1.4 concludes with the applied test and development scenarios.

#### 4.3.1.1 Proof of Concept

For the proof of the peak detection's concept the same strategy is applied analogously, as introduced in Section 4.2.1.1.

#### 4.3.1.2 Data Models and Algorithm Configuration

In the following the utilized data models are described, as well as the resulting *PR* objectives and the conclusions for the required model transposition.

**Data Model:**   To provide an universal and scalable data model as demanded in Section 4.3.1.1, the following approach is utilized with regard to the requirements in the *PR*'s introduction in Section 4.1.2: A set of *Variant Gaussian Random Fields* (*VGRFs*), a more general form of the *GFD* data model type as introduced in Section 3.2.3 to test the TF K-MEANS Algorithm, describes the distribution of the data items within a feature system of different variable ranges and magnitudes. A single *VGRF* can be defined as:

$$VGRF = \begin{pmatrix} offset_0, & range_0 \\ \dots & \dots \\ offset_i, & range_i \\ \dots & \dots \\ offset_n, & range_n \end{pmatrix} \qquad (4.15)$$

where the pair *offset_i* and *range_i* describes the field's distribution for the feature $i$, i.e. the start and the end $(offset + / - range)$ of an embedded normal distribution. In this way the field is restricted for each of the $n$ features, describing a $n$-dimensional ellipsoid within the entire feature space. Hence the data model can be described as a vector of *VGRF*s, each with a field *weight* such as:

$$description_{data\ model} = \begin{pmatrix} VGRF_0, & weight_0 \\ \dots & \dots \\ VGRF_m, & weight_m \end{pmatrix} \qquad (4.16)$$

Instantiating such a description as a data model with the size $k$, means firstly the random determination of all feature offsets and ranges for each of the $m$ *VGRF*s. Secondly, $n_{element}$ data items are assigned randomly, but weight-proportionally, to one of the fields and are initialized normally distributed with regard to the feature regions of this field. Figure 4.16 illustrates an instantiated data model of several fields with different feature regions.

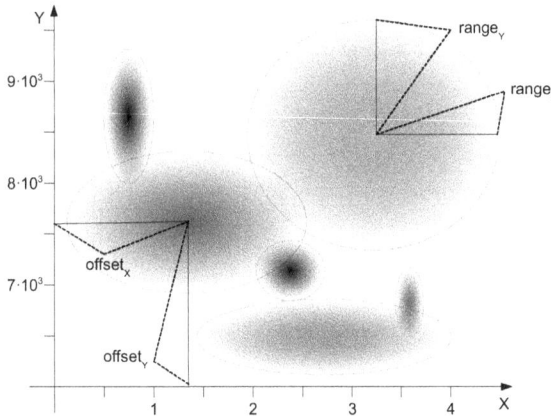

**Figure 4.16:** *Data Model Example based on VGRFs*; Shown is a data model of six Gaussian fields based on two features $X$ and $Y$. Prior to the data items' assignment, the fields are defined randomly by their center locations, e.g. *offset*$_{X,Y}$, their sizes, e.g. *range*$_{X,Y}$ and weights (not in the Figure). With regard to the field weights the data items are assigned randomly to one of the fields and their features are determined within the field's feature regions, forming normal distributions (gray ellipsoids). Note that the orders of magnitude and the offsets of $X$ and $Y$ are different and thus scaled down to allow a meaningful data model representation.

To create multiple data model descriptions with different feature ranges and field weights, the following matrix is required:

$$description_{data\ model\ space} = \begin{pmatrix} feature_{start} & feature_{end} \\ range^{rel}_{min} & range^{rel}_{max} \\ offset^{rel}_{min} & offset^{rel}_{max} \\ weight_{min} & weight_{max} \end{pmatrix} \quad (4.17)$$

where $feature_{start\,/\,end}$ defines the allowed range for each of the $n$ features and $range^{rel}_{min\,/\,max}$ restricts the relative ranges for each feature of the $VGRF$s. The same applies for the relative feature offsets $offset^{rel}_{min\,/\,max}$. Finally $weight_{min\,/\,max}$ describes the weight's range of all $VGRF$s.

In this way data models of arbitrary size and Gaussian field based feature distributions are provided. The data items are grouped in a defined number of patterns, which have random individual weights and form random ellipsoids within the feature space, while utilizing definable boundaries. If it is ensured, e.g. by repeated random $VGRF$ creations, that the patterns are distant enough to be distinguishable, this type of data model is suited to serve as test base for $PR$ tasks.

**Pattern Recognition Objectives:**   In consequence the objectives of the $PR$ are:

- the detection of the $VGRF$'s count,

- the determination of the $VGRF$'s offsets, ranges and weight,

- and the assignment of the data items to one of the detected field.

As the ambiguity of the relationships between original and detected clusters, i.e. the decision, which detected cluster is most affine to which original $VGRF$, would introduce an additional $MOO$ task, the following approach is chosen to describe measurable $PR$ objectives from the data items' point of view: After the assignment of each data item to a cluster, the *fragmentation* of the original $VGRF$s and the detected clusters should be minimal, i.e. $i)$ the data items of an original $VGRF$ should be assembled in the same detected cluster and $ii)$ vice versa, the members of a detected cluster should originate from only a single original $VGRF$. Figure 4.17 illustrates the cases of fragmentation on examples. Usually the result of a $PR$ process involves all shown cases.

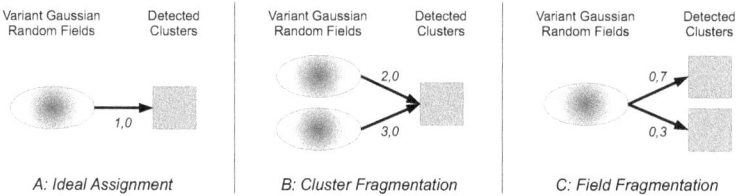

**Figure 4.17:** *PR Objectives based on Fragmentation Minimization;*
> **A: Ideal Assignment** - The ideal result of the $PR$ is the data items' assignment to clusters so, that the items of one original $VGRF$ are spread over only one single cluster, and no cluster contains items of different original $VGRF$s. Here a single field with the weight of 1.0 is assigned to a single cluster.
> **B: Cluster Fragmentation** - In this case two fields with the weights of 2.0 and 3.0 are assigned to one cluster, covering the original fields in a wrong merge cluster.
> **C: Field Fragmentation** - In this case a field with the weight of 1.0 is split into two clusters, exposing two wrong final patterns instead of one.

Hence the *PR*'s objective can be formulated as:

$$fragmentation_{VGRFs} = \sum_{k}^{VGRFs} frac_{rel}^{k}$$

$$fragmentation_{clusters} = \sum_{k}^{clusters} frac_{rel}^{k}$$

(4.18)

where $frac_{rel}^{k}$ stands for the single relative fragmentation of a *VGRF* or a cluster. To test the PD Algorithm only the data items are provided, so that the algorithm has to extract the field information from their features.

**Model Transposition:** As already stated in Section 2.1.1, large and high-dimensional data models with diverse feature domains impede or prevent an automated analysis, hence the DT Framework is utilized to transpose the models to a form, which allows for the application of the generalized PD Algorithm. For this purpose the same affinity functions for single values as in Formula (4.7) are applied, as all features of the tested data model belong to this domain. Furthermore, all features are treated with equal priorities.

To enhance the density concentrations of the sought clusters in the low-dimensional target space, the following optimal distance function is applied (compare Formula (2.8) in Section 2.3.2.1):

$$Func_{dist_{opt}}(aff) = (1 - aff)^2$$

(4.19)

### 4.3.1.3 Measuring Results and Performance

As explained in Section 4.2.1.1 the measurements focus is on the relative quality as difference from the ideal solution, and on the runtime. Table 4.6 shows an overview of the single measurements, detailed explanations can be found below.

| Measurement | Symbol | Description | Unit |
|---|---|---|---|
| Fragmentation of *VGRF*s | $fragmentation_{VGRFs}$ | deviation of the ideal field-cluster matching | - |
| Fragmentation of Clusters | $fragmentation_{clusters}$ | deviation of the ideal cluster-field matching | - |
| Total Fragmentation | $fragmentation_{total}$ | overall matching deviation (total quality) | - |
| DT Runtime | $t_{DT}$ | DT execution time | *seconds* |
| DT Runtime Slope | $slope_{DT}$ | relative DT runtime increment | *second/MSI* |
| *Clustering* Runtime | $t_{clustering}$ | *Clustering* execution time | *seconds* |
| Clustering Runtime Slope | $slope_{clustering}$ | relative clustering runtime increment | *second/MSI* |
| Total Runtime | $t_{total}$ | total execution time | *seconds* |

**Table 4.6:** *Measurements / Evaluations during* PD *Algorithm Testing;* Shown are the names and the symbols of the single measurements in combination with short descriptions and the according units.

The fragmentation of a single group, i.e. an original *VGRF* or a result cluster, is defined as:

$$fragmentation_{single\ group} = \frac{\sum_k^{groups} \left(frac^k\right)^2}{\left(\sum_k^{groups} frac^k\right)^2} \tag{4.20}$$

where $frac^k$ is the fraction of the single group's assignment to the complementary group $k$. For example the spreading of a $VGRF$ over two clusters $i$ and $j$ can be described as two fractions $frac^i$ and $frac^j$. The assignment of two $VGRF$s to a single cluster can be also expressed with two fractions in that manner. As the fields and the clusters can be fragmented simultaneously, fields and clusters may contain fractions of each other.

Consequently the assignment of a group to a single complementary group would result in a fragmentation value of 1 (no fragmentation - best case), the spreading over an infinite count of complementary groups would yield a value of 0 (maximal fragmentation - worst case). The fragmentation of all $VGRF$s and all clusters can be expressed as:

$$\begin{aligned} fragmentation_{VGRFs} &= \\ fragmentation_{clusters} &= \\ fragmentation_{groups} &= \frac{\sum_k^{groups} \left(fragmentation_{single\ group}^k \cdot weight^k\right)}{\sum_k^{groups} weight^k} \end{aligned} \tag{4.21}$$

taking into account the group weights (data item counts). Finally the total fragmentation of the $PR$'s result as overall quality measurement is defined as the average of both the field and the cluster fragmentation:

$$fragmentation_{total} = \frac{fragmentation_{fields} + fragmentation_{clusters}}{2} \tag{4.22}$$

The evaluation of the runtimes $t_{DT\,/\,clustering}$ of the DT Algorithm and the *Agglomerative Clustering*, as well as their according time complexities $slope_{DT\,/\,clustering}$, are performed similarly to the DT and TF runtime measurements (Sections 2.2.2 and 3.2.2).

### 4.3.1.4 Performance Evaluation

To validate the results and evaluate the performance of the PD Algorithm regarding quality and runtime, the following tests were performed:

Randomly created data models are processed by the $PR$ to detect the underlying $VGRF$s as density peaks and thus the sought patterns. The data models are varied step-wisely $i$) by the the number of Gaussian fields ($n_{fields}$) to alter the patterns to detect, and $ii$) by the number of data items ($n_{element}$) to enlarge the models. The algorithm's behavior is changed by a granularity parameter ($gain_{min}$), determining whether or not to recognize patterns as separated (Section 4.3.2.1).

The $VGRF$s for each data model are created in the following manner: First, the allowed value range and the order of magnitude of each feature are randomly determined ($feature_{start\,/\,end}$), the count of features is fixed ($n_{feature}$). In this scope the required $VGRF$s are set by determining their centers and expansions in each feature randomly ($range_{min\,/\,max}^{rel}$ and $offset_{min\,/\,max}^{rel}$) in addition to the field weights ($weight_{min\,/\,max}$). Subsequently the required count of data items are created and assigned randomly, but with respect to the field weights, to one of the $VGRF$, before determining the feature values within the scope of the assigned field, utilizing a normal distribution.

Table 4.7 shows the varied and constant model parameters (Section 4.3.1.2) in combination with the algorithm configurations and the measurements (Section 4.3.1.3).

| Scenario | Varied Parameter | Step Size | Fixed Parameter | Measurements |
|---|---|---|---|---|
| VGRF | $n_{fields} = 5..20$ | 1 | DT Framework: | $fragmentation_{VGRFs}$ |
| Pattern | $n_{element} = 1k..10k$ | $1k$ | $f_{SA} = 0.95$ | $fragmentation_{clusters}$ |
| Detection | $gain_{min} = 0.75..0.95$ | 0.05 | $n_{dimension} = 2$ | $fragmentation_{total}$ |
| | | | Data Model: | $t_{DT}$ |
| | | | $n_{feature} = 20$ | $t_{Clustering}$ |
| | | | $feature_{start/end} = 0.0/10^6$ | $t_{total}$ |
| | | | $range_{min/max}^{rel} = 0.0/0.5$ | $slope_{DT}$ |
| | | | $offset_{min/max}^{rel} = range_{max}^{rel}/1.0 - range_{max}^{rel}$ | $slope_{Clustering}$ |
| | | | $weight_{min/max} = 1.0/10.0$ | |

**Table 4.7:** *Test Scenarios for the* PD *Algorithm Testing*; Shown is configurations of the PD Algorithm test scenario, to analyze the influence of the data models' characteristics (Section 4.3.1.2) onto the quality and performance indicators (Section 4.3.1.3).

After the data models' transposition by the DT Framework (controlled by the $f_{SA}$ and $n_{dimension}$ parameters) and the clustering by the PD Algorithm, the qualities or results are evaluated (*fragmentation...*). While the algorithm operates, the runtimes and the slopes of each major step are measured ($t_{...}$ and *slope...*).

To reduce the effects of scattering, each test is instantiated and processed multiple, i.e. 10 times, utilizing different random seeds, providing measurements data, which are statistically useful.

The tests were carried out in parallel on a *Windows 7 Professional* workstation, hosting an *Intel Core* $i7 - 2760QM$ CPU with access to 12 GB of RAM. The results were collected automatically into detailed logs, which serve as the charts' database for the final analysis in Section 4.3.2.2.

## 4.3.2 Results

Presenting the results of TRACS-MOO's application to the *PR* issue, firstly Section 4.3.2.1 describes the developed algorithms and adaptations, and secondly 4.3.2.2 shows an overview of the performance analysis. Finally 4.3.2.3 summarizes and discusses the obtained insights.

### 4.3.2.1 The Peak Detection Algorithm

In the following the data preparation is introduced, which allows for the subsequently presented PD Algorithm as agglomerative clustering.

**Data Preparation:** As the other component of the PD Algorithm has already been introduced (DT Algorithm, Section 2), this Section focuses on the *PR* technique that enable the algorithm to detect the underlying structures, as described Section 4.1.2, following the concept shown there in Figure 4.7.

First, the model transposition, performed by the DT Algorithm, creates an uniform and low-dimensional equivalent of the provided data model (*VGRF*s based feature distribution, Section 4.3.1.2), where affine elements (data items) are positioned close to each other. Element affinity in this context signifies the similarity of features with regard to the *PR*'s objectives. This representation enables the subsequent agglomerative *CA* method to structure the given data model, forming groups of affine elements as the sought patterns.

Furthermore, the precalculated hierarchical data structure (element tree, Section 2.3.2.3), created and exported by the DT Algorithm, is utilized as base for the PD Algorithm, allowing it to involve additional information about coarse element groupings and shapes. The already available characteristics per tree branch $b$ are supplemented by the arithmetic center, analogue to the expected value:

$$\vec{\mu}^b = \frac{\sum_{b_{sub}} \vec{\mu}^{b_{sub}} \cdot weight^{b_{sub}}}{\sum_{b_{sub}} weight^{b_{sub}}} \qquad (4.23)$$

where $b_{sub}$ represents a subbranch (or leaf) of $b$ and $\vec{\mu}^{b_{sub}}$ is the previously calculated arithmetic center of this subbranch. Each subbranch is scaled by its *weight*, so that $\vec{\mu}^b$ is defined by the arithmetic, or gravity centers of all subbranches in a recursive manner.

Based on $\vec{\mu}^b$ a second characteristic is introduced as an expression for the density distribution within a branch, similar to the standard deviation:

$$\sigma^b = \frac{\sum_{b_{sub}} \left( |\vec{\mu}^b - \vec{\mu}^{b_{sub}}| \cdot weight^{b_{sub}} \right)}{\sum_{b_{sub}} weight^{b_{sub}}} \qquad (4.24)$$

which means, the smaller $\sigma^b$, the closer the encapsulated data items are to the arithmetic center $\vec{\mu}^b$, hence the larger is the average density gradient.

Utilizing these two new branch characteristics, the following metrics are defined to describe the branch $b$ with regard to the sought patterns:

$$flatness^b = \frac{\sigma^b}{dist^b_{maxleaf}}$$
$$wryness^b = \frac{|\vec{pos}^b - \vec{\mu}^b|}{dist^b_{maxleaf}} \qquad (4.25)$$
$$isolation^b = \frac{dist^b_{maxleaf}}{dist^{super}_{maxleaf}}$$

Here $dist^b_{maxleaf}$ is the former introduced *MLD* as radius of $b$ (Section 2.3.2.4), i.e. the smaller $\sigma^b$ in relation to the branch's extend, the smaller is the *flatness*$^b$, describing a high density gradient, which indicates a density peak. Furthermore, *wryness*$^b$ expresses the relative difference between the branch's geometric center $\vec{pos}^b$ and its center of gravity $\vec{\mu}^b$. A large wryness implies a flank of a peak, which appears larger than the current branch. A small wryness can mean, that the branch is equivalent to one the sought peaks. Finally *isolation*$^b$ describes the environment of $b$ by utilizing the radius $dist^{super}_{maxleaf}$ of the $b$'s super branch. The larger $b$ is in comparison to its super branch, the more the data items within the super branch are isolated from the rest, indicating a suitable pattern.

In conclusion the following *peak degree* defines how appropriate a branch $b$ can be regarded as a sought pattern, i.e. as a density peak:

$$deg^b_{peak} = \left( 1 - flatness^b \right) \left( 1 - wryness^b \right) \left( 1 - isolation^b \right) \qquad (4.26)$$

With respect to the three metrics defined in Formula (4.25) it means that, $i$) the steeper the density slope is toward the arithmetic center, $ii$) the more centered the density maximum is and $iii$) the more separated the branch is from the rest of the data, the more it can be regarded as a peak, i.e. as a sought cluster. The more a branch complies with all of these requirements, the more it can be considered as a pattern of the sought type.

These heuristics serve as base for the subsequent *CA* to identify branches or combinations of them as final clusters. As the element tree remains unchanged during the algorithm's processing and the calculations are based on the direct subbranches or leafs, these precalculations are done comparably fast and only once.

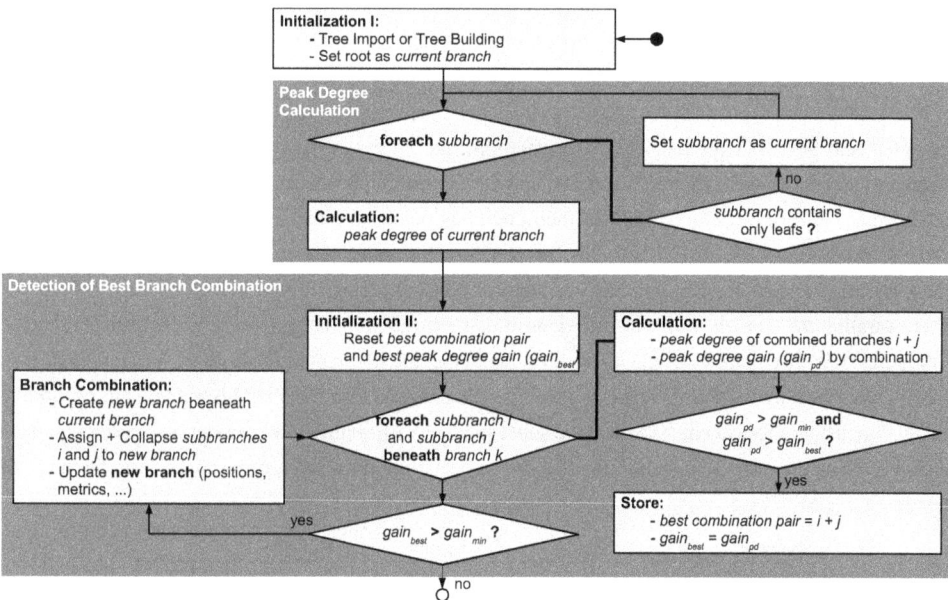

**Figure 4.18:** *Scheme of the* PD *Algorithm*;

**Initialization I:** The utilization of the tree building algorithms (Sections 2.3.2.3 and 2.3.2.4) or the import of an already created element tree provides the required hierarchical data representation. The algorithm starts at the tree's root.

**Peak Degree Calculation:** By processing the entire element tree in a bottom-up manner, the peak degree of each branch is calculated, based on the metrics of the subbranches.

**Detection of the Best Branch Combination:** After reinitializing the results of the previous iteration, each branch pair $i$ and $j$ beneath the same super branch $k$ is evaluated regarding the estimated peak degree gain ($gain_{pd}$) after a potential branch merge. If the best branch pair is found, whose $gain_{pd}$ is larger than a given threshold $gain_{min}$, the branch merge is performed by creating a new branch, assigning all leafs of $i$ and $j$ to it and constructing the required subbranches. Finally the merged branch is updated and its metrics and peak degree are calculated for the next iteration. If no branch pair yields a $gain_{pd}$ larger than the minimum, the algorithm terminates. The remaining branches represent the final clusters, and thus the result of the *PR* process.

**Agglomerative Clustering:** The base concept of the PD Algorithm is the iterative combination of branches of the provided element tree, until the peak degree (Formula (4.26)) can not be improved further. To evaluate a given branch pair $b^i$ and $b^j$ the *estimated gain* is introduced as the relative and weighted peak degree difference:

$$
gain^{b^i+b^j} = \frac{\dfrac{deg_{peak}^{b^i+b^j}}{deg_{peak}^{b^i}} weight^{b^i} + \dfrac{deg_{peak}^{b^i+b^j}}{deg_{peak}^{b^j}} weight^{b^j}}{weight^{b^i} + weight^{b^j}}
\tag{4.27}
$$

where $deg_{peak}^{b^i}$ and $deg_{peak}^{b^j}$ represent the previously calculated peak degrees (Formula (4.26)) of both branches and $weight^{b^i}$ and $weight^{b^j}$ describe the according branch weights, i.e. the leaf counts beneath the branches. The term $deg_{peak}^{b^i+b^j}$ stands for the estimated peak degree of a potentially merged branch based on $b^i$ and $b^j$, utilizing estimated characteristics and metrics.

Hence the PD Algorithm can be summarized as follows in Figure 4.18. The method requires an imported or previously constructed element tree (Sections 2.3.2.3 and 2.3.2.4). Initially the characteristics, i.e. the metrics and the peak degrees (Formulas (4.23) - (4.26)) are calculated, serving as base for the subsequent branch merging. In the following the algorithm detects iteratively the best branch pair, which $i)$ is beneath the same super branch, $ii)$ exposes the largest peak degree gain (Formula (4.27)) and $iii)$ has a potential peak degree larger than a given global threshold. Successively each best pair is merged until no further pairs comply with the requirements, and the algorithm terminates. The remaining branches are regarded as clustering result, as none of their combinations would improve the quality further with respect to the given criteria. In this way this $PR$ technique assembles branches of increasing sizes, treating large data amounts efficiently, to identify patterns of the sought type.

Illustrating the work flow of the PD Algorithm, Table 4.8 summarizes a test run, which is rendered in Figure 4.19: The features of the elements (data items) are distributed to various overlapping $VGRF$s of different sizes, thus forming a diverse and high-dimensional data model. This model is transposed to a low-dimensional equivalent, enabling the subsequent PD Algorithm to detect the now more separated analyzable patterns. The found clusters are mapped back to original feature space and form the final $PR$ result.

| Name | Symbol | Value |
|---|---|---|
| *Data Model* | | |
| *VGRF* count | $n_{field}$ | 20 |
| element count | $n_{element}$ | $10k$ |
| feature count | $n_{feature}$ | 20 |
| allowed feature range | $feature_{start/end}$ | $0.0/10^6$ |
| relative field sizes | $range_{min/max}^{rel}$ | $0.0/0.5$ |
| relative field offsets | $offset_{min/max}^{rel}$ | $range_{max}^{rel}/1.0 - range_{max}^{rel}$ |
| fields weights | $weight_{min/max}$ | $1.0/10.0$ |
| PD Algorithm | | |
| DT Framework: *SA* Factor | $f_{SA}$ | 0.95 (60 iterations) |
| DT Framework: target dimension count | $n_{dimension}$ | 2 |
| PD Algorithm: minimal peak degree gain | $gain_{min}$ | 0.85 |

**Table 4.8:** *Data Model Parameters and Algorithm Configuration for the Rendered Example in Figure 4.19*

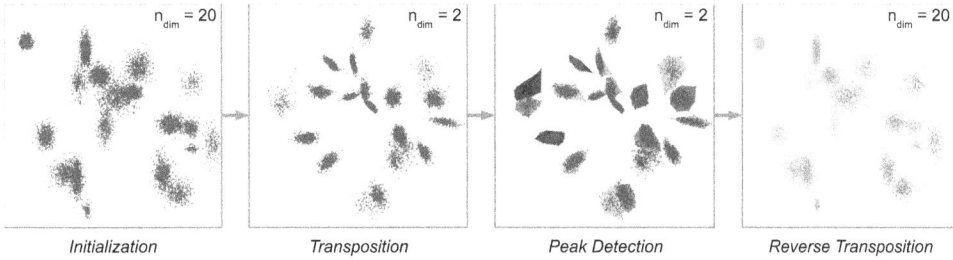

Initialization      Transposition      Peak Detection      Reverse Transposition

**Figure 4.19:** *Rendered* PD *Algorithm Example*;

> **Initialization:** The data model is initialized with 20 feature *VGRF*s and $10k$ elements (blue dots), assigned to them. The features are spread over different orders of magnitudes and regions, resulting in a diverse data model. Here only the first two (scaled) features ($x$- and $y$-axis) are shown.
>
> **Transposition:** The DT Framework transposes the data model to a system of two uniform axes, utilizing a non-linear optimal distance function to enhance the density concentrations. In consequence the patterns appear more clearly in the target space at the risk of covering small clusters.
>
> **Peak Detection:** Independent from the original feature domains and scalings the PD Algorithm detects the patterns as density peaks (colored polygons) in the transposition space.
>
> **Reverse Transposition:** After the data model's reverse mapping to the original feature space, the resulting clusters (colored dots) represent the identified *VGRF*s. The result yields a fragmentation degree of $0.96$ (Formula (4.22)), which is close to the optimal result.

### 4.3.2.2 Measurement Results

In the following representative measurements of the previously described *test scenario* (Section 4.3.1.4, Table 4.7) are illustrated.

The first charts in Figure 4.20 show the quality of peak detection (*fragmentation$_{total}$*, Formulas (4.20) - (4.22)), or short, the quality, in dependency of the model size. The number of elements ($n_{element}$) and the number of *VGRF*s were gradually increased, raising the difficulty to detect the correct patterns in adequate runtimes. The tests were performed several times, utilizing different minimal peak degree gains (*gain$_{min}$*, Formula (4.27)) to evaluate the parameter influences onto the quality.

First, for the most challenging case ($n_{element} = 10k$, $n_{field} = 20$) the charts expose a total quality (fragmentation) of $0.92$ for *gain$_{min}$* $= 0.85$, which is equivalent to an average of $4\%$ of inappropriate element assignments and inappropriate cluster contents each. Second, the quality is independent of the element count, as it remains constant for the tested model sizes. Third, the quality falls slightly from about $0.98$ to $0.92$ for increasing field counts.

Furthermore, the specified minimal peak degree gain determines the resulting quality significantly. In this case (in comparison to the known ideal results) a value of $0.85$ seems appropriate, as it's chart exposes the best overall quality. Nevertheless, in other test cases, other values may be more suitable, as for example the low quality of the tests with *gain$_{min}$* $= 0.95$ does not describe mandatory wrong results, but a finer pattern granularity, which may be more appropriate in other situations.

In conclusions it can be said, that the PD Algorithm is suitable to retrieve the covered *VGRF*s as sought patterns with only slight deviations, if an appropriate clustering granularity is chosen. Growing numbers of fields, i.e. a higher number of original patterns, cause declining qualities, as the overall amount of overlapping and the number of possible element assignments increase.

Following the test specification, Figure 4.21 shows the total runtime $t_{run}^{total}$ of the PD Algorithm during

Quality of Peak Detection

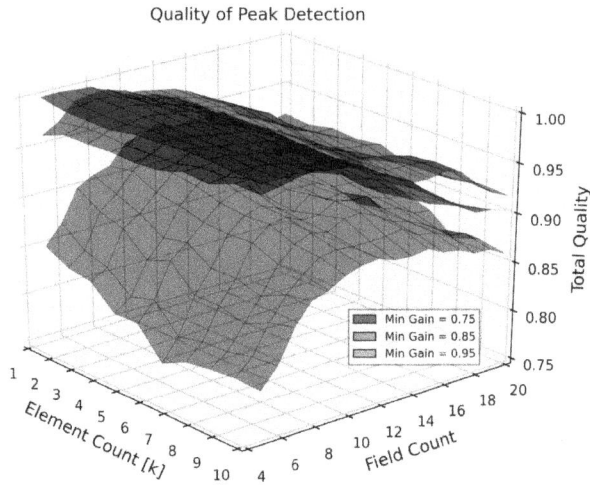

| Algorithm | Dimensions Count | SA Factor | Element Count | Field Count | Minimal Relative Gain |
|-----------|------------------|-----------|---------------|-------------|------------------------|
| PD Algorithm | $n_{dimension} = 2$ | $f_{SA} = 0.95$ | $n_{element} = 1k..10k$ | $n_{field} = 5..20$ | $gain_{min} = 0.75..0.95$ |

**Figure 4.20:** *Total* PD *Algorithm Quality of Results, dependent on Data Model Size and Gaussian Field Count.*

the *PR* task described above. The runtimes of both the DT Algorithm and the agglomerative clustering for all utilized values of $gain_{min}$ are integrated, thus representing the algorithm's runtime for all chosen granularities. The element count – field count layer contains a heatmap to clarify the model parameters' influence onto the runtime. The same applies for the chart's projections to both other layers.

It is obvious, that the runtime is independent of the field count, as it remains constant for growing numbers of fields, and increases linearly for growing element counts. In the most difficult case, i.e. $10k$ elements, whose features are distributed to 20 different *VGRF*s, the PD Algorithm terminates after about 175 seconds.

This runtime behavior proves the capability of the PD Algorithm to process much larger, prospective models, as in spite of the problem complexity of at least $O(n^2_{element})$, the runtime will only be doubled, if the model size (element count) is doubled.

To further investigate the runtime, Figure 4.22 shows the integrated runtimes $t_{DT/clustering}$ of both algorithm components, i.e. the DT Algorithm and the agglomerative clustering, for all field counts $n_{field}$ and granularities $gain_{min}$ in addition to their complexities $slope_{DT, clustering}$. Both the left $y$-axis for the component runtimes and the right $y$-axis for the runtime slopes expose logarithmic scales to cover the different orders of magnitude.

It is obvious, that the model transposition requires the majority of the runtime (30..150 seconds) in comparison to the agglomerative clustering (0.1..1.0 seconds), symbolized by the solid lines. A more important insight is, that the slopes of both runtimes, visualized as dotted lines, expose virtually constant values, i.e. $10^{-2} \frac{seconds}{element}$ for the DT Algorithm and $10^{-4} \frac{seconds}{element}$ for the clustering, which indicates linear runtime increments for all field counts and chosen clustering granularities.

Finally the crucial results, the total quality (*fragmentation$_{total}$*) and the total runtime ($t^{total}_{run}$) are sum-

Peak Detection Runtime

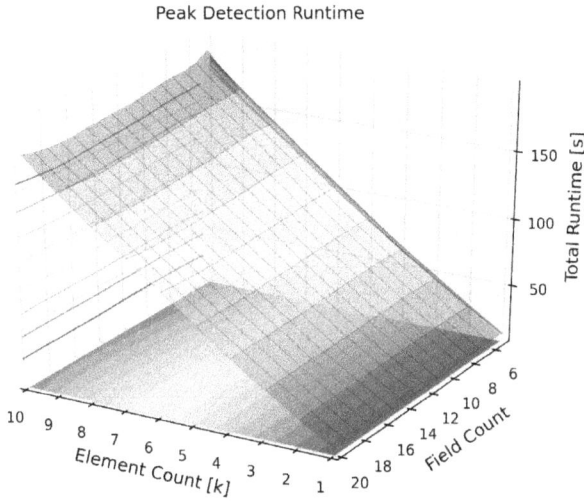

| Algorithm | Dimensions Count | SA Factor | Element Count | Field Count | Minimal Relative Gain |
|---|---|---|---|---|---|
| PD Algorithm | $n_{dimension} = 2$ | $f_{SA} = 0.95$ | $n_{element} = 1k..10k$ | $n_{field} = 5..20$ | $gain_{min} = 0.75..0.95$ |

**Figure 4.21:** *Total* PD *Algorithm Runtime, dependent on Data Model Size and Gaussian Field Count.*

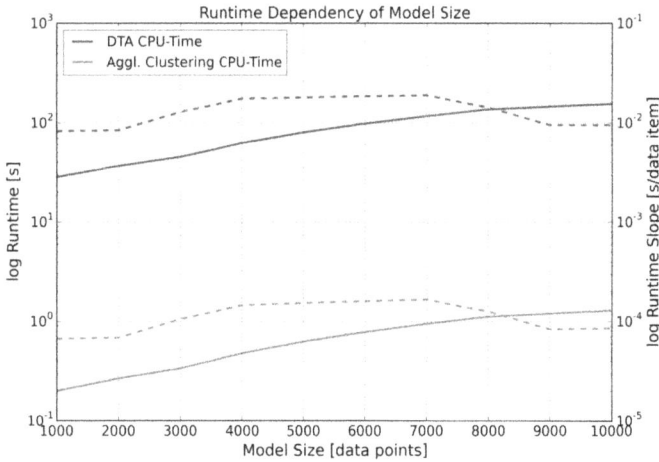

| Algorithm | Dimensions Count | SA Factor | Element Count | Field Count | Minimal Relative Gain |
|---|---|---|---|---|---|
| PD Algorithm | $n_{dimension} = 2$ | $f_{SA} = 0.95$ | $n_{element} = 1k..10k$ | $n_{field} = 5..20$ | $gain_{min} = 0.75..0.95$ |

**Figure 4.22:** *Single* PD *Algorithm Runtime, dependent on Data Model Size and Gaussian Field Count.*

marized in Figure 4.23. Here the quality is integrated for all field counts $n_{field}$ as well as for all utilized cluster granularities $gain_{min}$ and assigned to the right $y$-axis. The runtime chart is the sum of the single runtimes shown in Figure 4.22 and assigned to the left $y$-axis.

As conclusion regarding the *PR* quality it can be stated, that the fragmentation was almost constant between 0.94 and 0.95. According to the Formulas (4.20) - (4.22) this is equivalent to an average frag-

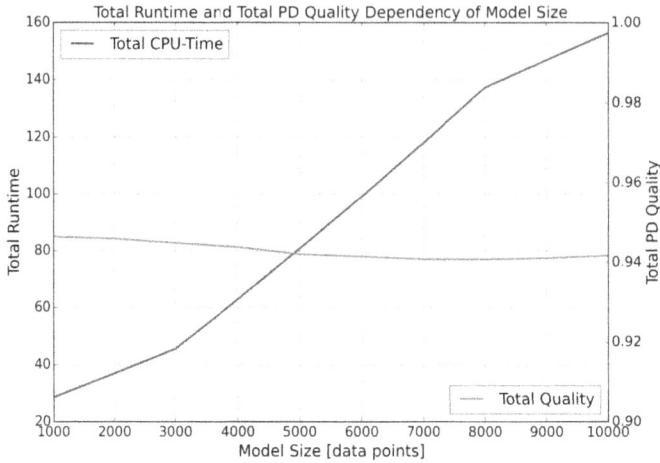

| Algorithm | Dimensions Count | SA Factor | Element Count | Field Count | Minimal Relative Gain |
|---|---|---|---|---|---|
| PD Algorithm | $n_{dimension} = 2$ | $f_{SA} = 0.95$ | $n_{element} = 1k..10k$ | $n_{field} = 5..20$ | $gain_{min} = 0.75..0.95$ |

**Figure 4.23:** *Total* PD *Algorithm Quality and Runtime, dependent on Data Model Size and Gaussian Field Count.*

mentation value of 3% for the Gaussian fields, i.e. 3% of their elements are assigned to another final cluster, and also 3% of the clusters content belongs originally to another field. The total runtime increased from 30 to 155 seconds and exposed a linear complexity, i.e. a constant slope for linear growing model sizes.

### 4.3.2.3 Results Summary

The *MDS* component introduced as the DT Framework (Section 2), was applied to a *MOO* task to validate and test its usability in this context. The approach follows the TRACS-MOO concept presented in Section 1.4, where it was postulated, that *MOO* tasks can be converted to a form, so that pareto-optimal solutions or the pareto-optimal front can be found by subsequent adapted *CA*s.

As illustrated in Section 4.1.2 the utilized *MOO* task is to detect underlying patterns within a provided set of data items with different features in varying value regions and orders of magnitude, usually referred to as *Pattern Recognition* (*PR*) in the field of *Knowledge Discovery in Databases* (*KDD*). The result should contain the number, shapes, sizes and locations of the underlying, unknown patterns, representing an abstraction of the data to enable general statements about regularities, repetitions or similarities within. The multi objectiveness is to find an appropriate tradeoff between the minimal pattern count and the maximal pattern density, as both extrema (one single pattern or one pattern per data item) provide no useful information.

The solution concept is to utilize the DT Framework to transpose the data model to a low-dimensional form, wherein the data items are positioned with regard to their feature affinities, suppressing their feature counts and diversities to enable methods of the *CA* to process the data efficiently. A newly introduced agglomerative clustering method utilizes the provided data tree structure from the DT Framework to identify the sought patterns in a bottom-up manner. For this purpose the algorithm detects and combines branch pairs iteratively, for which the largest gain of the defined pattern metrics (density gradient steepness, geometrical wryness and isolation from the rest of the data) can be expected. The *CA* terminates, if no branch pairs are available anymore, that comply with a given minimal improvement threshold, defining the granularity of the method. In this way a general purpose *PR* algorithm is provided, which is independent of the number, scalings and offsets of the data items features. The resulting PD Algorithm is described in Section 4.3.2.1.

The algorithm's analysis is based on artificial data models of arbitrary size and previously defined underlying patterns, providing known ideal solutions. The deviations from these known ideal clusterings serve as quality indicators. In this way and in combination with runtime measurements the performance of the PD Algorithm was evaluated. The analysis methods and model descriptions can be found in Section 4.2.1.

The measurement results in Section 4.3.2.2 prove, that the PD Algorithm is a suitable tool to retrieve the underlying patterns. The deviation from the ideal solution can be virtually neglected for all test cases, in particular for the most challenging ones (large numbers of data items and underlying patterns). The runtime increases linearly for growing data item counts and remains constant for growing field counts, representing a linear time complexity $O(n_{element})$ and enabling the algorithm for much larger data models. Furthermore, the tests show, that the model transposition requires much more runtime (155 seconds for $10k$ data items) than the actual clustering (1 second for $10k$ data items), i.e. after the typically one-time transformation, several runs of the comparably fast PD Algorithm can be performed with varying granularities to retrieve the desired pattern types.

Although the utilized testing strategy covers a broad range of *PR* task types, conclusions about the approach's generality are difficult. As already stated in Section 3.1.1 the final purpose of the sought patterns determines the characteristics (count, shapes, sizes, ..) of the clusters to detect, and thus the quality of result of each *CA*. Nevertheless, for applications which emphasize patterns of the Gaussian fields type, i.e. normal distributions in various regions and orders of magnitude for a large number of features, the PD Algorithm is an appropriate method with only one behavior defining parameter (the granularity) to process large numbers of data items efficiently.

## 4.4 Discussion

In the following several aspects of the TRACS-MOO's application to existing *MOO* problems are discussed:

*Prove of Concept:* The approach's application to the segmentation and pattern recognition issues serve as proof of the TRACS-MOO' concept, introduced in Section 1.4. Utilizing the approaches of the DT Framework and the TF K-MEANS Algorithm (Sections 2 and 3) abstracts the given tasks and requires only few adaptations and extensions to treat the different and the comparable complex tasks. Hence it could be shown that the TRACS-MOO is a suitable approach to handle such kind of *MOO* problems.

*Performance:* The tests of the SEGM Algorithm showed constant high qualities of results, but also partial non-linear time complexities. The BLC Algorithm as an adaptation for the abstract concepts (DT Framework and TF K-MEANS Algorithm) to an actual problem exhibited inconstant runtime slopes and thus long runtimes while processing prospectively large models. The generalized *MDS* component and the *CA* technique scaled virtually linearly as expected. Nevertheless, on the utilized test system the resulting SEGM Algorithm was able to treat models of up to $50k$ neurons almost ideally in acceptable time. With regard to the neural hardware systems of *FACETS* [Schemmel et al., 2010] or of the *HBP* [HBP, 2012] this model size is equivalent to 30 wafer ICs, thus making the algorithm suitable until further notice[7].

The PD Algorithm as application of the TRACS-MOO approach to the *PR* issue showed a virtually linear runtime behavior, enabling the method to process much larger data sets than during the tests. The majority of the runtime was required for the transposition step (DT Framework), which is usually performed only once. The subsequent generalized *CA* executed comparably fast (about 1 second for $10k$ data items), which allows for repeated executions to retrieve optimal results. The quality of result deviated with 3% in average only marginally from the ideal results.

*Analysis:* As descriptions of algorithms for similar problems do not provide sufficient data to enable reliable statements about qualities of results and runtimes[8], artificial models were processed. To evaluate the qualities of results, the models are designed in a manner, that the (one) ideal solution for each configuration is known, serving as reference for the algorithm results. For the purpose of runtime complexity measurements the models offer the possibility to scale their sizes arbitrarily. This strategy enables independent performance measurements and provides a general concept for other algorithm analyses. To enlarge the analysis base, similar algorithms or frameworks, e.g. concepts of the software aided hardware design (placement and routing tools) or other methods of the *KDD*, can be adapted to the given tasks, to also process the artificial models, providing results for algorithm comparisons.

In the case of the segmentation problem, the resulting SEGM Algorithm can be applied to existing neural networks, which are already mapped to comparable hardware, to validate the algorithm with biological based data and existing hardware, if such data is available. It is noteworthy, that the final mapping quality is unknown until the completion of the entire mapping process (including the routing and the parameter transformation steps), hence the considered placement quality here is mere an upper bound of the final quality. Analogously the PD Algorithm can be utilized to scan existing data sets for known patterns, if available, to be comparable to other methods of its kind.

*Enhancements:* As stated the BLC Algorithm as SEGM Algorithm component exhibits the highest time complexity, thus requiring a profound investigation to detect and improve runtime relevant content. If this is not possible, the concept of the balancing approach as introduced in Section 4.2.2.1 has to be questioned and redesigned if necessary. As the quality of result is close to ideal, it might be considered to apply (meta-) heuristics to decrease the SEGM Algorithm's runtime at the cost of quality. To accelerate the balancing method, it is conceivable to swap more than one element at a time, also at the costs of lower cluster qualities. Furthermore, stochastic approaches, e.g. random cluster reassignments in a dynamic area to quickly balance over-occupied clusters, as well as more suitable initial element assignments, can improve the runtime behavior at acceptable quality costs. Another more general approach to decrease the overall runtime is to involve additional data, provided by the models to enhance the initial element – cluster assignments, e.g. the until now neglected population information. Finally the abstraction of the balancing issue may lead to other already exiting approaches to equalize item – container assignments, e.g. solutions for the knapsack problem [Kellerer et al., 2004], whose adaptations and applications can also reduce the algorithm's time complexity.

As the PD Algorithm executes already comparably fast and its results are close to the ideal, it can be

---

[7]assuming strongly connected networks, i.e. neurons with maximal synaptic fan-in
[8]as far as evaluable during literature studies

taken into consideration $i)$ to extend its fields of application, i.e. extend the algorithm to detect patterns, which do not fit to the utilized metrics (e.g. non Gaussian fields), $ii)$ to increase the robustness to data noise or $iii)$ to enable hierarchical clustering, i.e. the detection of embedded clusters. In a wider scope, it is conceivable not only to retrieve affine data groups, but also to recognize correlations or other internal data dependencies.

   *Applications:* Because of the generality of both the developed SEGM Algorithm as well as the processed models, it can be supposed that the approach is applicable to other types of neural networks and neural hardware systems, which consist of computing cores and exhibit hierarchical structures. Moreover, it is probable that this algorithm is adaptable to other *MOO* tasks, that involve the determination of item – container assignments with respect to several optimization objectives and secondary constraints, e.g. modifications of the knapsack problem, or resource utilizations in the contexts of production planning or energy distribution.

   Furthermore, also the PD Algorithm can be utilized to process other *PR* tasks, that are based on normal distributions. Because of its short execution time, it is considerable to scan transposed data models multiple times with different parameter values, e.g. different granularities, to detect the most meaningful underlying patterns. In particular for unknown large data sets, this method may prove as useful to analyze the internal structures, as only several *PR* scans with different assumptions can reveal the sought patterns within [Bishop et al., 2006].

# 5 Summary and Final Discussion

With respect to the importance of *Multi-Objective Optimization* (*MOO*) in the context of the today's information processing and the limitations of current approaches to treat large and complex tasks in practical time and with little adjustment costs, this work proposes the concept of the *Transposition and Cluster Analysis based Multi-Objective Optimization* (TRACS-MOO), based on data model transformations and subsequent *Cluster Analysis*s (*CA*s) to solve *MOO* problems. The approach abstracts the transposition of large, high-dimensional and diverse data models to low-dimensional uniform equivalents within the generalized *Dimensional Transposition Framework* (DT Framework), which is optimized regarding data similarity conservation, i.e. the relations of the data items to each other are preserved, and low runtime complexity, i.e. the doubling of the model size also causes only a doubled runtime. The *CA* step is performed in one case by the *Tree Fusion k-Means Algorithm* (TF K-MEANS Algorithm), which is designed to group large numbers of data items to large numbers of clusters with also linear runtime complexity. In the other case a generalized *Agglomerative Clustering* is utilized, which is capable of handling unknown numbers of clusters of different sizes and shapes. Applying and adapting these components to two representative *MOO* examples illustrates and proves the usability of the TRACS-MOO concept, by solving these tasks with high qualities of results and generally low runtimes with linear time complexities. The abstract DT and TF components, as well as their application extensions are tested and analyzed by utilizing artificial, scalable databases, to determine valid parameter ranges, qualities of results and runtimes, and to enable repeatable tests.

The introduction in Section 1 outlines the fields of application of the *MOO* along with common approaches to treat those tasks. A subsequent summarizing of the current challenges clarifies, that conventional methods are pushed to their limits when processing large *MOO* tasks of many variables and optimization objectives. Hence the TRACS-MOO as a novel concept is proposed to approach these tasks by utilizing methods of the *CA* and the *Knowledge Discovery in Databases* (*KDD*). The concept is separated into three components, forming the subsequent sections.

As described in Section 2, the first component abstracts the required model transformation to enable the *MOO* tasks to be processable by *CA* techniques. Current challenging optimization tasks involve data models of large numbers of variables, target functions and secondary conditions, preventing the most *CA* methods to yield sufficient results in acceptable times. To transpose the models to low-dimensional uniform equivalents, the DT Framework, a generalized *Multi-Dimensional Scaling* (*MDS*) algorithm, is conceptualized and implemented. It is based on the pair-wise comparison of data items, which provides optimal distances, expressing the optimization objectives as item similarities. In combination with the current distances within the uniform target system, relative position changes (forces) to approach the optimal positions are applied. During several iterations, the integrated forces of all data items are calculated and the items are repositioned in an alternating manner, determining a so called equilibrium, in which all current distances are maximally close to their optima, thus representing the sought transformed model. As an increase of the data item count would cause a quadratic calculation effort increment of this transposition, several tree based heuristics are applied to reduce the time complexity at minimal transposition quality costs. During the development and for the framework's analysis a set of test scenarios were designed, to determine optimal algorithm parameters and to measure qualities of results and runtimes. For this purpose artificial models of arbitrary sizes, representing corner cases of conceivable problem classes, were utilized. It was shown that the DT Framework is able to transpose all applied model types with a sufficient high quality and a virtually linear time complexity. Thus, by supporting generalized data and function interfaces the approach is suitable to transpose arbitrary data models, and thus to transform a broad range of *MOO* tasks.

The second abstract component is represented by the TF K-MEANS Algorithm, described in Section 3. Following the TRACS-MOO's concept, the transposition of the data model via the DT Framework, optimizes the target objectives implicitly and enables *CA* methods to extract the results. The distance based expression of the data item similarities suggests a centroid-based approach to find dense data item sets as pareto-optimal results. Consequently one of the most common *CA* approaches, the *k-means* algorithm, is enhanced to the TF K-MEANS Algorithm, which focuses on the processing of large data sets with an only linear time complexity, while being *exact*, i.e. it yields the same results as the original technique. This is accomplished by the application of a tree-based preliminary clustering, which allows accelerated decisions about the data items' affiliations to clusters. The algorithm analysis was performed similar to the test approach for the DT Framework: artificial models of arbitrary sizes, applying extrema of data distribution types, were clustered in different test scenarios to explore the algorithm's configuration space and performance. The measurements yielded, that the TF K-MEANS Algorithm is able to cluster models with large numbers of items and target clusters with results identical to the original *k-means* algorithm, but with comparable low runtimes and virtually constant runtime slopes, i.e. linear time complexity.

Section 4 illustrates the application and adaptation of the TRACS-MOO's base, the DT Framework and subsequent *CA*s, e.g. the TF K-MEANS Algorithm, to current *MOO* problems. First, the *segmentation* issue is considered with its task to assign large numbers of data items to groups while targeting several optimization objectives and comply with secondary conditions. As representative scenario the *placement* task within the context of neural hardware[1] configurations was chosen. The task is to transfer a given neural network, including its topology and behavior defining parameters, to grids of neural hardware cores with limited capacities, connection restrictions and parameter range limitations. The TRACS-MOO approach is applied by utilizing the former introduced *MDS* and *CA* methods and extending them by a balancing algorithm to treat the secondary conditions, which could be accomplished with comparable little adaptation effort. Analyses, based on models with known theoretical ideal results, exposed high qualities of results, i.e. placements close to the ideal, and mixed time complexity types: the abstracted base components exhibited a linear, the balancing technique as extension a non-linear runtime complexity, thus restricting the processable data model size.

Second, the *Peak Detection* Algorithm (PD Algorithm) handles the general *Pattern Recognition* (*PR*) issue, which describes the task to find patterns, i.e. regularities, repetitions or similarities in unknown, potentially large and complex data sets. As example the detection of affine data item groups was chosen, representing a broad field of application within the scope of the *PR*. The task is to discover the count and characteristics of the underlying patterns, along with the assignment of each data item to one of these patterns to provide an abstraction of the data set, allowing for general statements regarding internal

---

[1] hardware, which consists of artificial neurons and synapses to simulate the behavior of neural networks

structures. For this purpose the DT Framework is utilized and combined with a tree-based agglomerative *CA* method, firstly to unify the data models, and secondly to detect regions of density peaks as sought patterns. During tests, which processed Gaussian field based data models, i.e. general artificial data sets of arbitrary sizes and with known internal patterns to compare the PD Algorithm's results against, it was shown, that this *PR* technique yields results very close to the ideal result at linear time complexity. In this way the application of the TRACS-MOO concept to a second task type with also little additional adaptation effort was illustrated.

With respect to the formulated theses in Section 1.4 the following conclusions can be drawn:

*Thesis 1:* *MOO* problems, including large, high dimensional and diverse data sets, can be transformed to low-dimensional, uniform equivalents by the *MDS* technique within the DT Framework. This allows common and enhanced *CA* methods, in this case the TF K-MEANS Algorithm and the PD Algorithm, to process these data efficiently. Efficiency in this context means a comparable high quality of result, low runtime complexity and little effort to adapt exiting, generalized components to the actual problem. All this is proven by test based algorithm investigations.

*Thesis 2:* Clustering results of transposed models represent pareto-optimal solutions for the original *MOO* problem, if suitable *CA* methods are applied and the clusters are appropriately interpreted. This concept is proven by the treatment of the segmentation and *PR* issue, both representing wide fields of application and yielding high quality of results with regard to known ideal solutions. The adjustment of the *MDS* and *CA* parameters and priorities to create different comparable solutions, traces the actual pareto front, marking the TRACS-MOO as a mixed form of the *scalarization* and the *decision* approaches.

*Thesis 3:* Similar to the conclusion of thesis 2, the approach exposes linear time complexities, if appropriate adaptations are applied. This behavior is shown by runtime measurements while processing data models of gradually increasing sizes. As far as evaluable this is superior to most of the known methods and enables the approach for much larger tasks, as the computed problem amount per time unit keeps constant.

*Thesis 4:* The processing of artificial models, covering the corner cases of general conceivable data model types, leads to virtually ideal results with regard to the known theoretical best solution, i.e. a further quality increasing is not possible, not even by conventional methods. Due to this and the generality of the algorithm tests, it can be assumed, but not proven, that the application of the TRACS-MOO concept to real data, or data already processed by other algorithms, will result in comparable or better solutions.

Reflecting the insights regarding the theses $1 - 4$, it can be stated, that the TRACS-MOO is a suitable approach to treat *MOO* problems with respect to high overall qualities of results as well as low runtimes and linear runtime complexities. It can be utilized to process large data sets and to explore pareto fronts.

In conclusion it can be shown by abstraction, iterative development and extensive testing, that the novel approach of TRACS-MOO is a viable and versatile technique for *MOO* in large and complex data models. By applying and enhancing existing methods of the *KDD*, it offers the potential to improve the present *MOO* solutions and performances or to enable their determination in the first place. This can be achieved by the following:

- *Adaptation to Pending Tasks of the MOO*: As the utilization of artificial data is well suited to evaluate the approach's performance, it provides only limited statements about its behavior if applied to problems based on real data. On the one hand this encapsulates already solved tasks, arising the question whether the TRACS-MOO achieves comparable or better results, which would substantiate its capability and versatility. On the other hand the (more) successful treatment of so far unsolved, or insufficiently solved *MOO* tasks would also be a worthwhile objective to prove the potential of the concept. For example the abstraction of the segmentation may lead to a more general approach for the *knapsack* problem, making the TRACS-MOO more comparable to, or combinable with algorithms for this general issue. Furthermore, the provision of the available interfaces would verify the data and control access design.

- *Analysis of Further Problem Classes*: The algorithm investigation considers problems, which are related to common *CA* tasks, and are as such are predestined to utilize cluster based techniques. In consequence it would be desirable to extend the approach to less suitable *MOO* problems. For example the adaptation to the class of function solvers would open a large field of potential application, as many *MOO* tasks can be described as a vector of (numerical) variables, a set of target functions to optimize and additional secondary conditions. The quantization of the entire decision space would result in a large transposable model, whose subsequent clustering may reveal the most dense variable sets as potential pareto front. In this way the exploration of more general and/or abstracted problem classes may improve the TRACS-MOO further and validate the concept, preparing it for the application to more problems to solve.

- *Realization of the Evaluated Enhancements*: The summary sections of the three previous chapters illustrate conceivable enhancements and further investigations of the DT Framework, the TF K-MEANS Algorithm and their application to existing *MOO* problems. This encompasses the utilization of known (meta-) heuristics for the purpose of quality or runtime improvements, parallelization approaches and internal structure optimizations as well as algorithm profilings to identify performance bottlenecks during redesigns. The realization of an appropriate subset of these proposals may lead to an increase of the overall performance of the TRACS-MOO implementation.

In this way a prospective strategy is proposed to further analyze, develop and apply the presented *MOO* concept.

# 6 Appendix

## 6.1 Dimensional Transposition Framework

### 6.1.1 Profiling Results

| Method Name | Time Costs | Total Tick Count | Ticks / Call | Calling Method | Total Callings | Ratio: Callings |
|---|---|---|---|---|---|---|
| CompareProperties | 49,10% | 2.683.616.643 | 92,63 | | | |
| | | | | Leaf.CompareProperties | 28.971.000 | 100,00% |
| CalculateSpecificWeight | 14,67% | 802.017.010 | 27,68 | | | |
| | | | | Leaf.CalculateSpecificWeight | 28.971.000 | 100,00% |
| Property.CompareTo | 11,52% | 629.655.988 | 1,09 | | | |
| | | | | CompareProperties | 579.420.000 | 100,00% |
| CalculateTotalForce | 9,15% | 500.228.393 | 17,27 | | | |
| | | | | CalculateForce | 28.971.000 | 100,00% |
| CalculateSquareDistance | 4,36% | 238.352.880 | 8,20 | | | |
| | | | | CalculateDistance | 29.062.000 | 100,00% |
| GetPropertyEnumerator | 3,43% | 187.490.666 | 1,62 | | | |
| | | | | CompareProperties | 57.942.000 | 50,00% |
| | | | | CalculateSpecificWeight | 5.794.2000 | 50,00% |
| CalculateForce | 2,07% | 113.235.156 | 1.952,33 | | | |
| | | | | CalculateForceUpward | 29.000 | 50,00% |
| | | | | CalculateForce | 29.000 | 50,00% |
| Leaf.CompareProperties | 1,47% | 80.338.982 | 2,77 | | | |
| | | | | CalculateForce | 28.971.000 | 100,00% |
| Leaf.CalcSpecificWeight | 1,40% | 76.545.019 | 2,64 | | | |
| | | | | CalculateSingleForce | 28.971.000 | 100,00% |
| CalcDistance | 1,29% | 70.428.592 | 2,42 | | | |
| | | | | UpdateInferiorSuperDist | 122.000 | 0,42% |
| | | | | CalculateSingleForce | 2.8971.000 | 99,48% |
| | | | | RepairTree | 30.000 | 0,10% |
| .. | .. | .. | .. | .. | .. | .. |

**Table 6.1:** EQ *Algorithm Example Profiling Log*;
  *Data Model*: 1.000 elements with 20 features, distributed in 10 gaussian fields of different sizes;
  *Algorithm*: DT Algorithm;
  All entries below 1% Time Costs were truncated; the left side represents the Time Costs of all involved methods, the right side shows the calling distribution of the appropriate method.

| Method Name | Time Costs | Total Tick Count | Ticks / Call | Calling Method | Total Callings | Ratio: Call-ings |
|---|---|---|---|---|---|---|
| CompareProperties | 36,57% | 66.617.105 | 97,77 | | | |
| | | | | Branch.CompareProperties | 38.839 | 5,70% |
| | | | | Leaf.CompareProperties | 642.504 | 94,30% |
| Property.CompareTo | 10,98% | 20.007.863 | 1,47 | | | |
| | | | | CompareProperties | 13.626.860 | 100.0% |
| CalculateSpecificWeight | 10,06% | 18.372.437 | 29,19 | | | |
| | | | | CalculateSpecificWeight | 629.455 | 100.0% |
| CalculateSingleForce | 8,39% | 629.455 | 24,27 | | | |
| | | | | CalculateForce | 629.455 | 100.0% |
| CalculateSingleForce | 4,83% | 8.801.504 | 6,90 | | | |
| | | | | CalculateDistance | 1.275.743 | 100.0% |
| RepairTree | 3,51% | 6.396.446 | 106.607,43 | | | |
| | | | | SingleTransposeStep | 58 | 96,67% |
| | | | | Run | 2 | 3,33 % |
| UpdateProperties | 3,39 % | 6.183.111 | 1.018,63 | | | |
| | | | | UpdateProperties | 6.070 | 100.0 % |
| CalculateForce | 2,87% | 5.277.387 | 31,64 | | | |
| | | | | CalculateForceUpward | 29.493 | 17,68% |
| | | | | CalculateForce | 137.294 | 82,32% |
| GetPropertyEnumerator | 2,15% | 3.922.957 | 2,15 | | | |
| | | | | CompareProperties | 936.422 | 51,26% |
| | | | | CalculateSpecificWeight | 890.540 | 48,74% |
| ReAssign | 1,74% | 3.176.228 | 23,07 | | | |
| | | | | RepairTree | 30.510 | 22,16% |
| | | | | ReAssign | 107.151 | 77,84% |
| CalculateDistance | 1,72% | 3.130.235 | 2,45 | | | |
| | | | | CalcDistance | 907.373 | 71,13% |
| | | | | CalculateSingleForce | 368.370 | 28,87% |
| CalcDistance | 1,46% | 2.668.080 | 2,62 | | | |
| | | | | UpdateClusterDistTIE | 75.635 | 2,25% |
| | | | | UpdateA2CAssignTIE | 368.370 | 7,43% |
| | | | | CheckAtomTrInEq | 40.557 | 3,98% |
| | | | | UpdateInferiorSuperDist | 368.370 | 28,87% |
| | | | | CalculateMaxLeafDistRe | 138.034 | 13,56% |
| | | | | PlaceAroundCenter | 2.528 | 0,25% |
| | | | | CalculateSingleForce | 261.085 | 25,64% |
| | | | | RepairTree | 30.510 | 3,00% |
| | | | | ReAssign | 363.370 | 35,69% |
| CompareAllProperties | 1,29% | 2.355.457 | 3,67 | | | |
| | | | | InsertionStep | 1.000 | 0,16% |
| | | | | FindMostAffinInferior | 12.049 | 1,88% |
| | | | | CalculateSingleForce | 629.455 | 97,97% |
| SingleTransposeStep | 1,09% | 1.994.204 | 34382,83 | | | |
| | | | | Run | 58 | 100.0% |
| .. | .. | .. | .. | .. | .. | .. |

**Table 6.2:** DT *Algorithm Example Profiling Log*;
*Data Model*: 1.000 elements with 20 features, distributed in 10 gaussian fields of different sizes;
*Algorithm*: DT Algorithm;
All entries below 1% Time Costs were truncated; the left side represents the Time Costs of all involved methods, the right side shows the calling distribution of the appropriate method.

## 6.1.2 Measurement Results

### 6.1.2.1 Transposition Quality Dependency of the Simulated Annealing Factor

Transposition Quality of EQ-Transposed DVD Models      Transposition Quality of DT-Transposed DVD Models

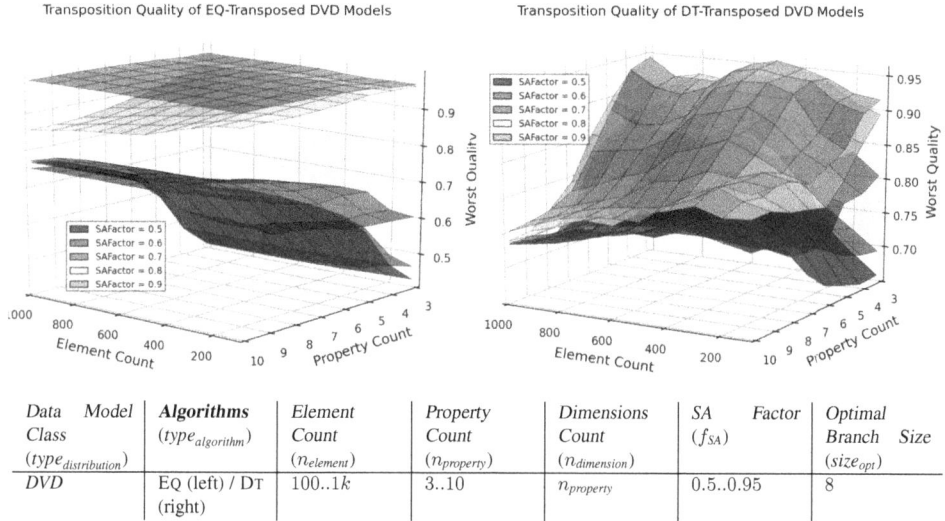

| Data Model Class $(type_{distribution})$ | **Algorithms** $(type_{algorithm})$ | Element Count $(n_{element})$ | Property Count $(n_{property})$ | Dimensions Count $(n_{dimension})$ | SA Factor $(f_{SA})$ | Optimal Branch Size $(size_{opt})$ |
|---|---|---|---|---|---|---|
| DVD | EQ (left) / DT (right) | $100..1k$ | $3..10$ | $n_{property}$ | $0.5..0.95$ | 8 |

**Figure 6.1:** *Worst Quality of Result ($quality_{worst}$) of the* EQ *and* DT *Algorithms applied to DVD Data Models, dependent on Element and Property Count, as well as SA Factor.*

Transposition Quality of EQ-Transposed CVD Models      Transposition Quality of DT-Transposed CVD Models

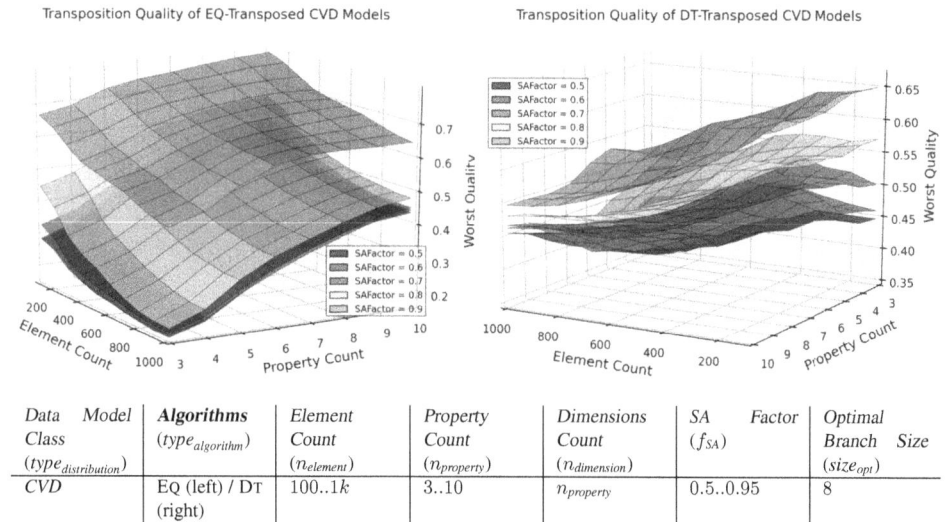

| Data Model Class $(type_{distribution})$ | **Algorithms** $(type_{algorithm})$ | Element Count $(n_{element})$ | Property Count $(n_{property})$ | Dimensions Count $(n_{dimension})$ | SA Factor $(f_{SA})$ | Optimal Branch Size $(size_{opt})$ |
|---|---|---|---|---|---|---|
| CVD | EQ (left) / DT (right) | $100..1k$ | $3..10$ | $n_{property}$ | $0.5..0.95$ | 8 |

**Figure 6.2:** *Worst Quality of Result ($quality_{worst}$) of the* EQ *and* DT *Algorithms applied to CVD Data Models, dependent on Element and Property Count, as well as SA Factor.*

## 6.1.2.2 Transposition Quality Dependency of the Branch Size

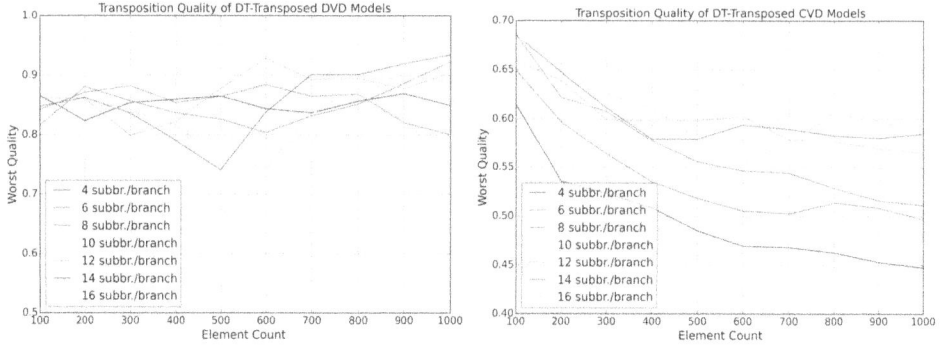

| Data Model Class ($type_{distribution}$) | Algorithms ($type_{algorithm}$) | Element Count ($n_{element}$) | Property Count ($n_{property}$) | Dimensions Count ($n_{dimension}$) | SA Factor ($f_{SA}$) | Optimal Branch Size ($size_{opt}$) |
|---|---|---|---|---|---|---|
| DVD (left) / CVD (right) | DT | 100..1k | 3..10 (averaged) | $n_{property}$ | $f_{SA}^{best} = 0.9$ | 4..16 |

**Figure 6.3:** *Worst Quality of Result ($quality_{worst}$) of the* DT *Algorithm applied to DVD and CVD Data Models, dependent on Element Count and Branch Size.*

## 6.1.2.3 Transposition Quality Dependency of the Dimension Count

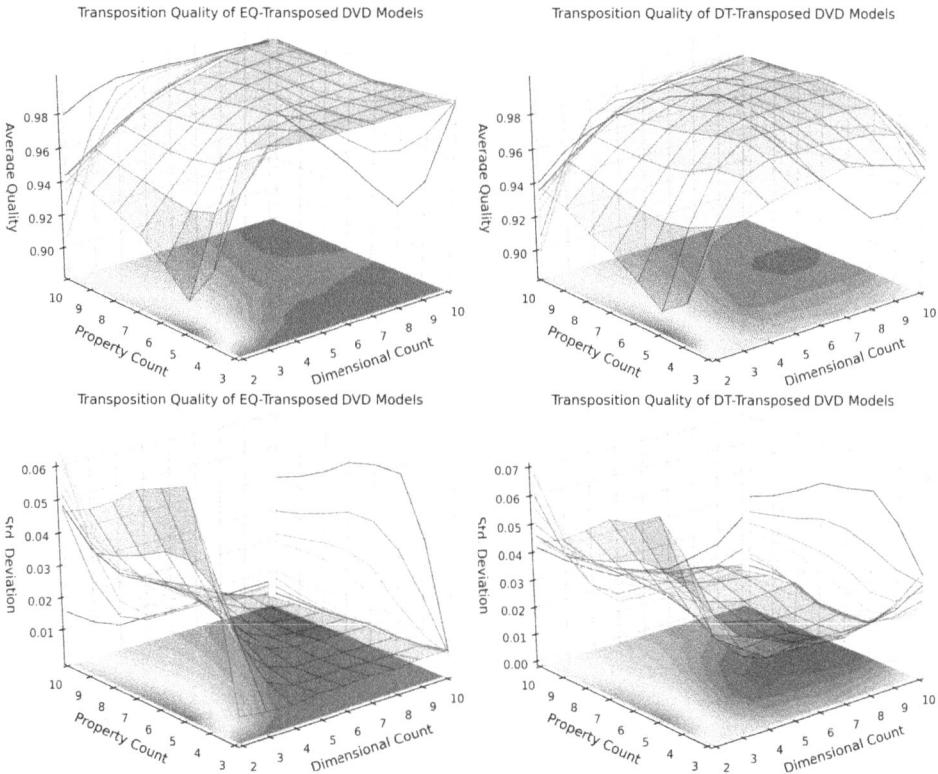

Transposition Quality of EQ-Transposed DVD Models

Transposition Quality of DT-Transposed DVD Models

Transposition Quality of EQ-Transposed DVD Models

Transposition Quality of DT-Transposed DVD Models

| Data Model Class ($type_{distribution}$) | Algorithms ($type_{algorithm}$) | Element Count ($n_{element}$) | Property Count ($n_{property}$) | Dimensions Count ($n_{dimension}$) | SA Factor ($f_{SA}$) | Optimal Branch Size ($size_{opt}$) |
|---|---|---|---|---|---|---|
| DVD | EQ (left) / DT (right) | $1k$ | $3..10$ | $2..10$ | $f_{SA}^{best} = 0.9$ | $size_{opt}^{best} = 16$ |

**Figure 6.4:** *Average Quality of Result ($quality_{average}$, top) and its Standard Deviation ($quality_{SD}$, bottom) of the* EQ *and* DT *Algorithms applied to DVD Data Models, dependent on Dimension and Property Count.*

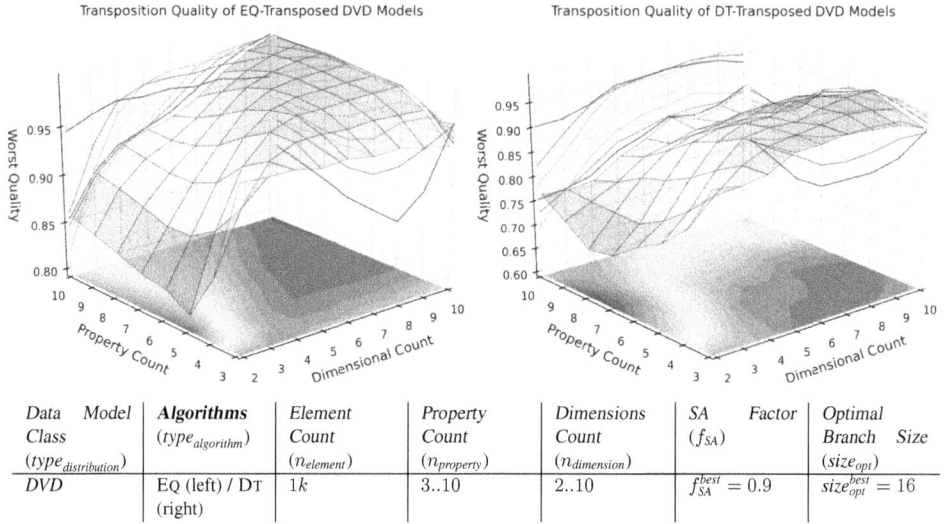

| Data Model Class $(type_{distribution})$ | Algorithms $(type_{algorithm})$ | Element Count $(n_{element})$ | Property Count $(n_{property})$ | Dimensions Count $(n_{dimension})$ | SA Factor $(f_{SA})$ | Optimal Branch Size $(size_{opt})$ |
|---|---|---|---|---|---|---|
| DVD | EQ (left) / DT (right) | $1k$ | $3..10$ | $2..10$ | $f_{SA}^{best} = 0.9$ | $size_{opt}^{best} = 16$ |

**Figure 6.5:** *Worst Quality of Result ($quality_{worst}$) of the* EQ *and* DT *Algorithms applied to DVD Data Models, dependent on Dimension and Property Count.*

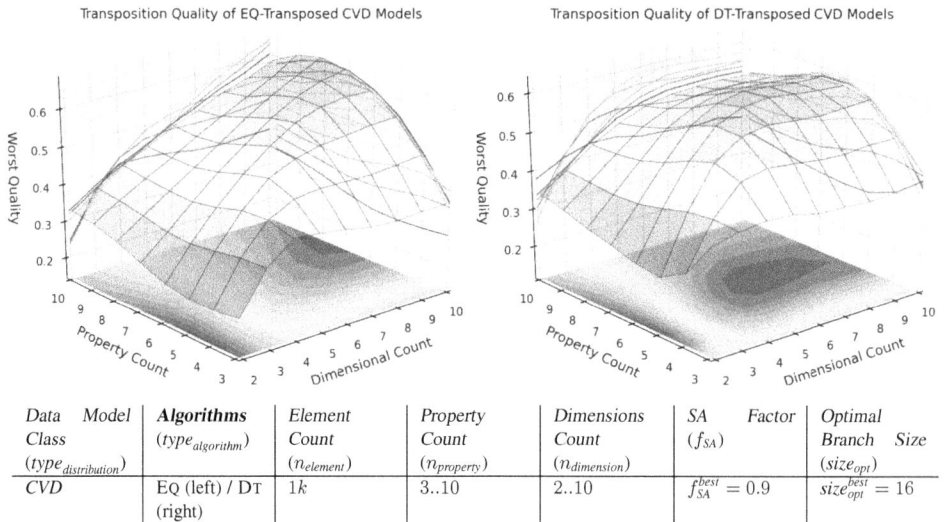

| Data Model Class $(type_{distribution})$ | Algorithms $(type_{algorithm})$ | Element Count $(n_{element})$ | Property Count $(n_{property})$ | Dimensions Count $(n_{dimension})$ | SA Factor $(f_{SA})$ | Optimal Branch Size $(size_{opt})$ |
|---|---|---|---|---|---|---|
| CVD | EQ (left) / DT (right) | $1k$ | $3..10$ | $2..10$ | $f_{SA}^{best} = 0.9$ | $size_{opt}^{best} = 16$ |

**Figure 6.6:** *Worst Quality of Result ($quality_{worst}$) of the* EQ *and* DT *Algorithms applied to CVD Data Models, dependent on Dimension and Property Count.*

## 6.1.2.4 Runtime Dependency of the Data Model Size

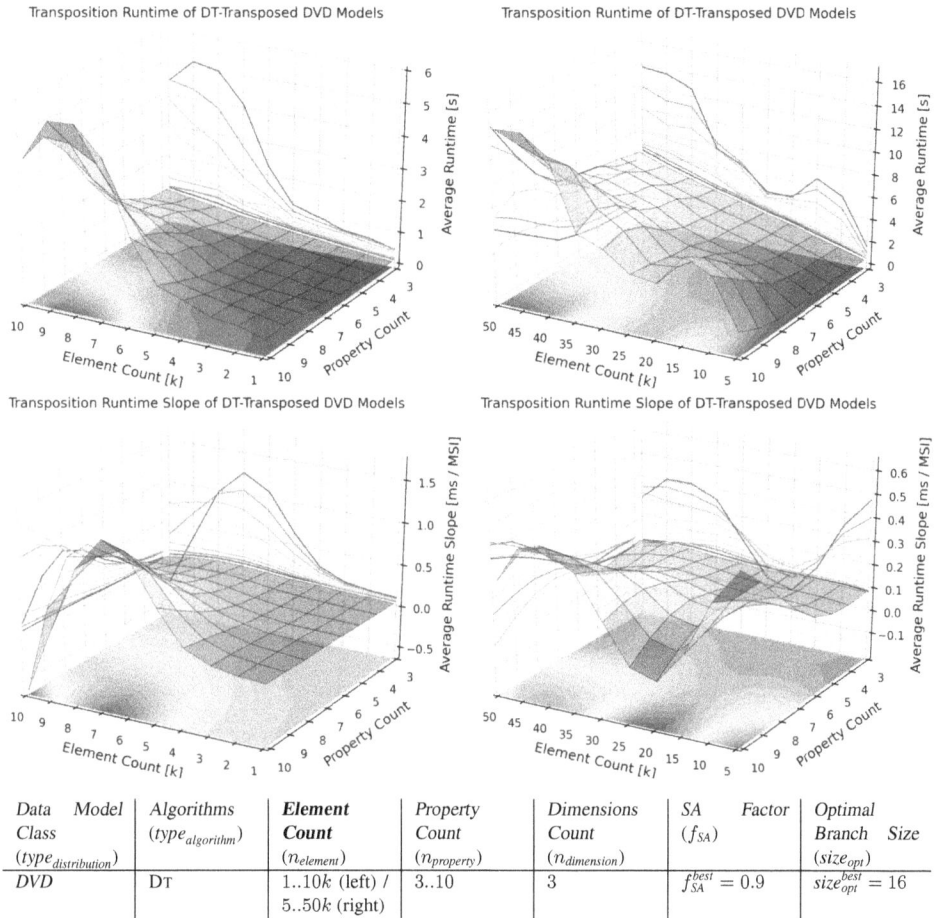

| Data Model Class ($type_{distribution}$) | Algorithms ($type_{algorithm}$) | **Element Count** ($n_{element}$) | Property Count ($n_{property}$) | Dimensions Count ($n_{dimension}$) | SA Factor ($f_{SA}$) | Optimal Branch Size ($size_{opt}$) |
|---|---|---|---|---|---|---|
| DVD | DT | 1..10k (left) / 5..50k (right) | 3..10 | 3 | $f_{SA}^{best} = 0.9$ | $size_{opt}^{best} = 16$ |

**Figure 6.7:** *Average Runtime ($t_{run}$ top) and its Slope ($slope_{runtime}$, bottom) of the* DT *Algorithm applied to DVD Data Models, dependent on Element and Property Count.*

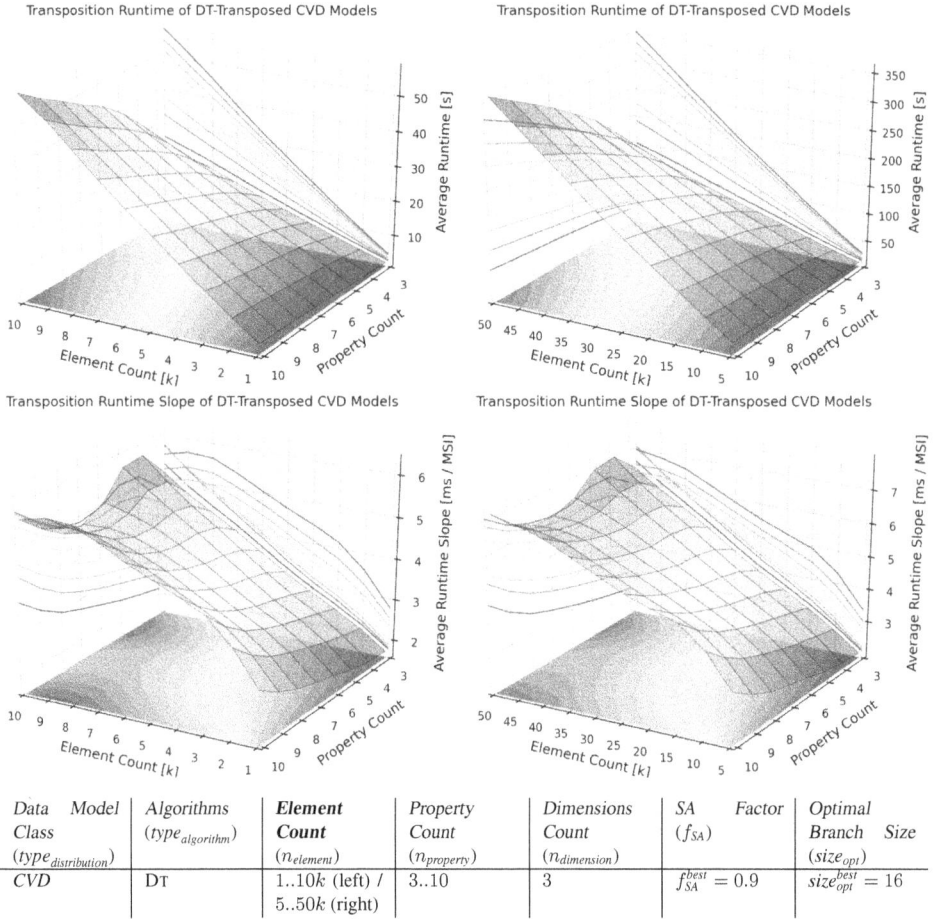

| Data Model Class ($type_{distribution}$) | Algorithms ($type_{algorithm}$) | **Element Count** ($n_{element}$) | Property Count ($n_{property}$) | Dimensions Count ($n_{dimension}$) | SA Factor ($f_{SA}$) | Optimal Branch Size ($size_{opt}$) |
|---|---|---|---|---|---|---|
| CVD | DT | 1..10$k$ (left) / 5..50$k$ (right) | 3..10 | 3 | $f_{SA}^{best} = 0.9$ | $size_{opt}^{best} = 16$ |

**Figure 6.8:** *Average Runtime ($t_{run}$ top) and its Slope ($slope_{runtime}$, bottom) of the* DT *Algorithm applied to CVD Data Models, dependent on Element and Property Count.*

## 6.2 *Tree Fusion k-Means* Algorithm

### 6.2.1 Profiling Results

| Method Name | Time Costs | Total Tick Count | Calling Methods | Total Callings | Ratio: Callings |
|---|---|---|---|---|---|
| CheckAtom | 60,76% | 117.274.278 | | | |
| | | | UpdateA2CAssignments | 57.000 | 100,0% |
| UpdateMinA2CDistances | 16,31% | 31.470.388 | | | |
| | | | PerformTrInEq | 57 | 100,0% |
| CalculateSquareDistance | 10,41% | 20.084.400 | | | |
| | | | CalculateDistance | 1.951.365 | 100,0% |
| UpdateClusterDistances | 5,40% | 10.417.071 | | | |
| | | | PerformTrInEq | 57 | 100,0% |
| CalculateDistance | 2,92% | 5.626.896 | | | |
| | | | UpdateClusterDistances | 1.524.066 | 78,10% |
| | | | UpdateA2CAssignments | 216.703 | 11,11% |
| | | | CheckAtom | 210.596 | 10,79% |
| UpdateA2CAssignments | 2,22% | 4.277.702 | | | |
| | | | PerformTrInEq | 57 | 100,0% |
| InitializeMinA2CDistances | 1,58% | 3.051.257 | | | |
| | | | PerformTrInEq | 1 | 100,0% |
| .. | .. | .. | .. | .. | .. |

**Table 6.3:** TRIN K-MEANS *Algorithm Example Profiling Log*;
Data Model: *ED*, 10.000 elements onto 500 clusters;
*Algorithm*: TRIN K-MEANS Algorithm by [Elkan, 2003];
All entries below 1% Time Costs were truncated; the left side represents the Time Costs of all involved methods, the right side shows the calling distribution of the appropriate method.

| Method Name | Time Costs | Total Tick Count | Calling Methods | Total Callings | Ratio: Callings |
|---|---|---|---|---|---|
| CalculateSquareDistance | 13,52% | 14.067.450 | | | |
| | | | CalculateDistance | 1.574.000 | 100,0% |
| CheckAtom | 11,72% | 12.190.477 | | | |
| | | | UpdateA2CAssignments | 981.608 | 100,0% |
| Filter | 10,94% | 11.380.055 | | | |
| | | | PerformTrInEq | 57 | 0,02% |
| | | | StepDownOrCollect | 307.670 | 99,98% |
| CalculateA2CMetrics | 10,84% | 11.275.571 | | | |
| | | | UpdateMinMaxA2CDist. | 791.358 | 100,0% |
| StepDownOrCollect | 6,98% | 7.255.354 | | | |
| | | | Filter | 295.407 | 100,0% |
| UpdateA2CAssignments | 6,76% | 7.029.984 | | | |
| | | | PerformTrInEq | 3.143 | 100,0% |
| CalculateMinMaxA2CDist. | 5,00% | 5.208.630 | | | |
| | | | CalculateA2CMetrics | 429.383 | 100,0% |
| CalculateDistance | 4,98% | 5.184.488 | | | |
| | | | UpdateClusterDistances | 92.152 | 5,85% |
| | | | UpdateA2CAssignments | 549.574 | 34,92% |
| | | | CheckAtom | 219.273 | 13,93% |
| | | | UpdateInferiorSuperiorDist. | 31.724 | 2,02% |
| | | | CalculateMaxLeafDistance | 35.584 | 2,26% |
| | | | UpdateMetrics | 216.310 | 13,74% |
| | | | CalculateCB2EBDistance | 429.383 | 27,28% |
| CalculateCB2EBDistance | 4,68% | 4.865.566 | | | |
| | | | CalculateA2CMetrics | 429.383 | 100,0% |
| UpdateMinMaxA2CDist. | 4,24% | 4.410.300 | | | |
| | | | Filter | 323.912 | 100,0% |
| ValidateClusters | 4,12% | 4.284.546 | | | |
| | | | Filter | 323.912 | 100,0% |
| UpdateMetrics | 3,34% | 3.476.624 | | | |
| | | | PerformTreeUpdate | 57 | 0,04% |
| | | | UpdateMetrics | 144.723 | 99,96% |
| IsUp2Date | 3,11% | 3.231.260 | | | |
| | | | UpdateMinMaxA2CDist. | 876.247 | 100,0% |
| UpdateBestMaxA2CDist. | 3,05% | 3.169.186 | | | |
| | | | Filter | 323.912 | 100,0% |
| UpdateMinA2CDistances | 1,12% | 1.166.981 | | | |
| | | | PerformTrInEq | 3.143 | 100,0% |
| UpdateChangedClusters | 1,08% | 1.127.739 | | | |
| | | | PerformTrInEq | 3.143 | 98,22% |
| | | | PerformTreeUpdate | 57 | 1,78% |
| .. | .. | .. | .. | .. | .. |

**Table 6.4:** TF K-MEANS *Algorithm Example Profiling Log*;
*Data Model: ED*, 10.000 elements onto 500 clusters;
*Algorithm*: TF K-MEANS Algorithm;
All entries below 1% Time Costs were truncated; the left side represents the Time Costs of all involved methods, the right side shows the calling distribution of the appropriate method.

## 6.2.2 Measurement Results

### 6.2.2.1 Runtime Dependency of the Data Model Type and Size

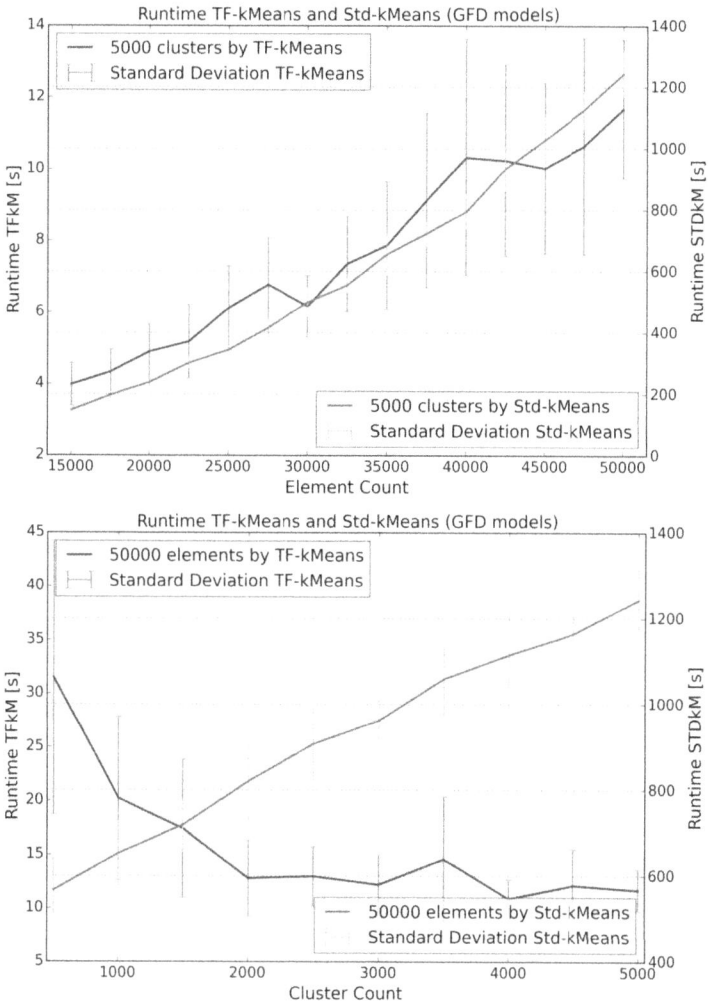

| Algorithms (*type*$_{algorithm}$) | Data Model Class (*type*$_{distribution}$) | Element Count (*n*$_{element}$) | Cluster Count (*n*$_{cluster}$) | Dimensions Count (*n*$_{dimension}$) |
|---|---|---|---|---|
| STD (blue) / TF (green) | *GFD* | 15*k*..50*k*   (top), 50*k* (bottom) | 5*k* (top), 500..5000 (bottom) | 2 |

**Figure 6.9:** *Runtimes ($t_{total}$) of the* TF *and the* STD K-MEANS *Algorithms applied to GFD Data Models, dependent on Element and Cluster Count.*

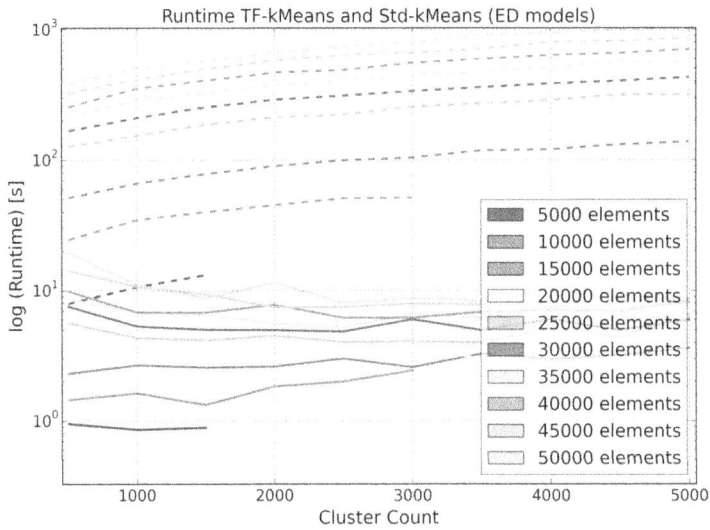

**Figure 6.10:** *Runtimes ($t_{total}$) of the* TF *and the* STD K-MEANS *Algorithms applied to ED Data Models, dependent on Cluster Count*

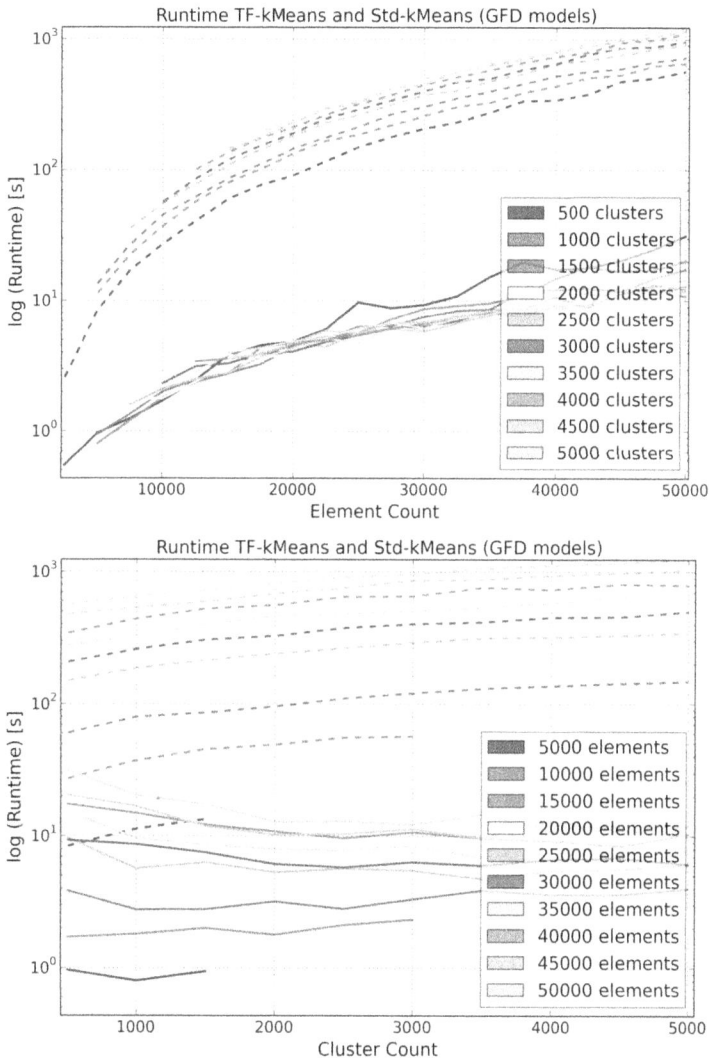

**Figure 6.11:** *Runtimes ($t_{total}$) of the* TF *and the* STD K-MEANS *Algorithms applied to GFD Data Models dependent, on Element (top) and Cluster Count (bottom)*

## 6.2.2.2 Speedup Dependency of the Data Model Type and Size

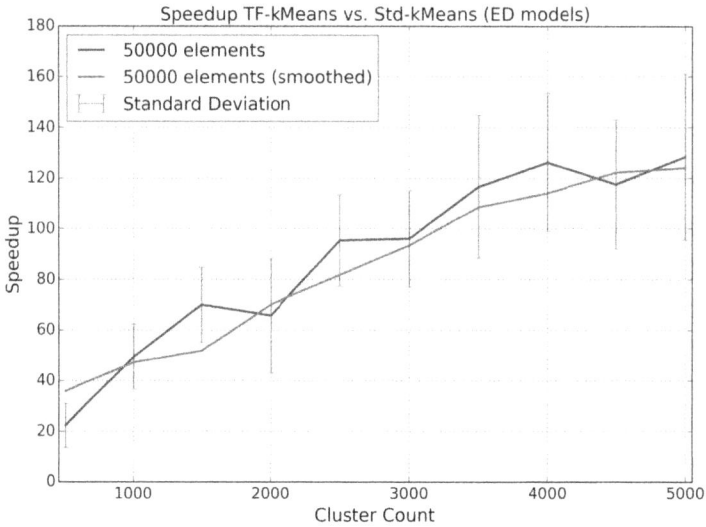

| Algorithms | Data Model Class | Element Count | Cluster Count | Dimensions Count |
|---|---|---|---|---|
| $(type_{algorithm})$ | $(type_{distribution})$ | $(n_{element})$ | $(n_{cluster})$ | $(n_{dimension})$ |
| STD / TF | ED | $50k$ | 500..5000 | 2 |

**Figure 6.12:** *Speedup ($q_{speedup}$) of the* TF *vs. the* STD K-MEANS *Algorithm applied to ED Data Models, dependent on Cluster Count*

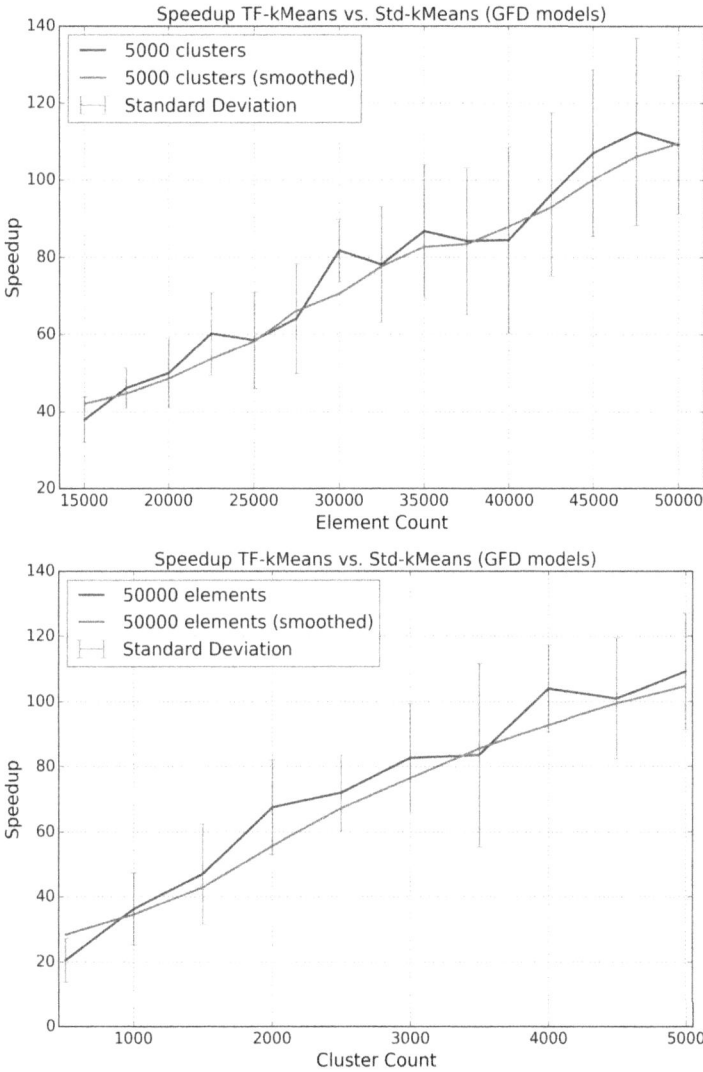

**Figure 6.13:** *Speedup ($q_{speedup}$) of the* TF *vs. the* STD K-MEANS *Algorithm applied to GFD Data Models, dependent on Element (top) and Cluster Count (bottom)*

| Algorithms | Data Model Class | Element Count | Cluster Count | Dimensions Count |
|---|---|---|---|---|
| ($type_{algorithm}$) | ($type_{distribution}$) | ($n_{element}$) | ($n_{cluster}$) | ($n_{dimension}$) |
| STD / TF | GFD | $5k..50k$ | $500..5000$ | 2 |

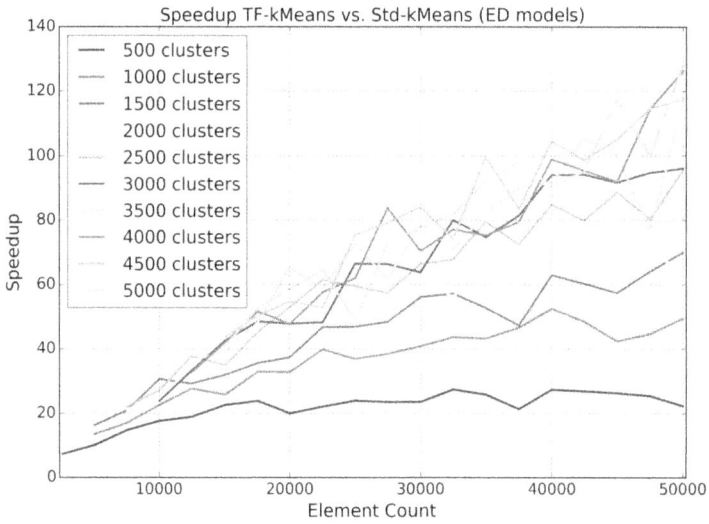

Figure showing "Speedup TF-kMeans vs. Std-kMeans (ED models)" with y-axis "Speedup" (0 to 140) and x-axis "Element Count" (10000 to 50000). Legend: 500 clusters, 1000 clusters, 1500 clusters, 2000 clusters, 2500 clusters, 3000 clusters, 3500 clusters, 4000 clusters, 4500 clusters, 5000 clusters.

| Algorithms $(type_{algorithm})$ | Data Model Class $(type_{distribution})$ | Element Count $(n_{element})$ | Cluster Count $(n_{cluster})$ | Dimensions Count $(n_{dimension})$ |
|---|---|---|---|---|
| STD / TF | ED | $5k..50k$ | $500..5000$ | 2 |

**Figure 6.14:** *Speedup ($q_{speedup}$) of the* TF *vs. the* STD K-MEANS *Algorithm applied to ED Data Models, dependent on Element and Cluster Count*

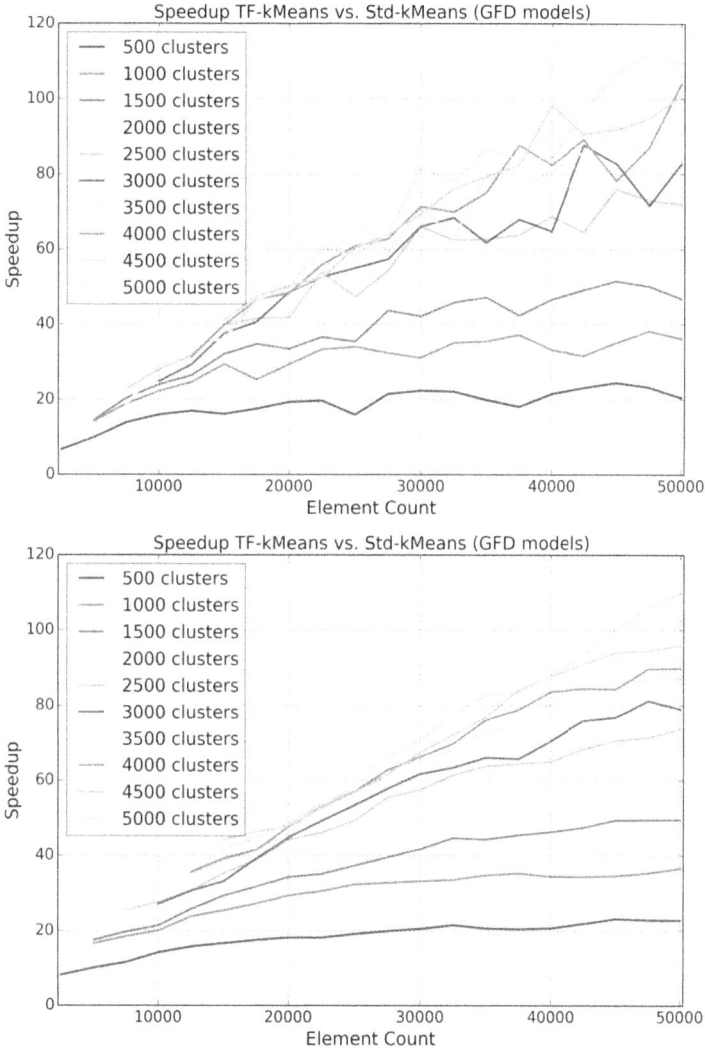

| Algorithms | Data Model Class | Element Count | Cluster Count | Dimensions Count |
| $(type_{algorithm})$ | $(type_{distribution})$ | $(n_{element})$ | $(n_{cluster})$ | $(n_{dimension})$ |
|---|---|---|---|---|
| STD / TF | GFD | $5k..50k$ | $500..5000$ | 2 |

**Figure 6.15:** *Speedup ($q_{speedup}$, averaged (top) and smoothed (bottom)) of the* TF *vs. the* STD K-MEANS *Algorithm applied to GFD Data Models, dependent on Element Count*

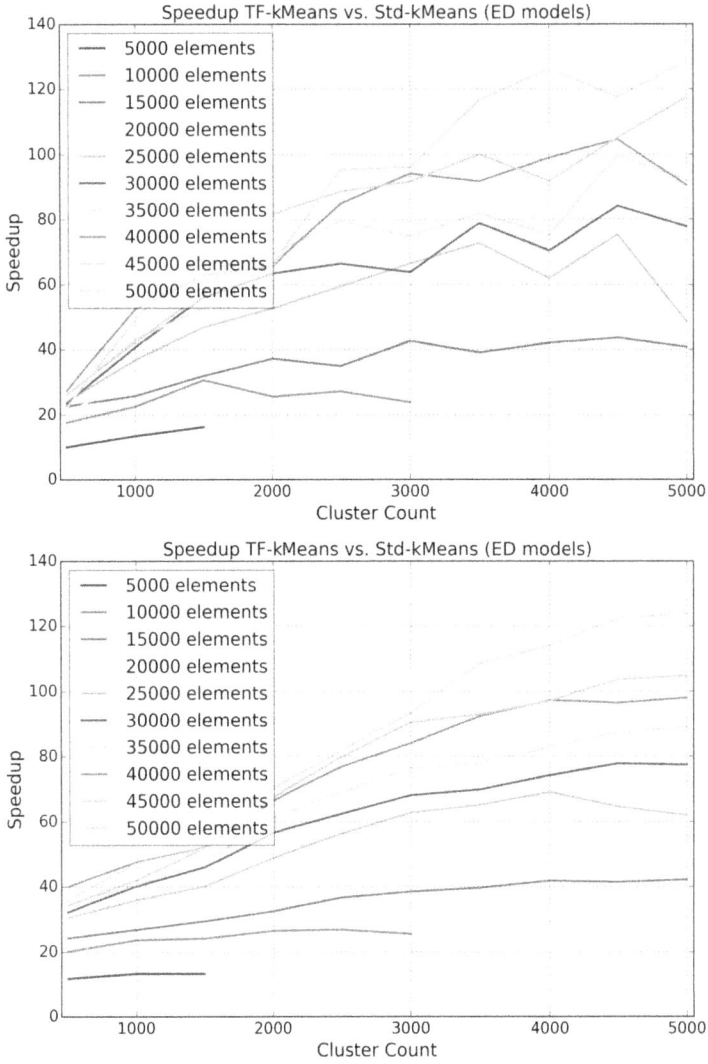

Figure 6.16: Speedup ($q_{speedup}$, averaged (top) and smoothed (bottom)) of the TF vs. the STD K-MEANS Algorithm applied to ED Data Models, dependent on Cluster Count

| Algorithms ($type_{algorithm}$) | Data Model Class ($type_{distribution}$) | Element Count ($n_{element}$) | Cluster Count ($n_{cluster}$) | Dimensions Count ($n_{dimension}$) |
|---|---|---|---|---|
| STD / TF | ED | 5k..50k | 500..5000 | 2 |

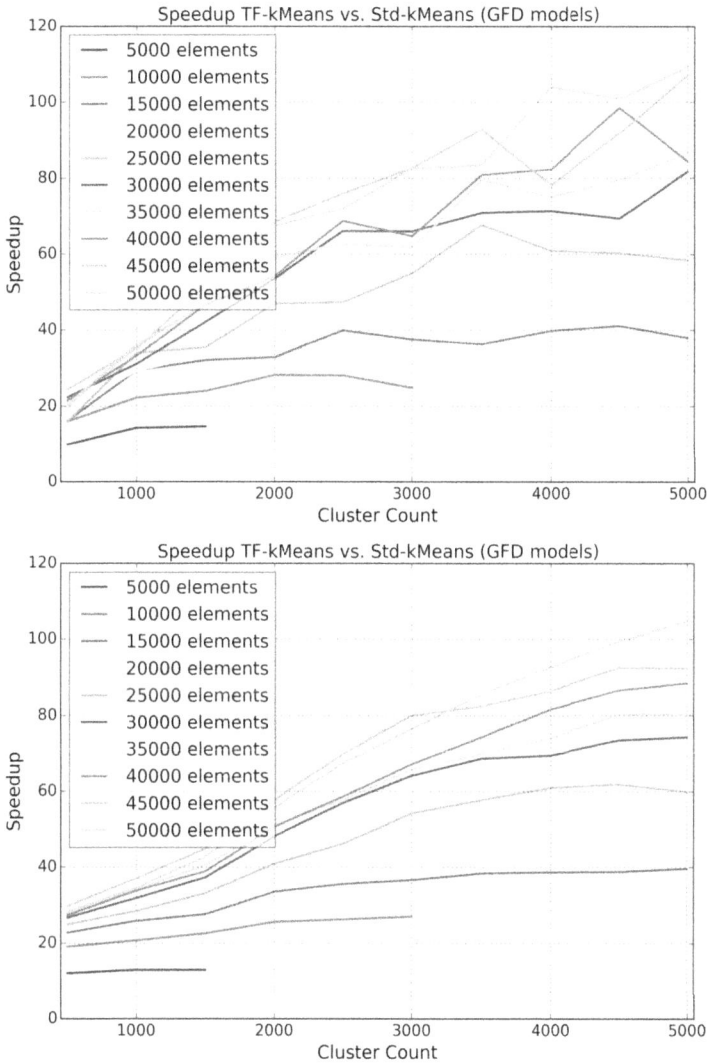

| Algorithms | Data Model Class | Element Count | Cluster Count | Dimensions Count |
|---|---|---|---|---|
| $(type_{algorithm})$ | $(type_{distribution})$ | $(n_{element})$ | $(n_{cluster})$ | $(n_{dimension})$ |
| STD / TF | GFD | $5k..50k$ | $500..5000$ | 2 |

**Figure 6.17:** *Speedup ($q_{speedup}$, averaged (top) and smoothed (bottom)) of the TF vs. the STD K-MEANS Algorithm applied to GFD Data Models, dependent on Cluster Count*

### 6.2.2.3 Speedup and Runtime Slope Dependency of Large Scale Data Models

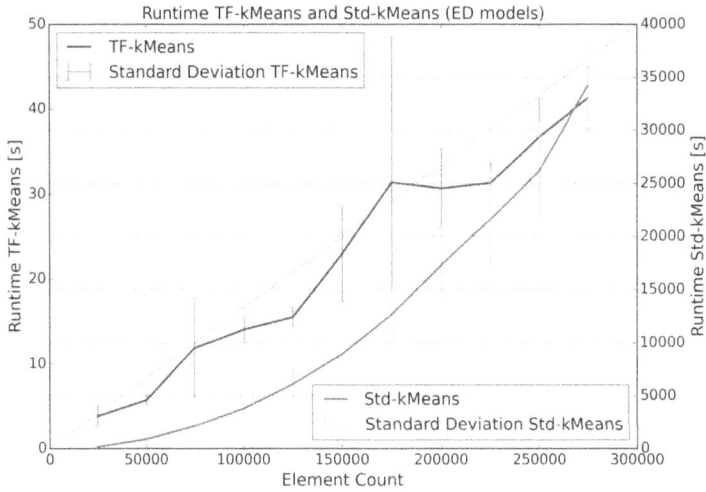

| Algorithms | Data Model Class | Element Count | Cluster Count | Dimensions Count |
|---|---|---|---|---|
| $(type_{algorithm})$ | $(type_{distribution})$ | $(n_{element})$ | $(n_{cluster})$ | $(n_{dimension})$ |
| STD / TF | ED | $5k..50k$ | $n_{element}/10$ | 2 |

**Figure 6.18:** *Runtimes ($t_{total}$) of the* TF *and the* STD K-MEANS *Algorithms applied to ED Data Models, dependent on Element and Cluster Count*

### 6.2.2.4 Speedup Dependency of the Dimension Count

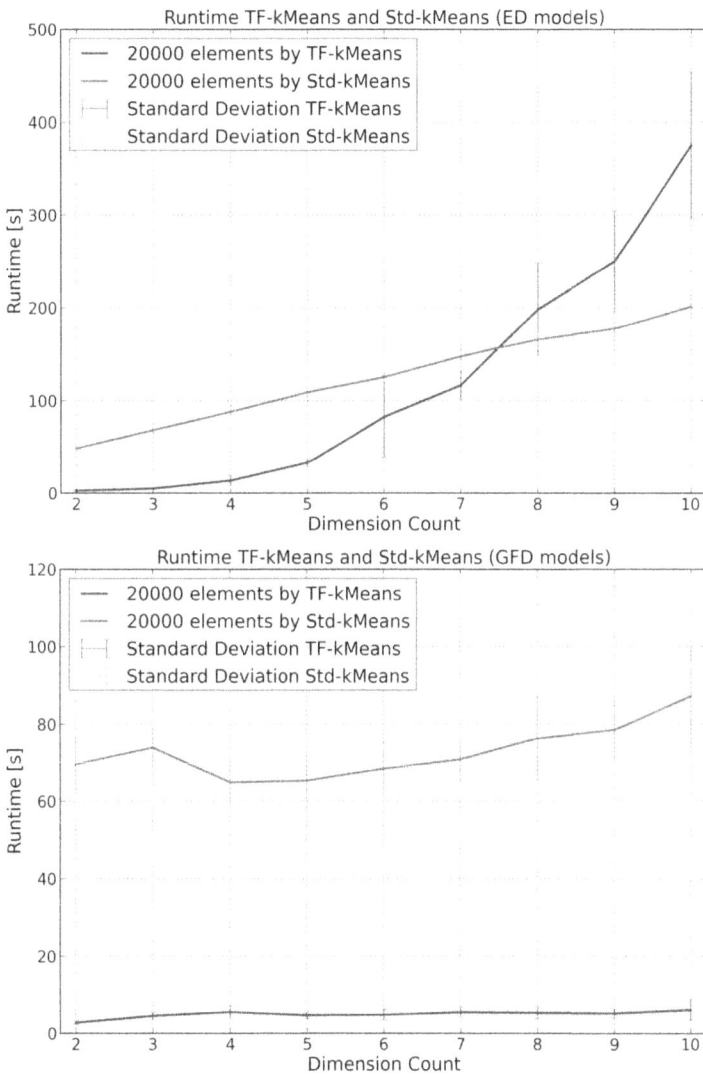

| Algorithms | Data Model Class | Element Count | Cluster Count | Dimensions Count |
|---|---|---|---|---|
| $(type_{algorithm})$ | $(type_{distribution})$ | $(n_{element})$ | $(n_{cluster})$ | $(n_{dimension})$ |
| STD (green) / TF (blue) | ED (top), GFD (bottom) | $20k$ | $1k$ | $2..10$ |

**Figure 6.19:** *Runtimes ($t_{total}$) of the* TF *and the* STD K-MEANS *Algorithms applied to ED and GFD Data Models, dependent on Dimensions Count*

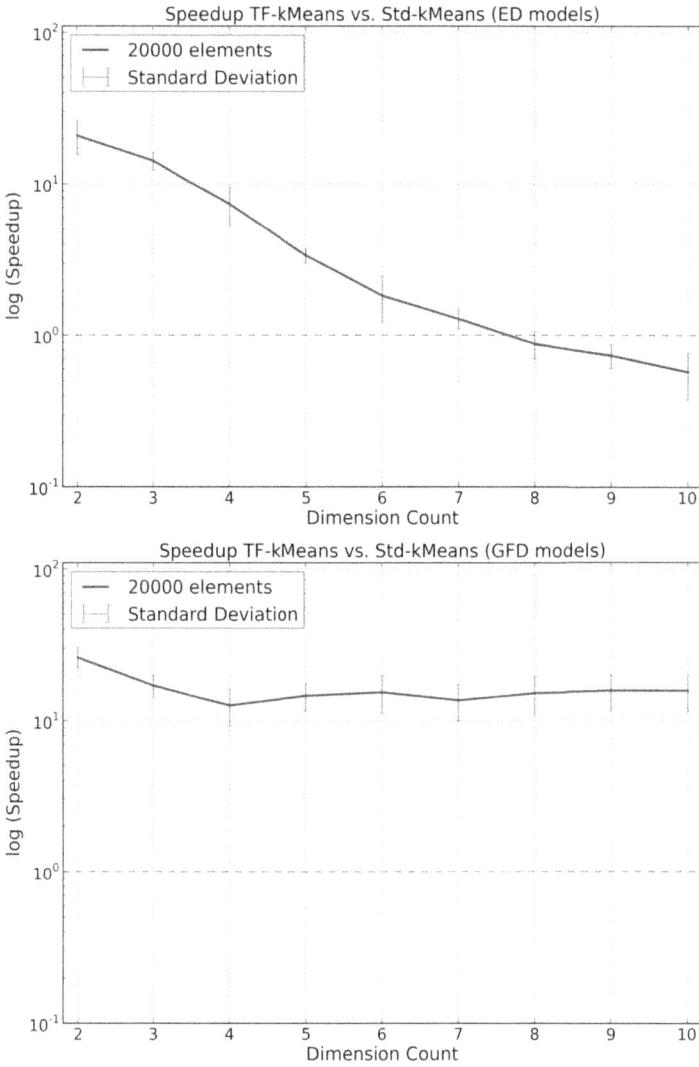

**Figure 6.20:** *Speedup ($q_{speedup}$) of the TF vs. the STD K-MEANS Algorithm applied to ED and GFD Data Models, dependent on Dimensions Count*

| Algorithms ($type_{algorithm}$) | Data Model Class ($type_{distribution}$) | | Element Count ($n_{element}$) | Cluster Count ($n_{cluster}$) | Dimensions Count ($n_{dimension}$) |
|---|---|---|---|---|---|
| STD / TF | ED (top), (bottom) | GFD | $20k$ | $1k$ | $2..10$ |

## 6.2.2.5 Speedup Dependency of the Branch Size

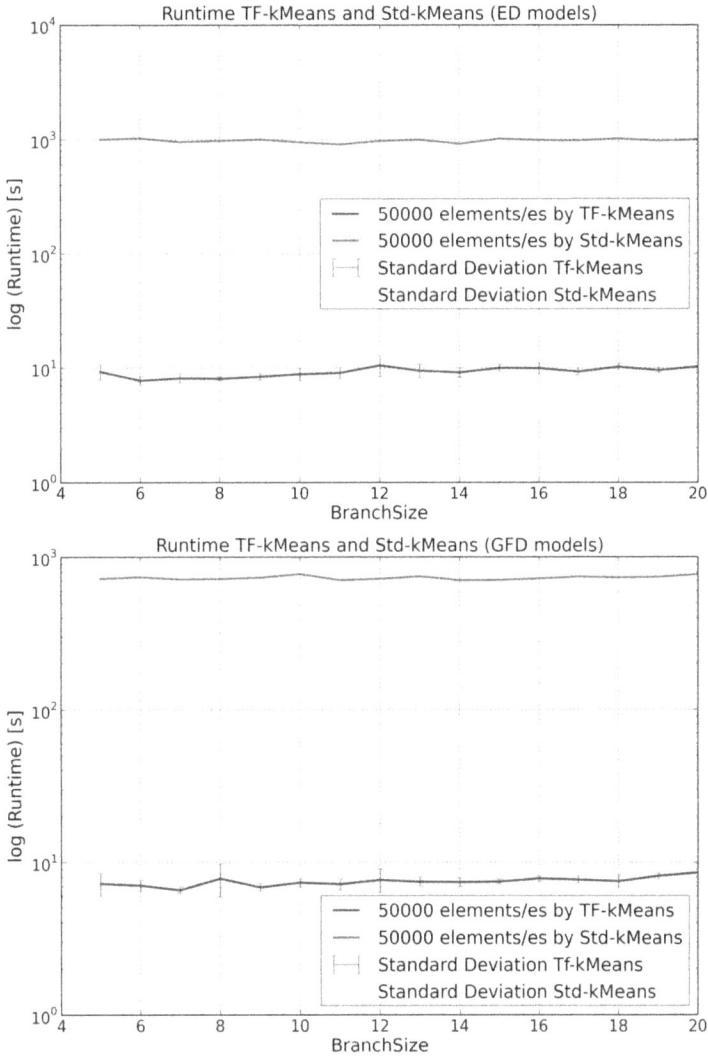

**Figure 6.21:** *Runtimes ($t_{total}$) of the* TF *and the* STD K-MEANS *Algorithms applied to ED and GFD Data Models, dependent on Branch Size*

| Algorithms ($type_{algorithm}$) | Data Model Class ($type_{distribution}$) | Element Count ($n_{element}$) | Cluster Count ($n_{cluster}$) | Dimensions Count ($n_{dimension}$) | Branch Size ($size_{opt}$) |
|---|---|---|---|---|---|
| STD / TF | ED (top), GFD (bottom) | 10k | 1k | 2 | 4..20 |

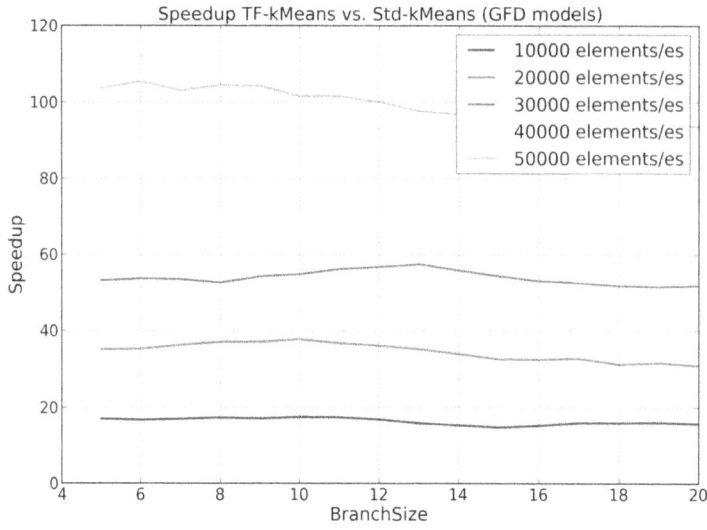

| Algorithms | Data Model Class | Element Count | Cluster Count | Dimensions Count | Branch Size |
|---|---|---|---|---|---|
| $(type_{algorithm})$ | $(type_{distribution})$ | $(n_{element})$ | $(n_{cluster})$ | $(n_{dimension})$ | $(size_{opt})$ |
| STD / TF | GFD | $1k..10k$ | $n_{element}/10$ | 2 | $4..20$ |

**Figure 6.22:** *Speedup ($q_{speedup}$) of the* TF *vs. the* STD K-MEANS *Algorithm applied to GFD Data Models, dependent on Element and Cluster Count, as well as Branch Size*

## 6.2.2.6 Efficiency Dependency of the Data Model Size

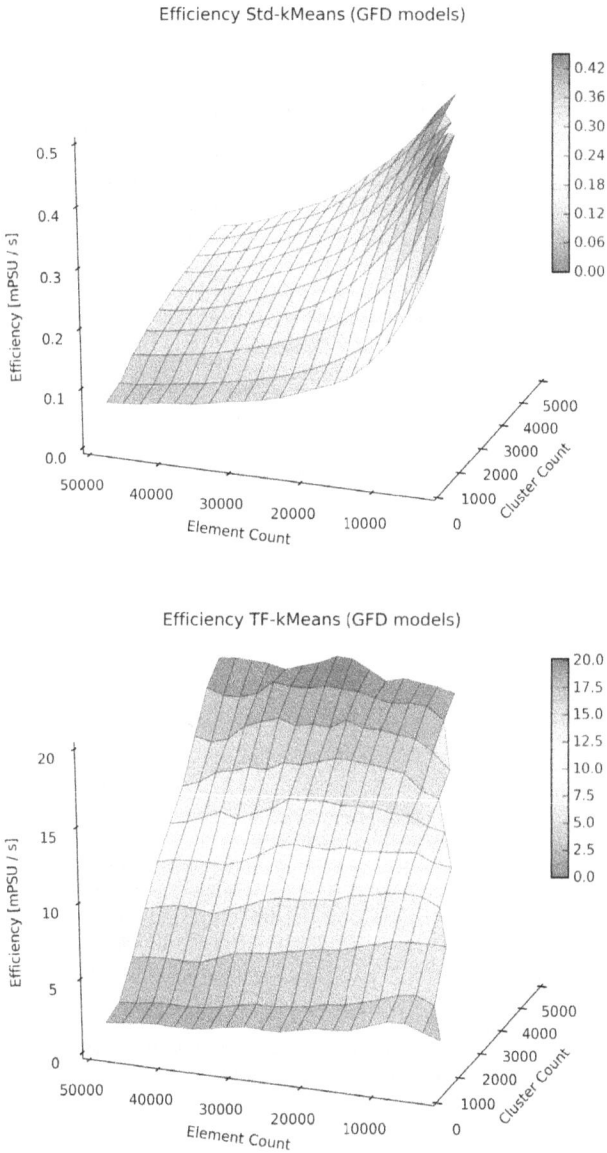

### Efficiency Std-kMeans (GFD models)

### Efficiency TF-kMeans (GFD models)

| Algorithms | Data Model Class | Element Count | Cluster Count | Dimensions Count |
|---|---|---|---|---|
| $(type_{algorithm})$ | $(type_{distribution})$ | $(n_{element})$ | $(n_{cluster})$ | $(n_{dimension})$ |
| STD (top) / TF (bottom) | GFD | $5k..50k$ | $500..5000$ | $2$ |

**Figure 6.23:** *Efficiency ($q_{efficiency}$) of the* STD *and the* TF K-MEANS *Algorithms applied to GFD Data Models, dependent on Element and Cluster Count;* Efficiency is defined as *PSU* per *TU*, see Formula (3.3).

## 6.3 Applications

### 6.3.1 Segmentation

#### 6.3.1.1 Data Models and Algorithm Configuration

**Self Adaptation** : As the property value ranges are unknown, the following approach adapts the values to the DT Framework's operational range $(0, 1)$, utilizing a linear scaling. For a given range $(value_{min}, value_{max})$ should apply:

$$m \cdot value_{min} + n = 0$$
$$m \cdot value_{max} + n = 1$$

(6.1)

where $m$ and $n$ represent the linear scaling. During the equilibration process (Section 2.3.2.2) the range's boundaries are updated as

$$value_{min} = min(value_{min}, value^i)$$
$$value_{max} = max(value_{max}, value^i)$$

(6.2)

where $value^i$ is the value of the current of element $i$. Hence the scaling can be calculated:

$$m = \frac{1}{value_{max} - value_{min}}$$
$$n = -m \cdot value_{min}$$

(6.3)

and applied to $value^i$ to provide the adapted property value:

$$^{scl}value^i = m \cdot value^i_{min}$$

(6.4)

**Specific Weight** : To enable the DT Framework to create hierarchical structures the following expression replaces the former static *weight* of an element or branch (see also Sections 2.3.2.1 and 2.3.2.2):

$$weight^* = weight \left(1 - \left(1 - aff^{pot}_{FanOut, FanIn}\right) \cdot \left(1 - \left(\frac{scaling_{base}}{sca}\right)^{pot}\right)\right)$$

(6.5)

Here $aff_{FanOut, FanIn}$ stands for the combined *Fan* affinity of the current comparison. The parameters *pot* and *sca* allow the degree of hierarchical substructuring.

# Acronyms

**MOO** *Multi-Objective Optimization*

**CA** *Cluster Analysis*

**PCA** *Principal Component Analysis*

**MDS** *Multi-Dimensional Scaling*

**MAO** *Mathematical Optimization*

**EVA** *Evolutionary Algorithms*

**TRACS-MOO** *Transposition and Cluster Analysis based Multi-Objective Optimization*

**DT Framework** *Dimensional Transposition Framework*

**DT Algorithm** *Dimensional Transposition Algorithm*

**EQ Algorithm** *Equilibration Algorithm*

**CVD** *Continuous Value Distribution*

**DVD** *Distinct Value Distribution*

**MSI** *Model Size Increment*

**PC** *Property Comparison*

**SA** *Simulated Annealing*

**TB** *Tree Building*

**MLD** *Maximal Leaf Distance*

**IA** *Inferior Affinity*

**TM** *Tree Maintenance*

**HBR** *Hierarchical Branch Representation*

**TB-EQ** *Tree Based Equilibration*

**SUB** *Substitution*

**INS** *Insertion*

**TF K-MEANS Algorithm** *Tree Fusion k-Means Algorithm*

**STD K-MEANS Algorithm**  *Standard k-Means Algorithm*

**TRIN K-MEANS Algorithm**  *Triangular Inequality based k-Means Algorithm*

***MLIF***  *Minimal Leaf Interspace Filtering*

$LI_{min}$  *Minimal Leaf Interspace*

$LI_{max}$  *Maximal Leaf Interspace*

***TF***  *Tree Fusion*

***CDE***  *Cluster Distance Estimation*

***PSU***  *Problem Size Unit*

***TU***  *Time Unit*

***ED***  *Equal Distribution*

***GFD***  *Gaussian Fields Distribution*

**SEGM Algorithm**  SEGMENTATION ALGORITHM

**SISEGM Algorithm**  *Single Segmentation Algorithm*

**BLC Algorithm**  *Balancing Algorithm*

***PR***  *Pattern Recognition*

***KDD***  *Knowledge Discovery in Databases*

**PD Algorithm**  *Peak Detection* Algorithm

***VGRF***  *Variant Gaussian Random Field*

# Bibliography

Acharya, T. and Ray, A. K. (2005). *Image Processing: Principles and Applications*. Wiley & Sons, Inc., Hoboken, Bew Jersey.

Agarwal, A., Phillips, J. M., and Venkatasubramanian, S. (2010). A unified algorithmic framework for multi-dimensional scaling. *arXiv preprint arXiv:1003.0529*.

Aggarwal, C. C., Wolf, J. L., Yu, P. S., Procopiuc, C., and Park, J. S. (1999). Fast algorithms for projected clustering. *ACM SIGMOD Record*, 28:61–72.

Agrawal, R., Gehrke, J., Gunopulos, D., and Raghavan, P. (2005). Automatic subspace clustering of high dimensional data. *Data Mining and Knowledge Discovery*, 11:5–33.

Andrews, N. O. and Fox, E. A. (2007). Recent developments in document clustering. Technical report, Department of Computer Science, Virginia Tech, Blacksburg, VA 24060.

Arkhangel'skii, A. V. (1990). *General Topology I: Basic Concepts and Constructions Dimension Theory (Encyclopaedia of Mathematical Sciences)*. Springer-Verlag.

Arthur, D., Manthey, B., and Roeglin, H. (2009). k-means has polynomial smoothed complexity. In *Proceedings of the 50th Symposium on Foundations of Computer Science (FOCS)*.

Arthur, D. and Vassilvitskii, S. (2007). k-means++: the advantage of careful seeding. *Proceedings of the eighteenth annual ACM-SIAM symposium on Discrete algorithms*, pages 1027–1035.

Bae, S.-H., Qiu, J., and Fox, G. (2012). High performance multidimensional scaling for large high-dimensional data visualization. *IEEE TRANSACTION OF PARALLEL AND DISTRIBUTED SYSTEM,*.

Ball, G. H. and Hall, D. J. (1965). Isodata, a novel method of data anlysis and pattern classification. Technical report, Standford Research Instiute.

Bar-On, D., Wolter, S., van de Linde, S., Heilemann, M., Nudelman, G., Nachliel, E., Gutman, M., Sauer, M., and Ashery, U. (2012). Super-resolution imaging reveals the internal architecture of nano-sized syntaxin clusters. *Journal of Biological Chemistry*, 287(32):27158–27167.

Bellman, R. E. (1957). Dynamic programming. In *Proceedings of the National Academy of Science uf the USA*. Princeton University Press.

Bentley, J. L. (1990). K-d trees for semidynamic point sets. *Proceeding SCG '90 Proceedings of the sixth annual symposium on Computational geometry*, pages 187–197.

Berkhin, P. (2006). A survey of clustering data mining techniques. *Grouping Multidimensional Data*, pages 25–71.

Betzig, E., Hell, S. W., and Moerner, W. E. (2014). How the optical microscope became a nanoscope.

Bezdek, J. C., Ehrlich, R., and Full, W. (1984). Fcm: The fuzzy c-means clustering algorithm. *Comupter and Geosciences*, 10:191–203.

Bishop, C. M. et al. (2006). *Pattern recognition and machine learning*, volume 1. springer New York.

Boehm, C., Kailing, K., Kriegel, H.-P., and Kroeger, P. (2004). Density connected clustering with local subspace preferences. In *Data Mining, IEEE International Conference on*.

Borg, I. and Groenen, P. J. (2005). *Modern multidimensional scaling: Theory and applications*. Springer Series in Statistics.

Bradley, P. S., Fayyad, U. M., and Reina, C. A. (1998). Scaling clustering algorithms to large databases. In *Proceedings of the 4th International Conference on Knowledge Discovery & Data Mining*.

Brandes, U. and Pich, C. (2007). Eigensolver methods for progressive multidimensional scaling of large data. In *Graph Drawing*, pages 42–53. Springer.

Bronstein, A. M., Bronstein, M. M., and Kimmel, R. (2005). Generalized multidimensional scaling: A framework for isometry-invariant partial surface matching. *Proceedings of the National Academy of Sciences of the United States of America*, 103:1168–1172.

Brüderle, D., Petrovici, M., Vogginger, B., Ehrlich, M., Pfeil, T., Millner, S., Grübl, A., Wendt, K., Müller, E., Schwartz, M.-O., de Oliveira, D., Jeltsch, S., Fieres, J., Schilling, M., Müller, P., Breitwieser, O., Petkov, V., Muller, L., Davison, A., Krishnamurthy, P., Kremkow, J., Lundqvist, M., Muller, E., Partzsch, J., Scholze, S., Zühl, L., Mayr, C., Destexhe, A., Diesmann, M., Potjans, T., Lansner, A., Schüffny, R., Schemmel, J., and Meier, K. (2011). A comprehensive workflow for general-purpose neural modeling with highly configurable neuromorphic hardware systems. *Biological Cybernetics*, 104:263–296.

Can, F. and Ozkarahan, E. A. (1990). Concepts and effectiveness of the cover-coefficient-based clustering methodology for text databases. *ACM Transactions on Database Systems*, 15:483–517.

Caramia, M. (2006). *Effective resource management in manufacturing systems: optimization algorithms for production planning*. Springer.

Cheng, H.-M., Groeger, P., Hartmann, A., and Schlierf, M. (2014). Bacterial initiators form dynamic filaments on single-stranded dna monomer by monomer. *Nucleic acids research*, page gku1284.

C#.NET, M. (2001).

Coello, C. C., Lamont, G. B., and van Veldhuizen, D. A. (2007). *Evolutionary Algorithms for Solving Multi-Objective Problems*. Springer Science & Business Media.

de Amorim, R. C. and Mirkin, B. (2012). Minkowski metric, feature weighting and anomalous cluster initializing in k-means clustering. *Pattern Recognition*, 45:1061–1075.

de Silva, V. and Tenenbaum, J. B. (2003). Global versus local methods in nonlinear dimensionality reduction. *NIPS*, pages 721–728.

Deb, K. (2001). *Multi-Objective Optimization Using Evolutionary Algorithms*. John Wiley & Sons.

Dhillon, I. S. and Modha, D. S. (2000). A data-clustering algorithm on distributed memory multiprocessors. *Lecture Notes in Computer Science*, 1759:245–260.

Ding, C. and He, X. (2004). K-means clustering via principal component analysis. In *Proceeding ICML '04 Proceedings of the twenty-first international conference on Machine learning*, page 29.

Dunn, J. C. (1974). A fuzzy relative of the isodata process and its use in detecting compact well-separated clusters. *Journal of Cybernetics*, pages 32–57.

Ehrgott, M. (2005). *Multicriteria Optimization*. Springer, 2. ed. edition.

Ehrlich, M. and Schüffny, R. (2013). Neural schematics as a unified formal graphical representation of large-scale neural network structures. *Frontiers in neuroinformatics*, 7.

Elkan, C. (2003). Using the triangle inequality to accelerate k -means. In *Proceedings of the Twentieth International Conference on Machine Learning*.

Eschrich, S., Ke, J., Hall, L. O., and Goldgof, D. B. (2003). Fast accurate fuzzy clustering through data reduction. *IEEE Transactions on Fuzzy Systems*, 11:262–270.

Ester, M., Kriegel, H.-P., Sander, J., and Xu, X. (1996). A density-based algorithm for discovering clusters in large spatial databases with noise. In *International Conference on Knowledge Discovery and Data Mining*.

Estivill-Castro, V. (2002). Why so many clustering algorithms. *ACM SIGKDD Explorations Newsletter*, 4:65.

Frahling, G. and Sohler, C. (2006). A fast k-means implementation using coresets. In *Proceedings of the twenty-second annual symposium on Computational geometry*.

Guha, S., Rastogi, R., and Shim, K. (1998). Cure: an efficient clustering algorithm for large databases. In *International Conference on Management of Data*.

Hartigan, J. A. (1975). *Clustering algorithms*. New York, NY : Wiley.

HBP (2012). *Human Brain Project*. https://www.humanbrainproject.eu/de.

Heintzman, N. D., Stuart, R. K., Hon, G., Fu4, Y., Ching, C. W., Hawkins, R. D., Barrera, L. O., Calcar, S. V., Qu, C., Ching, K. A., Wang, W., Weng, Z., Green, R. D., Crawford, G. E., and Ren, B. (2007). Distinct and predictive chromatin signatures of transcriptional promoters and enhancers in the human genome. *Nature Genetics*, 39:311–318.

Hestenes, M. R. (1958). Inversion of matrices by biorthogonalization and related results. *Journal of the Society for Industrial and Applied Mathematics*, 6:51–90.

Hilbert, D. (1904). Grundzüge einer allgemeinen theorie der linearen integralgleichungen, mathematisch-physikalische klasse. *Nachrichten von der Königlichen Gesellschaft der Wissenschaften zu Göttingen*.

Houle, M. E., Kriegel, H.-P., Kroeger, P., Schubert, E., and Zimek, A. (2010). Can shared-neighbor distances defeat the curse of dimensionality? In *Proceedings of the 22nd International Conference on Scientific and Statistical Database Management*.

Huang, Z. (1997). Clustering large data sets with mixed numeric and categorical values. In *Proceedings of the 1st Pacific-Asia Conference on Knowledge Discovery and Data Mining*.

Huang, Z. (1998). Extensions to the k-means algorithm for clustering large data sets with categorical values. In *Data Mining and Knowledge Discovery 2*. Kluwer Academic Publishers.

Inaba, M., Katoh, N., and Imai, H. (1994). Applications of weighted voronoi diagrams and randomization to variance-based k-clustering. *Proceedings of the tenth annual symposium on Computational geometry*, pages 332–339.

Indyk, P. and Motwani, R. (1998). Approximate nearest neighbors: towards removing the curse of dimensionality. *Proceeding STOC '98 Proceedings of the thirtieth annual ACM symposium on Theory of computing*, pages 604–613.

Inequality, T. (2014).

Jaeger, T. F. (2008). Categorical data analysis: Away from anovas (transformation or not) and towards logit mixed models. *Journal of Memory and Language*, 59:434–446.

Jain, A., Murty, M., and Flynn, P. (1999). Data clustering: A review. *ACM Computing Surveys*, 31:264–323.

Jain, A. K. (2008). Data clustering: 50 years beyond k-means. In *International Conference of Pattern Recognition*.

Jain, A. K., Duin, R. P. W., and Mao, J. (2000). Statistical pattern recognition: A review. *Pattern Analysis and Machine Intelligence, IEEE Transactions on*, 22(1):4–37.

Jiang, D., Tang, C., and Zhang, A. (2004). Cluster analysis for gene expression data: A survey. *IEEE Transactions on Knowledge and Data Engineering*, 16:1370–1386.

Jolliffe, I. (2005). *Principal Component Analysis. Encyclopedia of Statistics in Behavioral Science*. Wily Online Library.

Kailing, K., Kriegel, H.-P., and Kroeger, P. (2004). Density-connected subspace clustering for high-dimensional data. In *Proceedings of the 2004 SIAM International Conference on Data Mining*.

Kanungo, T., Mount, D. M., Netanyahu, N. S., Piatko, C. D., Silverman, R., and Wu, A. Y. (2002). An efficient k-means clustering algorithm: Analysis and implementation. *IEEE Transactions On Pattern Analysis and Machine Intelligence*, 24:881–892.

Kellerer, H., Pferschy, U., and Pisinger, D. (2004). *Knapsack problems*. Springer.

Kirkpatrick, S. (1984). Optimization by simulated annealing: Quantitative studies. *Journal of statistical physics*, 34(5-6):975–986.

Kittler, J., Hatef, M., Duin, R. P., and Matas, J. (1998). On combining classifiers. *Pattern Analysis and Machine Intelligence, IEEE Transactions on*, 20(3):226–239.

Kriegel, H.-P., Kroeger, P., and Zimek, A. (2009). Clustering high-dimensional data: A survey on subspace clustering, pattern-based clustering, and correlation clustering. *ACM Transactions on Knowledge Discovery from Data (TKDD)*, 3:1:1 – 1:58.

Kruskal, J. B. (1956). On the shortest spanning subtree of a graph and the traveling salesman problem. *Proceedings of the American Mathematical Society*, 7:48–50.

Kruskal, J. B. (1964). Multidimensional scaling by optimizing goodness of fit to a nonmetric hypothesis. *Psychometrika*, 29:1–27.

Lathauwer, L. D., Moor, B. D., and Vandewalle, J. (2000). A multilinear singular value decomposition. *SIAM Journal on Matrix Analysis and Applications*, 21:1253–1278.

Liu, Y., Zhang, D., Lu, G., and Ma, W.-Y. (2007). A survey of content-based image retrieval with high-level semantics. *Pattern Recognition*, 40(1):262–282.

Lloyd, S. P. (1982). Least squares quantization in pcm. *IEEE Transactions on Information Theory*, 28:129–137.

Lukashin, A. V., Lukashev, M. E., and Fuchs, R. (2003). Topology of gene expression networks as revealed by data mining and modeling. *Bioinformatics*, 15:1909–1916.

Mahajan, M., Nimbhorkar, P., and Varadarajan, K. (2009). The planar k-means problem is np-hard. *Lecture Notes in Computer Science*, 5431:274–285.

Marler, R. and Arora, J. (2004). Survey of multi-objective optimization methods for engineering. *Structural and Multidisciplinary Optimization*, 26(6):369–395.

Mather, P. M. (1999). *Computer Processing of Remotely-Sensed Images*. Wiley & Sons, Inc., Hoboken, Bew Jersey, second edition edition.

McCallum, A., Nigam, K., and Ungar, L. H. (2000). Efficient clustering of high-dimensional data sets with application to reference matching. In *Proceeding KDD '00 Proceedings of the sixth ACM SIGKDD international conference on Knowledge discovery and data mining*.

Miettinen, K. (1999). *Nonlinear Multiobjective Optimization*. Springer Science & Business Media.

Miettinen, K., Ruiz, F., and Wierzbicki, A. (2008). Introduction to multiobjective optimization: Interactive approaches. In Branke, J., Deb, K., Miettinen, K., and Słowiński, R., editors, *Multiobjective Optimization*, volume 5252 of *Lecture Notes in Computer Science*, pages 27–57. Springer Berlin Heidelberg.

Ng, R. T. and Han, J. (2002). Clarans: A method for clustering objects for spatial data mining. *IEEE TRANSACTIONS ON KNOWLEDGE AND DATA ENGINEERING*, 14:1003–1016.

Oehler, K. (1995). Combining image compression and classification using vector quantization. *Pattern Analysis and Machine Intelligence, IEEE Transactions*, 17:461–473.

Parsons, L., Haque, E., and Liu, H. (2004). Subspace clustering for high dimensional data: a review. *ACM SIGKDD Explorations Newsletter - Special issue on learning from imbalanced datasets*, 6:90–105.

Pearson, K. (1901). On lines and planes of closest fit to systems of points in space. *Philosophical Magazine*, 11:559–572.

Pelleg, D. and Moore, A. (1999). Accelerating exact k-means algorithms with geometric reasoning. In *Research Showcase*.

Pelleg, D. and Moore, A. (2000). X-means: Extending k-means with efficient estimation of the number of clusters. *Proceedings of the Seventeenth International Conference on Machine Learning*, pages 727–734.

Philbin, J. (2007). Object retrieval with large vocabularies and fast spatial matching. In *IEEE Conference on Computer Vision and Pattern Recognition*.

Pohekar, S. and Ramachandran, M. (2004). Application of multi-criteria decision making to sustainable energy planning - a review. *Renewable and Sustainable Energy Reviews*, 8(4):365–381.

Pulido, G. and Coello Coello, C. (2004). Using clustering techniques to improve the performance of a multi-objective particle swarm optimizer. In Deb, K., editor, *Genetic and Evolutionary Computation GECCO 2004*, volume 3102 of *Lecture Notes in Computer Science*, pages 225–237. Springer Berlin Heidelberg.

Punj, G. and Stewart, D. W. (1983). Cluster analysis in marketing research: Review and suggestions for application. *Journal of Marketing Research*, 20:134–148.

Rabiner, L. (1989). A tutorial on hidden markov models and selected applications in speech recognition. *Proceedings of the IEEE*, 77(2):257–286.

Rakesh, A., Johannes, G., Dimitrios, G., and Prabhakar, R. (2005). *Automatic Subspace Clustering of High Dimensional Data*, volume 11 of *Data Mining and Knowledge Discovery*. Springer Netherlands.

Roman, S. (2007). *Advanced Linear Algebra*. Graduate Texts in Mathematics. Springer.

Roweis, S. T. and Saul, L. K. (2000). Nonlinear dimensionality reduction by locally linear embedding. *Science*, 290:2323–2326.

Schemmel, J., Brüderle, D., Grübl, A., Hock, M., Meier, K., and Millner, S. (2010). A wafer-scale neuromorphic hardware system for large-scale neural modeling. In *Proceedings of the 2010 IEEE International Symposium on Circuits and Systems (ISCAS)*, pages 1947–1950.

Schemmel, J., Grübl, A., Meier, K., and Muller, E. (2006). Implementing synaptic plasticity in a VLSI spiking neural network model. In *Proceedings of the 2006 International Joint Conference on Neural Networks (IJCNN)*. IEEE Press.

Schwartz, S. H. and Bilsky, W. (1990). Toward a theory of the universal content and structure of values: Extensions and cross-cultural replications. *Journal of Personality and Social Psychology*, 58:878–891.

Shapiro, L. and Stockman, G. (2001). *Computer Vision*. Prentice Hall.

Shepard, R. N. (1980). Multidimensional scaling, tree-fitting, and clustering. *Science*, 210:390–398.

Snyman, J. (2005). *Practical mathematical optimization: an introduction to basic optimization theory and classical and new gradient-based algorithms*, volume 97. Springer.

Soni, N. and Ganatra, A. (2012). Categorization of several clustering algorithms from different perspective: A review. *International Journal of Advanced Research in Computer Science and Software Engineering*, 2:63–68.

Steinbach, M., Karypis, G., and Kumar, V. (2000). A comparison of document clustering techniques. In *KDD workshop on text mining*.

Steinbach, M., Tan, P.-N., Kumar, V., Klooster, S., and Potter, C. (2003). Discovery of climate indices using clustering. In *International Conference on Knowledge Discovery and Data Mining*.

Taboada, H. and Coit, D. W. (2007). Data clustering of solutions for multiple objective system reliability optimization problems. *Quality Technology & Quantitative Management Journal*, 4(2):35–54.

ter Braak, C. J. F. (1986). Canonical correspondence analysis: A new eigenvector technique for multivariate direct gradient analysis. *Ecology*, 67:1167–1179.

Theiler, J. P. and Gisler, G. (1997). Contiguity-enhanced k-means clustering algorithm for unsupervised multispectral image segmentation. In *Algorithms, Devices, and Systems for Optical Information Processing*.

Thomopoulos, S., Bougoulias, D., and Chin-Der, W. (1995). Dignet: an unsupervised-learning clustering algorithm for clustering and data fusion. *IEEE Transactions on Aerospace and Electronic Systems*, 31:21–38.

Topcuoglu, H., Hariri, S., and Wu, M.-Y. (2002). Performance-effective and low-complexity task scheduling for heterogeneous computing. *Parallel and Distributed Systems, IEEE Transactions*, 13:260–274.

Torgerson, W. S. (1952). Multidimensional scaling: I. theory and method. *Psychometrika*, 17:401–419.

Vattani, A. (2011). k-means requires exponentially many iterations even in the plane. *Discrete Computational Geometry*, 45:596–616.

Wendt, K. (2007). Mapping- und konfigurationstool fÃ¼r das facets-design-framework. Master's thesis, Technical University of Dresden.

Wilkinson, J. H. (1965). *The Algebraic Eigenvalue Problem*. Clarendon Press Oxford.

Willett, P. (1988). Recent trends in hierarchic document clustering: A critical review. *Information Processing & Management*, 24:577–597.

Williams, H. P. (2013). *Model Building in Mathematical Programming*. John Wiley & Sons.

Xu, X., Ester, M., Kriegel, H.-P., and Sander, J. (1998). A distribution-based clustering algorithm for mining in large spatial databases. In *Data Engineering, 1998. Proceedings., 14th International Conference on*, pages 324–331. IEEE.

Yehuda, K., Carmel, L., and Harel, D. (2002). Ace: A fast multiscale eigenvectors computation for drawing huge graphs. In *Information Visualization*.

Zhang, T., Ramakrishnan, R., and Livny, M. (1996). Birch: An efficient data clustering method for very large databases. In *SIGMOD*.

www.ingramcontent.com/pod-product-compliance
Lightning Source LLC
Chambersburg PA
CBHW081102220326
41598CB00038B/7202